MATHEMATICAL TECHNIQUES in GIS

SECOND EDITION

Peter Dale

MATHEMATICAL TECHNIQUES in GIS

SECOND EDITION

CRC Press
Taylor & Francis Group
Boca Raton London New York

CRC Press is an imprint of the
Taylor & Francis Group, an **informa** business

CRC Press
Taylor & Francis Group
6000 Broken Sound Parkway NW, Suite 300
Boca Raton, FL 33487-2742

First issued in paperback 2019

ISBN-13: 978-1-4665-9554-5 (hbk)
ISBN-13: 978-0-367-86846-8 (pbk)

Library of Congress Cataloging-in-Publication Data

Dale, Peter.
   [Introduction to mathematical techniques used in GIS]
   Mathematical techniques in GIS / Peter Dale. -- Second Edition.
      pages cm
   Revised edition of: Introduction to mathematical techniques used in GIS.
   Includes bibliographical references and index.
   ISBN 978-1-4665-9554-5 (Hardcover : acid-free paper)
   1. Geography--Mathematics. 2. Geographic information systems. I. Title.

G70.23.D35 2014
910.285--dc23

2014001339

Visit the Taylor & Francis Web site at
http://www.taylorandfrancis.com

and the CRC Press Web site at
http://www.crcpress.com

Contents

Preface to Second Edition

Many people wishing to make use of geographic information systems (GIS) start from a limited mathematical background. In *Mathematical Techniques in GIS*, Second Edition, the text as before focuses on those who are unfamiliar with mathematics and need to understand the principles behind the manipulation of spatial data. The new text adds additional material. The first nine chapters lay out the basic foundations and introduce the reader to the relevant techniques and shorthand notations that frequently occur in mathematics; the remaining five chapters build on the earlier material. These later chapters place particular emphasis on the use of the techniques in GIS and geomatics. Throughout the text there are a number of examples shown in boxes and there is a summary at the end of each chapter listing all the important ideas that have been introduced.

Preface to First Edition

This book has been written for nonmathematicians who wish to understand some of the assumptions that underlie the manipulation and display of geographic information. It assumes very little basic knowledge of mathematics but moves rapidly through a wide range of data transformations, outlining the techniques involved. Many of these are precise, building logically from certain underlying assumptions; others are based on statistical analysis and the pursuit of the optimum rather than the perfect and definite solution.

Mathematics has its own form of shorthand that often gets in the way of understanding what is going on. For those who are unfamiliar with mathematical notation this can be daunting; but it cannot be avoided. It can in many cases be kept to a minimum and in what follows, the derivation of some of the formulae is placed in boxes that can be digested at leisure without interrupting the narrative. But at the end of the day, compromise has had to be made and as the text progresses there is an increasing use of symbols.

This spirit of compromise is most apparent in the selection of topics discussed. Many things have had to be left out—indeed every chapter could be expanded to a full book and most would require several volumes in order to do their subject justice. *Introduction to Mathematical Techniques Used in GIS* is therefore a book that allows the reader to get started and then to turn to the many more informative texts that are available.

The text begins with an introduction to geographic data but soon focuses on the "where" rather than the "what." It assumes that the data have been measured and refrains from discussing the techniques of measurement science, other than to recognize that measurement is prone to error. Pure mathematics, even when dealing with vague concepts, provides precise answers that can be verified by anyone. Even statistical analysis uses processes that can be programmed into a computer to give a consistent answer, even when the underlying assumptions are not met or the hypothesis has been incorrectly formulated. The apparent exactness of an answer does not mean that it is correct. To understand the output from, for example, a geographic information system, one needs to understand the quality of the data that are entered into the system, the algorithms behind the data processing, and the limitations of the graphic displays.

This book deals with only part of the bigger picture. It focuses on the basic mathematical techniques, building the whole of mathematics in a series of steps that are the foundations for a deeper understanding. It seeks to lay the foundations for the more complex forms of manipulation that arise in the handling of spatially related data.

The technology behind geographic information systems (GIS) allows such data to be gathered, processed, and displayed. The power and appeal of such systems often lie in their graphical output, the maps that they create. Users of GIS need to understand the quality of that output so that they can advise others on the integrity of their results. The issue is not a matter of which buttons to push but rather of the quality of the information that has been produced. Quality means "fitness for purpose" and "safety in use."

This book therefore looks at some of the fundamentals and provides an introduction to spatial data manipulation through which users of GIS may come to understand whether what they do results in what can genuinely be described as a "quality product." It has been copy edited for an American market, hence the spelling of words such as "meter" for the English "metre" and "center" for "centre."

The Author

Peter Dale trained as a land surveyor and worked for seven years in Uganda before entering the academic world. He ultimately became a professor in land information management at the University College London. He is an Honorary President of the International Federation of Surveyors and was awarded an OBE in recognition of his services to surveying. He is now retired and lives in a remote area of Scotland. Peter Dale can be contacted via e-mail at: peter.f.dale@btinternet.com.

List of Tables

List of Illustrations

List of Boxes

List of Examples

1 Characteristics of Geographic Information

1.1 GEOGRAPHIC INFORMATION AND DATA

It used to be said that geography was about "maps," as distinct from "chaps." Without doubt today it is about both, and a lot more besides. Ultimately, geography is about making sense of the world around us and this is done by observing, measuring, and processing data about the environment and then presenting the information either as text or pictorially. In particular, it is concerned with why things are where they are.

In recent years, much has been said and written about geographic information systems or GIS, which are tools that can help the process of understanding. Although there are various interpretations of what is meant by the acronym "GIS," the majority of people would accept that it includes a computer system of hardware and software that can be used to record, manage, integrate, manipulate, analyze, and display data that are spatially referenced to the Earth. The term *spatially referenced* means that their location can be described by measured quantities. *Data* are basic facts that can somehow or other be measured and turned into information.

Information is the commodity that is used by people when they make decisions. Too many facts can be confusing—there may be many different possible routes from one's home to the nearest shopping mall, each of which has its own quality of road surface, slopes, twists, turns and intersections, street lamps, drain covers, and so forth. All these facts about each route can be measured and recorded but all that the average user really wants to know is which is the shortest route. This is a piece of information that can be extracted from the basic data.

The term *shortest* is ambiguous since it could mean shortest in terms of time or shortest in terms of distance; these are not necessarily the same thing. The types of data that need to be collected depend on the use to which the information is to be put. The required output determines the required input and the manner in which the data may need to be processed. One can of course start with a set of data and see what sense can be made of all the facts and figures. Frequently, the most effective way to do this is through pictures, especially maps and graphs. Advocates of the use of GIS often quote the 19th century case in London where the location of cases of cholera were plotted on a map, which then showed clearly that they formed a cluster around an infected well whose water had become contaminated.

When processing data, two golden rules always apply.

1. Bad data plus good processing gives rise to unreliable information.
2. Good data plus bad processing also gives rise to unreliable information.

If data are to be converted into good quality information, then both the data and the way in which they are processed must be of good quality, that is, they must be "fit for purpose" or "safe in use." In the discussions that follow we will focus on the basic mathematical principles underlying how data are processed and not on the technical aspects of measurement or how the data are acquired.

1.2 CATEGORIES OF DATA

Data come essentially in two forms—*categorical* and *numerical*. As their name suggests, categorical data are those that are placed in a category or codified according to a classification system. Such data are sometimes referred to as *nominal data* and have no numerical value as such. Whether a piece of fruit is an apple or a pear or something else depends on the object itself and the way in which fruits have been classified. For many objects there are internationally and scientifically recognized standards for classification though even then there is the occasional dispute over whether some new discovery belongs as a subset of an existing class or whether it represents a totally new species.

With some data, categorization is less scientific, for instance, when designating the type of land use at a particular location. Although within each country there may be a national land use classification system, it does not mean that all those who record land use abide by it and it certainly does not follow that every country uses the same system. A building may be used in several different ways with, for example, the basement as a gymnasium, the ground floor as a shop, the next floor as commercial offices, and the top floor as a residential accommodation. In spite of national guidelines, investigators may still disagree as to how the use of the building should be categorized. It is, however, not the aim of this book to analyze the problems of data classification but rather to note that it is an issue that intimately affects the quality of data.

Once data have been categorized they can be subjected to comparison without being quantified. Thus, the data can be placed in a rank so that a is said to be more than b, which is more than c, and so forth. Such data are described as *ordinal,* an example of which is a list of preferences (area a is a nicer place to live than area b, etc.). Various statistical tests exist to process and analyze the differences between ranks or sequences of ordinal data but these too will not be discussed here.

Once the data have been categorized it is often necessary to indicate their magnitude. This may be done through the use of *discrete* or *continuous* variables. A discrete variable is one that can only take distinct values while a continuous variable is one that changes only gradually, allowing any intermediate values. Some data can only be measured in terms of whole numbers (called *integers*—such as the number of children in a family) while other items can be measured on a continuous scale (such as the height of each child). One can, of course, talk about the average family size being in the decimal system (see Chapter 2), 2.54 children, even though it is impossible to have 54 out of 100 parts of a child. Such a figure is useful for some practical purposes especially when associated with an estimate of its reliability, as discussed in Chapter 12.

Discrete variables are precise and are often expressed as whole numbers or integers (0, 1, 2, 3, etc.). More particularly, they can take a succession of distinct values

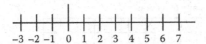

FIGURE 1.1　A scale bar.

at set intervals along a scale for which there are no intermediate values. Such data are often referred to as *interval data* (see Figure 1.1).

The data may be positive or negative but such items can only be compared quantitatively on the basis of the differences between them. Only when the values are *absolute* can valid conclusions be drawn about their relative sizes. One can say that a family with four children has twice as many youngsters as a family with two children because "zero children" is an absolute point of reference. One should not, however, say that a temperature of 16°C is twice as hot as a temperature of 8°C as zero on the centigrade scale is an arbitrarily chosen point.

The highest level of measurement is the *ratio scale*, which differs from the interval scale in that it relates to absolute zero (in the case of temperature this is approximately –273°C). Absolute temperature, length, and breadth are examples of measures on a ratio scale. They are *continuous variables* in that they are not restricted to integer forms but can take any value whatsoever from zero upward. The numerical quantity used to express the measurement of a continuous variable, such as the length of a line or the area of a field, presupposes a standard unit of measure. The numerical value represents the ratio between the quantity measured and the unit of measurement (e.g., the meter or "metre").

Geographical data have one particular characteristic that distinguishes them from all other forms of data, namely, location. Graphical data can be plotted on a map and be represented by points, lines, and areas. From a theoretical perspective, a point on its own has no dimension, a line has one dimension (length), and an area has two and a volume three. In practice, a point on a map is a blob or very small area while a line has thickness and also direction. Each has a category (the "what") representing some attribute or attributes associated with it, and each has a location ("where"). Examples of how the "what" may be categorized as points, lines, and areas when used by cartographers are shown in Table 1.1.

To define the location of any point there must be some reference to which the point can be related. The most common reference system uses a rectangular grid made up of squares of a standard size. For absolute position (as distinct from relative

TABLE 1.1

Points, Lines, and Areas on Maps

Feature	Points	Lines	Areas
Physical objects	Corner of building	Road network	Planning zone
Statistical values	Sampling point	Isoline	Layer tints
Areas	Central point	Boundary line	Polygon
Surfaces	Height point	Contour	Hill shading
Text	House numbers	Street names	District names

Northings or *y* Direction

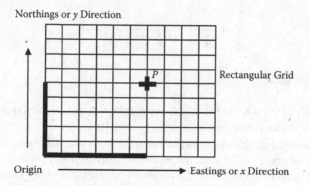

FIGURE 1.2   Rectangular Cartesian coordinates.

position), the grid must have a point of *origin* from which measurements are taken. Points may then be located so far east (or to the right) of the origin and so far north (or up the page), using a standard unit of measure. The two distances are called the *coordinates* or more particularly the *rectangular Cartesian coordinates* (named after the French mathematician Descartes).

In Figure 1.2, the rectangular grid coordinates of *P* are (x, y) relative to the origin and are shown here as $(6, 5)$. The idea can be extended to three dimensions by adding the height above the origin. Although in some countries the *x*-direction is taken as being to the north or upward, here we will follow the convention that the direction across the page is the *x*-direction while up the page is described as the *y*-direction (x, y, or "in the door and up the stairs," which is the opposite of many computer software graphics packages that measure the position of points from the top of the screen downward). Height is then in the *z*-direction. For any point on a three-dimensional object the coordinates would be (x, y, z). For two-dimensional ("flat Earth") displays, $z = 0$.

Simple mathematical techniques can be used to analyze the locations of points that have been referenced to a rectangular grid. Sometimes it is useful to use a non-rectangular or skewed grid, for example, when trying to show three dimensions on a flat piece of paper (Figure 1.3). Data manipulation is slightly more complicated in these circumstances although the underlying principles are the same.

An alternative way of measuring the location of a point is through the use of *polar coordinates* (Figure 1.4). These describe points by their distance from an origin and their direction relative to some reference line. The direction is known as the *bearing*

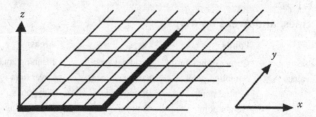

FIGURE 1.3   Nonrectangular or skewed grid.

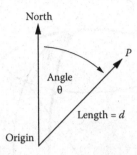

FIGURE 1.4 Polar coordinates.

and is normally measured clockwise from the north (or up the page). In Figure 1.4, P has polar coordinates (θ, d) relative to the origin, d being the distance, and θ (the Greek letter "theta") being the direction or bearing from the north.

Angles and distances are examples of measures on the ratio scale. Distances are normally expressed as a ratio, for instance, between the amount of space between two points in comparison with a standard length. Under the *Systeme International* (SI) the standard unit of length is known as the *metre* and was once defined as the distance between two marks on a bar of platinum kept at constant temperature in Paris. It is now defined by stipulating that the speed of light is 299,792,458 meters (or "metres") per second. As we will show later, trigonometrical formulae allow polar coordinates (θ, d) to be converted into Cartesians (eastings and northings or x and y) and vice versa.

Angles are a ratio between the amount of turning and a complete turn. They may be expressed as a proportion of either 360 degrees—written as 360° with each degree being subdivided into 60 minutes (60') and each minute into 60 seconds (60"); or 400 grads (where 100 grads equates with a quarter turn, with submeasurements being expressed as decimals) or 2π (two pi) radians where "pi" is the ratio between the diameter of a circle and its circumference.

Angular measures are important in surveying where positions may be expressed as if the Earth were a sphere using what are known as *spherical coordinates*. The *latitude* of a point is its angular distance north or south of the equator and is often represented by the Greek letter "phi" or ø. The *longitude* of a point is an angular measure east or west of the Greenwich standard meridian: it is normally represented by the Greek letter "lambda" or λ (see Figure 1.5).

The altitude or height of any point is measured as a distance above a reference level or surface, such as a mathematical shape that best approximates to the size and shape of the Earth. It is not normally given a Greek letter and hence the coordinates of points are either expressed as (ø, λ) or as (ø, λ, H). The use of Greek letters is common in mathematics. The full alphabet is given in Table 1.2.

For more accurate work, rather than assuming that the Earth is a perfect sphere, its shape is taken to be an *ellipsoid* (an ellipse rotated on its shorter axis creating a squashed sphere) as discussed in Chapter 4. However, for many practical purposes, the Earth can be regarded as a sphere. The word *accurate* as used here relates to nearness to the truth. The word *precision* will be used to refer to the exactness with

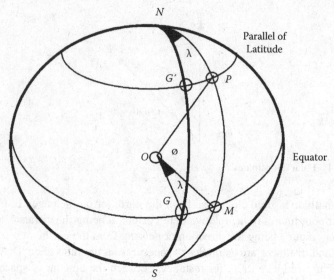

GG´ is the Greenwich Meridian of Origin (Zero Longitude)
GM is the Equator (Zero Latitude)
O is the center of the spherical Earth
The angle GOM is the *longitude* of P = λ
The angle MOP is the *latitude* of P = ø

FIGURE 1.5   Latitude and longitude.

which a value is expressed, whether the value is right or wrong. The word *precision*
is often used to describe the number of decimal places that represent the quantity—
for example, a distance expressed as 2.105 m (where "m" means meters) is more precise than 2 m though the latter may be nearer to the truth and hence more accurate.

On a flat surface, a straight line represents the shortest distance between two
points. On a curved surface such as a sphere, a so-called straight line bends with the
surface while a line of sight is even more bent because the light is refracted in the

TABLE 1.2
The Greek Alphabet

Letter	Name	Letter	Name	Letter	Name
A α	alpha	I ι	iota	P ρ	rho
B β	beta	K κ	kappa	Σ σ	sigma
Γ γ	gamma	Λ λ	lambda	T τ	tau
Δ δ	delta	M μ	mu	Y υ	upsilon
E ε	epsilon	N ν	nu	Φ ø	phi
Z ζ	zeta	Ξ ξ	xi	X χ	chi
H η	eta	O ο	omicron	Ψ ψ	psi
Θ θ	theta	Π π	pi	Ω ω	omega

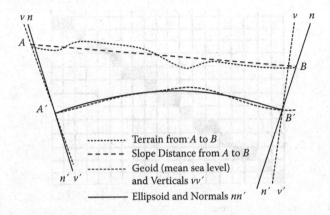

FIGURE 1.6 Straight-line distances.

atmosphere. We will not deal with the consequences of the latter effect. In Chapter 11, we will consider how to transform measurements on curved surfaces onto a plane through map projections. Before doing so, measurements taken on the surface of the Earth need to be adjusted so that they fit on a mathematical surface known as the *spheroid* or *ellipsoid* of reference that is the best approximation to mean sea level. The surface defined by the mean level of the seas, assuming that they extend under the mountains, is known as the *geoid*. The difference in length between what is measured and what is calculated is important in *geodesy*, which is the scientific study of the size and shape of the Earth. Figure 1.6 illustrates how the vertical as shown by a plumb line (here the lines *vv'*) may not in practice point at the center of the Earth (the lines *nn'* are perpendicular to the mathematical surface *A'B'*) due, for example, to it being pulled aside by nearby mountains. We will discuss some of these matters later but for now we will focus on a flat Earth and two-dimensional representations.

Geographic information systems handle lines either as vectors or rasters. A *vector* is a quantity that represents both direction and distance. A polar coordinate as described above is an example of a vector quantity. Every straight line has a direction and length (such as *A* to *B* in Figure 1.7) while a curved line may be considered as being made up of a series of short lines or vectors. In Chapter 8, there is further discussion on vectors.

On the other hand, a line can be considered as a series of points adjacent to each other, each point being of small but finite size (Figure 1.8). A television screen or a dot matrix printer produces what to the eye may appear like smooth curved lines but

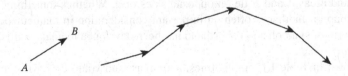

FIGURE 1.7 Lines as vectors.

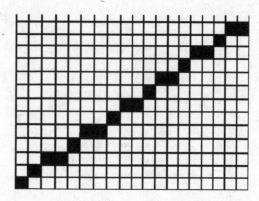

FIGURE 1.8   A line as a raster image.

which in practice are a series of points on a grid. Such a representation is called a *raster image* and each small square is known as a *pixel*.

Raster imaging can be used to show lines or areas. Raster data are simple for a computer to handle, but require relatively large amounts of computer storage. Given an area 20 cm by 20 cm and a grid cell size or resolution of 100 dots to the centimeter (i.e., one tenth of a millimeter) there will need to be a storage capacity for four million bits of information. For each vector it is necessary to record only the coordinates of the start and end points of straight-line sections.

One of several advantages of pixelation is that each pixel can be allocated a number or set of numbers that indicate the characteristics of the point concerned. Those numbers may represent the color of the point, such as the proportion of red, green, and blue (referred to as *RGB*) to be used on a color television; or they may, for instance, represent the various wavelengths in the electromagnetic spectrum that have been picked up by a scanner, for example, in remote sensing. The values can either be analyzed through "number crunching" on a computer or they can be displayed visually as an image that the human brain can analyze and interpret.

A line on a flat surface may be regarded as an item in its own right; alternatively, it may be regarded as the division between two areas: one on the left and the other to the right of the line. *Topology* involves the study of *adjacency*, that is, what lies beside a given area, *containment* (what is contained within it), and *connectivity* (how lines or areas are connected to other lines or areas).

Thus, in Figure 1.9, the area *B* is adjacent to the area *A* while the area *C* is contained within the area *A*. Point *P* is connected to point *Q*. The fact that the line *PQ* may be straight or curved is of no significance; what is important is that there is a division and areas *A* and *B* lie on opposite sides of it. Whether something belongs to one group or another is often an important consideration in mathematics and a specific form of symbolism or shorthand has been developed to analyze this, known as *set theory*.

As shown in Table 1.1, points, lines, and areas (and volumes) have size, shape, and location and also a classification. They each form part of a set of information. A set, also called a *class*, is a collection of related objects that can be treated as an

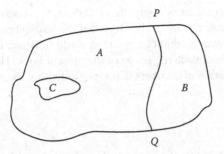

FIGURE 1.9 Topology.

entity in its own right. A set may be finite in size, such as the letters in an alphabet, or infinite, such as the set of integers that has no limit. A set is often identified by a pair of curved brackets such as (all human beings) or if they are in a specific sequence, angled brackets <letters in the Roman alphabet>. The symbol "∈" is used to indicate that an object is a member of a specified set {"a" ∈ < Roman alphabet >} while ∉ indicates that the object is not a member {"6" ∉ < Roman alphabet >}.

"A ⊂ B" shows that set A is a subset of set B (and "A ∉ B" that A is not a subset) while "A ⊃ B" means that set A includes set B. "A ∪ B" is the *union* of sets A and B, that is, it is the combination of set A and set B, the symbol "∪" sometimes being referred to as *cup* while "∩," sometimes referred to as *cap*, is the *intersection* of A and B. See Figure 1.10.

We will not develop the ideas of topology nor the processes of classifying data, although we will be concerned with the outcomes of such classification when, in Chapters 12 and 13, we touch on elementary statistical techniques commonly used in geomatics and GIS. Throughout this book we will focus on two and three dimensions (length, breadth, and height, or latitude, longitude, and altitude) and how relevant data may be manipulated. Time is, of course, a further dimension and strictly

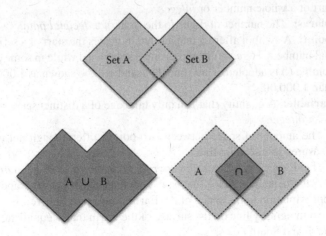

FIGURE 1.10 Union and intersection of data sets A and B.

speaking we should consider not only the (x, y) or (x, y, z) coordinates of a point but also the (x, y, z, t) where t is time. Also, from a mathematical perspective, there is no inherent reason why we should not consider a world in which there are many more dimensions but this is outside the scope of the present book. Here, we will concentrate on the manipulation of numbers that represent either location or the quantities found at a location.

SUMMARY

Absolute value: The value of a quantity for which there is an absolute zero and hence the quantity can be measured irrespective of its relation to other values. It is also used in mathematics as the magnitude of a number without regard to whether it is positive or negative.

Accuracy: The nearest to truth or correctness of a quantity. Distinguished from *precision*, which relates to the exactness of a quantity regardless of whether it is nearer to the truth.

Adjacency: A term used in *topology* for areas that are side by side.

Bearing: The direction of an object, normally quoted as an angle measured clockwise relative to the direction of north.

Cartesian coordinates: Numbers that indicate the distance from an origin and in directions parallel to two or three fixed lines known as the axes.

Categorical data: Data that are identified by their category or class.

Connectivity: A term used in *topology* to describe which lines or areas are connected to each other.

Containment: A term used in *topology* to describe objects within a given area or volume.

Continuous variable: A quantity that can take any intermediate value and is not restricted, for instance, to being an *integer*.

Data: Raw facts that have not been processed into *information*.

Decimal: A number that relates to a tenth part (or hundredth or thousandth, etc.), part of a whole number or *integer*.

Decimal places: The number of digits to the right of a *decimal point*.

Decimal point: A symbol after an integer that indicates the start of a series of decimal numbers. Here, a full stop (.) or period is used while in some countries a comma (,) is adopted. Thus, one thousand will be shown as 1,000.00 rather than 1.000,00.

Discrete variable: A quantity that can only take one of a distinct set of values such as an *integer*.

Distance: The amount of space between two points, often though not necessarily measured in a straight line.

Eastings: The distance in a *Cartesian coordinate system* east from an *origin*.

Ellipsoid: A mathematical shape that is, in effect, a squashed sphere and is a better approximation to the shape of the Earth than a pure sphere.

Equator: An imaginary line on the surface of the Earth that is equidistant from the North and South Poles.

Geodesy: The scientific study of the size and shape of the Earth.

Geoid: The shape of the Earth based on mean sea level and its imagined extension under or over land.

GIS: Geographic Information System (GIS) that is used to record, analyze, manipulate, and display information, which relates to some location.

Great circle: A line on the surface of a sphere where the plane that contains the line passes through the center of the sphere.

Information: Data that have been processed and presented in a form that permits decisions to be made.

Integer: A whole number such as 1, 2, 3, 4, and so on.

Intersection: Those elements of two or more data sets that overlap.

Interval data: A set of *discrete variables* that take values at set intervals.

Latitude: The angular distance of a point north or south of the *equator.*

Longitude: The angular distance of a point on the Earth's surface that is east or west of a *meridian* of origin, usually taken as the meridian through Greenwich in England.

Meridian of longitude: A *great circle* on the Earth's surface that passes through the North and South Poles and a given point.

Nominal data: Data that are identified only by their class or category, such as apples or pears.

Northings: The distance in a *Cartesian coordinate system* north from an *origin.*

Numerical data: Data that are expressed in terms of numbers.

Ordinal data: Data that can be arranged in a sequence, such as preferences.

Origin: A fixed point that is the start point for a coordinate system from which *eastings* and *northings* can be measured.

Parallel of latitude: A *small circle* on the Earth's surface along the line of which all points have the same *latitude.*

Pixel: The smallest area on a display screen that can be uniquely identified.

Polar coordinates: A coordinate system based on the measurements of *bearing* and *distance*, the latter being either the straight-line distance or, on a sphere, the distance along a *great circle.*

Precision: The exactness of a quantity, for instance, the number of *decimal places* to which the quantity is expressed.

Raster: A rectangular pattern of parallel lines that divides a screen up into a series of *pixels.*

Ratio scale data: Data that are *continuous* and *absolute.*

Rectangular coordinates: *Cartesian coordinates* in which the axes are at right angles.

Small circle: A line on the surface of a sphere where the plane that contains the line does not pass through the center of the sphere.

Spherical coordinates: The location of points based on the angular measurements of *latitude* and *longitude.*

Topology: The study of geometric properties and relationships that are not related to their size or shape.

Union: The combination of two data sets.

Vector: A quantity that has direction as well as magnitude. (The term is also used for a matrix that has only one row or one column—see Chapter 7.)

2 Numbers and Numerical Analysis

2.1 THE RULES OF ARITHMETIC

The mathematics that underpins all geographical analysis involves the application of rules, most of which are straightforward. Mathematics makes use of symbols, the most basic of which are shown in Table 2.1. We will add and explain other symbols later. In this chapter we will consider arithmetic, which is concerned with numerical calculations such as adding, subtracting, multiplying, and dividing.

The whole of arithmetic is based essentially on seven axioms, as shown in Box 2.1. Outside arithmetic, these axioms may not apply, for instance, when two raindrops running down a windowpane come together to make one raindrop so that in symbolic form: $1 + 1 = 1$. Furthermore, computer programmers often write "$N = N + 1$," meaning "Take the number in the box labeled N, add one to that number and put it back in the box labeled N"; although partially an arithmetic operation, the use of the "=" sign has a different meaning from that which we are considering here.

Axiom 1 in Box 2.1 states that adding a to b has the same result as adding b to a or in symbolic form, $a + b = b + a$, for instance, $2 + 5 = 5 + 2$. This applies to basic arithmetic but elsewhere the sequence of operations is important. If you rotate a dice forward and then sideways it will finish in a different position than if you rotate it sideways and then forward (see Figure 2.1). This illustrates how in some circumstances the sequence of operations can be important and we will discuss this further in Chapter 7 in the context of what are called matrices.

In this chapter, we will deal with simple arithmetic for which the axioms in Box 2.1 are fundamental. They are all necessary but they are not quite sufficient. Consider the calculation $2 + 3 * 4$. A pocket calculator will show $2 + 3$ equals 5; then enter 4 and multiply to give the answer 20. On the other hand, $3 * 4$ equals 12. Adding 2 gives 14. The same sum done in a different order gives a different answer.

Hence, we must have rules of priority. The simplest way to handle this is to place brackets around the groups that are together. Thus, in the first case, we have $(2 + 3) * 4$, while in the second case, we have calculated $2 + (3 * 4)$. We must distinguish between these two cases. This leads to Rule 2.1 in Box 2.2.

TABLE 2.1
Standard Symbols

Add	+	Less than	<	Equal	=
Subtract	–	More than	>	Nearly equal	≈
Multiply	*	Less or equal	≤	Not equal	≠
Divide	/	More or equal	≥		

BOX 2.1   AXIOMS OF ARITHMETIC

AXIOM 1: THE COMMUTATIVE LAW

For any numbers a and b,

> a plus b has the same value as b plus a.

Also,

> a multiplied by b has the same value as b multiplied by a.

Put into symbols,

$$a + b = b + a$$
$$a * b = b * a$$

The latter may also be written as

$$ab = ba$$

AXIOM 2: THE ASSOCIATIVE LAW

For any three numbers a, b, c,

$$(a + b) + c = a + (b + c)$$

Also,

$$(ab)c = a(bc)$$

AXIOM 3: THE DISTRIBUTIVE LAW

For any numbers a, b, c, then

$$a * (b + c) = a * b + a * c$$

or

$$a(b + c) = ab + ac$$

AXIOM 4

There is a number called **zero** (0) that for any a

$$a + 0 = a$$

(Historically, there was a debate as to whether 0 was actually a number. The Romans, for example, used the symbols I, V, X, L, C, D, and M to represent 1, 5, 10, 50, 100, 500, 1000 in the decimal system but had no zero.)

AXIOM 5

There is also a number 1 such that

$$a * 1 = a$$

AXIOM 6

For every value of *a* there is a number *d* such that

$$a + d = 0$$

(This then introduces the whole range of negative numbers.)

AXIOM 7

Provided *c* does *not* equal zero ($c \neq 0$), then

$$\text{if } c * a = c * b \text{ then } a = b$$

(Similarly, if $a + c = b + c$, then $a = b$ although this also applies if $c = 0$ but not when *c* is infinitely large.)

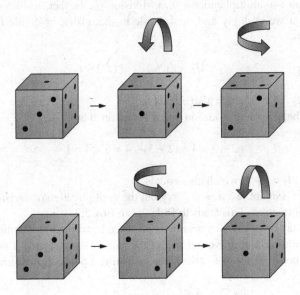

FIGURE 2.1   Rotating a dice (forward + sideways or sideways + forward).

BOX 2.2   RULES OF ARITHMETIC

RULE 2.1

Place items that are together in brackets and deal with what is inside the brackets first.

RULE 2.2

If there are no brackets or when what is inside the brackets has been evaluated, then deal with multiplication or division before addition or subtraction.

RULE 2.3

$$(positive) * (positive) = + = (positive)/(positive)$$

$$(positive) * (negative) = - = (positive)/(negative)$$

$$(negative) * (positive) = - = (negative)/(positive)$$

$$(negative) * (negative) = + = (negative)/(negative)$$

RULE 2.4

To add or subtract fractions they must share a common denominator.

Where there are no brackets, then we have to decide which comes first—addition (+), subtraction (–), multiplication (*), or division (/). In fact, it does not normally matter whether we multiply and then divide in that order, or divide first and then multiply. Thus,

$$3 * 4/2 = (3 * 4)/2 = 12/2 = 6$$

while $3 * (4/2) = 3 * 2$, which also equals 6.

The same happens with addition and subtraction. Thus,

$$2 + 3 - 4 = (2 + 3) - 4 = 5 - 4 = 1$$

while $2 + (3 - 4) = 2 - 1$, which also equals 1.

When there is doubt, we should carry out the multiplication or division before the addition or subtraction. This leads to Rule 2.2 in Box 2.2.

Numbers may be positive or negative and when handling these, simple rules also apply. Thus, adding a negative number is the same as subtracting a positive while subtracting a negative number is the same as adding a positive. Put another way:

$$4 + (-3) = 4 - 3 = 1$$

while $4 - (-3) = 4 + 3 = 7$.

Multiplication and division follow Rule 2.3 in Box 2.2.

Ordinary numbers not only come as positive or negative but also as integers or real numbers. An *integer* is a whole number such as 1, 2, 3, or 4. *Real* numbers are basically what happens in between and are often expressed in *decimal* parts (i.e., in submultiples of 10). Real numbers are either rational or irrational. A *rational* number is one that can be expressed as the ratio of two integers, for example, the decimal number $1.125 = 1 + (125/1000) = 1 + (1/8)$. *Irrational* numbers are ones that cannot be expressed in this way such as the number which multiplied by itself gives the value 2, known as the square root of 2 and shown symbolically as $\sqrt{2}$. Another example of an irrational number is π, which is the ratio between the circumference of a circle and its diameter. π has been calculated to millions of decimal places with the sequence of numbers never being repeated.

A *fraction* is a number that is expressed in the form of two integers such as (I/J) where I and J are whole numbers; for example, $\frac{1}{3} = 1/3$; $\frac{5}{8} = 5/8$; and $\frac{13}{25} = 13/25$ are all fractions. I is known as the *numerator* (the number that indicates how many there are) and J is the *denominator* (giving the denomination or category of the fraction). Fractions can be converted into decimal parts simply by dividing the denominator into the numerator; thus, 5 and 8 are integers while 5/8 is a fraction that can be expressed as the decimal number 0.625 ($= 625/1000$).

Numbers can also be *complex* or *imaginary* in that they have two parts, the first of which is real and the second a real number multiplied by the square root of (-1), often referred to as i, for instance, $(2 + 3i)$. Complex numbers obey special rules that contradict some of the rules listed above, notably that when multiplied by themselves they can produce a negative answer, contrary to Rule 2.3 in Box 2.2. This cannot happen in normal arithmetic. Complex numbers are particularly useful when handling vectors (which will be discussed in Chapter 8) but will not be dealt with here. For now, we will treat all numbers as integers or as real numbers.

Many integers are the product of smaller integers each of which is a *factor* of the bigger number. Thus, the factors of 14 are two and seven since $(2 * 7) = 14$. Positive numbers that cannot be expressed as the product of at least two smaller integers (apart from 1) are said to be *prime* numbers. Thus, 1, 2, 3, 5, 7, 11, 13, 17, 19, 23, 29, 31, 37, 41, 43, and 47 are all the prime numbers under 50. None of these numbers can be "factorized," that is, they have no factors other than 1.

Factorizing is important when dealing with fractions as it permits simplification. By removing all the common factors from the numerator and the denominator, we can, for example, simplify $\frac{561}{2431}$ to $\frac{187*3}{187*13}$ which reduces to $\frac{3}{13}$. Finding the *highest common factor* (here 187) can greatly simplify subsequent data processing.

Conversely, we may wish to combine two factors such as adding $\frac{7}{16}$ to $\frac{13}{40}$. Here, we need to find the *lowest common denominator*, which is the smallest number into which the two denominators can divide. In this case, it would be 80 since that is the smallest integer number that can be divided by both 16 and 40. There are, of course, higher numbers such as ($16 * 40$ or 640) but 80 is the smallest number that is divisible by both. Hence, by expressing the first fraction as $\frac{5*7}{5*16}$ or $\frac{35}{80}$ and the second as $\frac{2*13}{2*40}$ or $\frac{26}{80}$, the two fractions can be added; $\frac{7}{16} + \frac{13}{40} = \frac{35}{80} + \frac{26}{80} = \frac{61}{80}$. This leads to Rule 2.4 in Box 2.2.

Alternatively, fractions can be expressed as decimal numbers and then processed accordingly. A decimal number consists of an integer, a decimal point, and a series of fractions of the number 10. In some countries, notably in parts of mainland Europe, a comma is used instead of a decimal point while a full stop (or period) is used to separate out thousands. Thus, one million plus twenty-three hundredths would be written as 1.000.000, 23 but in this text, we will use the comma to separate the thousands and a full stop as the decimal separator so that our number will be shown as 1,000,000.23.

Decimals are easier to add or subtract than fractions. Thus, in the example above, $\frac{35}{80} = 0.4375$ while $\frac{26}{80} = 0.3250$ giving a sum of 0.7625, which is the same as $\frac{61}{80}$ expressed as a decimal.

Decimal numbers use a sequence of numerals that were derived from the Arabic system, with numbers represented as the symbols 0, 1, 2, 3, 4, 5, 6, 7, 8, 9. In Arabic the actual shapes of the symbols used for these numbers are different (، ١ ٢ ٣ ٤ ٥ ٦ ٧ ٨ ٩) but they are based on the number 10. In the decimal system, the number 8642 represents

$$8 * (10 * 10 * 10) + 6 * (10 * 10) + 4 * (10) + 2 * 1.$$

By using indices as discussed later, we can write $10 * 10 * 10$ as 10^3, $10 * 10$ as 10^2, 10 as 10^1 where the "3," "2," and "1" are called the *indices*. 1 can be written as 10^0 (in fact, any number whose index is 0 will equal 1). This means that

$$8642 = 8 * 10^3 + 6 * 10^2 + 4 * 10^1 + 2 * 10^0$$

Continuing the sequence downward, $(1/10) = 10^{-1}$, $(1/100) = 10^{-2}$, and the like, so that 0.325, for example, represents $3 * (1/10) + 2 * (1/100) + 5 * (1/1000)$, which may be written as:

$$0.325 = 3 * 10^{-1} + 2 * 10^{-2} + 5 * 10^{-3}$$

2.2   THE BINARY SYSTEM

Although the decimal numbering system is convenient for human beings, there are useful alternatives. The most common is the binary system, because it is appropriate for computers. The binary system works only with the numbers 0 and 1 (*binary digits or bits*) and increases by multiples of 2 rather than 10 (Example 2.1).

Early computers used to store numbers in 8 bits, a "bit" being simply "0 or 1," "off or on," or "no charge or with charge." Each storage box within the computer system could hold 8 of these bits at a time, giving any number from 0 and 255, since $256 = 2^8$. This is known as a *byte*. Now many computers use at least 64 bits. 2^{64}, which is "2" multiplied by "2" 64 times, represents a very large number.

EXAMPLE 2.1: THE BINARY SYSTEM

2 in decimals = **10** = $1 * 2^1 + 0 * 2^0$ in binary (since $2^0 = 1$)
3 in decimals = **11** = $1 * 2^1 + 1 * 2^0$ in binary
4 in decimals = **100** = $1 * 2^2 + 0 * 2^1 + 0 * 2^0$
8 in decimals = **1000** = $1 * 2^3 + 0 * 2^2 + 0 * 2^1 + 0 * 2^0$
12 in decimals = **1100** = $1 * 2^3 + 1 * 2^2 + 0 * 2^1 + 0 * 2^0$
14 in decimals = **1110** = $1 * 2^3 + 1 * 2^2 + 1 * 2^1 + 0 * 2^0$
16 in decimals = **10000** = $1 * 2^4 + 0$
31 in decimals = **11111** = $1 * 2^4 + 1 * 2^3 + 1 * 2^2 + 1 * 2^1 + 1 * 2^0$
32 in decimals = **100000** = $1 * 2^5 + 0$, and so on.
255 in decimals = **11111111** = $1 * 2^7 + 1 * 2^6 + 1 * 2^5 + 1 * 2^4 + 1 * 2^3 + 1 * 2^2$
$$+ 1 * 2^1 + 1 * 2^0$$
$$= 128 + 64 + 32 + 16 + 8 + 4 + 2 + 1 \text{ in decimals}$$

and consists of 8 bits or 1 byte.

In the binary system, the rules for addition, subtraction, multiplication, and division are similar to those for decimal numbers. For example, when adding two decimal numbers, work from the right to the left and "carry one" if over the base number 10 (Example 2.2).

The same procedure is followed with the binary system where increments go from 1 to 2 (= 10) to 4 (= 100) to 8 (= 1000) to 16 (= 10000), and so on, doubling each time and adding a zero, as shown in Example 2.3.

Similar procedures are followed for subtraction, multiplication, and division. A pocket calculator, for example, turns the decimal number that is typed in on the keypad into a binary number, processes it in that form and then returns the answer in decimals (which is why sometimes answers such as 79.99999999999999 appear instead of 80).

EXAMPLE 2.2: DECIMAL (BASE 10) ADDITION

To add 83 + 649:   649 = $6 * 10^2 + 4 * 10^1 + 9 * 1$. Then, to add
$+ 83 = 0 * 10^2 + 8 * 10^1 + 3 * 1$

Start on the right (singles) column. Add

$$(3 + 9 = 12 = 1 * 10^1 + 2 * 10^0).$$

Keep **2** and carry one over to the tens column.
Add the tens (= **4 + 8 + 1** carried over = **13** = $1 * 10^2 + 3 * 10^1$).
Keep the **3** and carry one over to the hundreds column.

$$(6 + 0 + 1 \text{ carried over} = 7) \text{ finishing with}$$

$$\text{Total} = 7 * 10^2 + 3 * 10^1 + 2 * 10^0 = 732$$

EXAMPLE 2.3: BINARY (BASE 2) ADDITION

To add the binary numbers **111** + **101** (Using **bold** for binary numbers):
Given:

$$\mathbf{111}\ (\text{in decimals } 1 * 2^2 + 1 * 2^1 + 1 * 2^0 = 7)$$

Add: **101** (in decimals $1 * 2^2 + 0 * 2^1 + 1 * 2^0 = 5$)
Start on the right (singles or 2^0) column

$$\mathbf{1 + 1 = 10}\ (\mathbf{1} \text{ lot of } 2^1 \text{ plus } \mathbf{0} \text{ lot of } 2^0)$$

Giving $+\mathbf{0} * 2^0$ (and carry one lot of 2^1)
Repeat on the center (twos or 2^1) column

$$\mathbf{(1 + 0) + (1} \text{ carried over}) = \mathbf{1 + 1 = 10}\ (\mathbf{1} \text{ lot of } 2^2 + \mathbf{0} \text{ lot of } 2^1)$$

Giving $+\mathbf{0} * 2^1 + \mathbf{0} * 2^0$ (carry $\mathbf{1} * 2^2$)
Repeat on the left (2^2) column

$$\mathbf{(1 + 1) + (1} \text{ carried over}) = \mathbf{10 + 1 = 11}\ (\mathbf{1} \text{ lot of } 2^3 \text{ plus } \mathbf{1} \text{ lot of } 2^2)$$

Giving $+\mathbf{1} * 2^2 + \mathbf{0} * 2^1 + \mathbf{0} * 2^0$ (carry $\mathbf{1} * 2^3$)
Add the **1** lot of 2^3 carried over

Total $= 1 * 2^3 + 1 * 2^2 + 0 * 2^1 + 0 * 2^0$ = 1100

$$(= 8 + 4 + 0 + 0 = 12 \text{ in decimals})$$

Hence, 111 + 101 = 1100.

Although numbers such as 11010101001010101 may be ideal for computers, they are tedious for human beings so in what follows we will stick to the decimal system.

2.3 SQUARE ROOTS

A number that is multiplied by itself is said to be the *square* of that number. For example, 3 * 3 is the square of three (since it is the area of a square that is three units by three units) and usually written as $3^2 = 9$. Thus, 9 is the square of three while 3 is the square root of 9, which is normally written as $\sqrt{9}$. Similarly, a number may be cubed and have a cube root, for instance, $4^3 = 64$ while $\sqrt[3]{64} = 4$.

Squares and square roots are of particular importance when dealing with coordinate systems. According to the theorem of Pythagoras (a proof of which is given in Chapter 3), in any right-angled triangle the square of the diagonal length is equal to the sum of the squares on the two shorter sides. Thus, if we have the rectangular Cartesian coordinates for two points A and B and the difference between them in the x-direction or eastings is E and the difference in the y-direction or northings is N, then (see Figure 2.2) the square of the distance from A to B is equal to $E * E + N * N$ or $(E^2 + N^2)$. Put another way, $AB = \sqrt{(E^2 + N^2)}$.

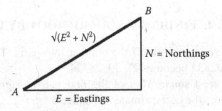

FIGURE 2.2 The distance from A to B.

In the very special case when $E = 3$ units and $N = 4$ units (or vice versa) then AB is exactly equal to 5 since $3^2 + 4^2 = 25 = 5^2$. A similar relationship exists with $5^2 + 12^2 = 169 = 13^2$. Normally, however, the square root will be an irrational number and not an integer. Some numbers are *perfect squares* (1, 4, 9, 16, etc.) in that their square roots are exact numbers. Most numbers do not have an exact square root; hence, any system that generates square roots can only produce an answer that is sufficient for the needs in hand.

The easiest way to find the square root of a number is to use a pocket calculator! If nothing other than a pencil and paper is available, then the value can be obtained by repeated attempts at long division until an answer to the appropriate level of significant numbers is achieved. It is done by *iteration* in which one gets closer and closer to the answer until sufficient significant figures have been reached.

Example 2.4 shows the calculation of a square root using multiplication and division. It finishes up with a number that is an approximation to the true value that, like $\sqrt{2}$, is irrational and the numbers after the decimal point could go on forever. We therefore need to define a suitable level of accuracy and precision beyond which there is no merit in going. A similar problem arises in geomatics where although the calculations may be precise and accurate, even when handling uncertainty, the data that are fed into formulae may be unreliable. The fact that a distance may be calculated in meters and expressed as 123.165284 does not mean that it is accurate to a one-thousandth part of a millimeter. Such a number may be calculated to six *decimal places* (six figures after the decimal point), but it does not mean that it is reliable in terms of what is really on the ground to that degree of precision.

The extent to which such a number is reliable can be expressed in terms of its *significant figures*. This is the number of figures after which we can write zeros for all the difference it will make. If the measurement is only reliable to the nearest ten meters, we could say that rather than being 123.165284, the distance is 120 meters and is reliable to two significant figures; to the nearest five significant figures the distance would be expressed as 123.17. In this case, there has been *rounding up* in that the fifth figure (here the number 6 in ".165284") has become 7 and the subsequent numbers have been ignored. The number 123.17 is closer to 123.165284 than is 123.16. If the original figure had been 123.164284, then it would have been *rounded down* and the best estimate of the number to five significant figures would have been 123.16.

There is, of course, a potential ambiguity when, for example, rounding 123.5 to three significant figures. Should the answer be 123 or 124? Some people would say "always round down," some "always round up," while others would go for the answer

EXAMPLE 2.4: FINDING A SQUARE ROOT BY ITERATION

1. The first stage is to mark off the number into pairs. Thus, a number such as 27392834 becomes 27, 39, 28, 34.
2. Take the nearest square root of the first pair of significant figures (here 27). Here the best estimate is 5 (since 5 * 5 = 25). Add zeros for the remaining pairs. The first trial number is 5000. Then, working to three decimal places:
3. Calculate (1/2) * (Trial + number/trial)

$$= 0.5 * (5000 + 27392834/5000)$$

$$= 0.5 * (5000 + 5478.567) = 5239.284$$

4. Repeat using 5239.284 as the best estimate

$$= 0.5 * (5239.284 + 27392834/5239.284)$$

$$= 0.5 * (5239.284 + 5228.354) = 5233.819$$

5. Repeat using 5233.819 as the best estimate

$$= 0.5 * (5233.819 + 27392834/5233.819)$$

$$= 0.5 * (5233.819 + 5233.814) = 5233.816$$

To three decimal places, 5233.816 is as close to the square root of our original number as we can get. The solution has thus converged quite rapidly from 5000 through 5239.284 through 5233.819 to 5233.816.

An explanation of why this works is given in Chapter 6, Section 6.4 on linearization. It is known as *Newton's method* after the mathematician and scientist Sir Isaac Newton.

that is an even number because it can be halved. The convention used here will be that (0, 1, 2, 3, 4) round down to 0 while (5, 6, 7, 8, 9) round up to 10. Thus, 123.5 would be rounded up to 124.

2.4 INDICES AND LOGARITHMS

Longhand calculations are tedious but there is another way than that shown in Example 2.4, whereby we can approach the problem of calculating square roots. It is based on indices. When we write the square of the number a as a^2, the number "2" is said to be the *index* of a or that a has been raised to the *power* of 2. The cube of a is a^3 with a being raised to the power of 3. If we multiply a by itself five times, then we obtain a^5—for example—2^5 or "two to the power of 5" equals 2 * 2 * 2 * 2 * 2 or 32. For any number a then if its square ($a * a$) is multiplied by its cube ($a * a * a$) we obtain $a * a * a * a * a = a^2 * a^3 = a^5$. Note that $5 = 2 + 3$. As a general rule, we add the indices when multiplying the same base number raised to separate powers—see Rule 2.5 in Box 2.3.

BOX 2.3 FURTHER RULES OF ARITHMETIC

RULE 2.5: THE LAW OF INDICES

When multiplying any number raised to the power of m (a^m) by itself raised to the power of n (a^n), then we add the indices so that $a^m * a^n = a^{(m+n)}$.

RULE 2.6

Any number raised to the power of zero equals 1

$$a^0 = 1$$

Any number raised to the power of 1 is itself

$$a^1 = a$$

If we divide a to the fifth by a squared, that is, $(a * a * a * a * a)/(a * a)$ or a^5/a^2, then the answer is $(a * a * a)$ or a^3. Here, $5 - 2 = 3$. Rule 2.5 applies if n is negative as well as if it is positive. Hence, we can write a^{-n} to mean either "divide by a^{+n}" or to mean $1/(a^n)$. A particular consequence of this is that if we divided a^n by itself then

$$(a^n/a^n) = a^n * a^{-n} = a^{(n-n)} = a^0 = 1.$$

As we stated earlier, any number raised to the power of zero equals 1. Similarly, if we divide $a^{(n+1)}$ by a^n then

$$(a^{(n+1)}/(a^n)) = a^{(n+1)} * a^{-n} = a^{(n+1-n)} = a^1 = a.$$

This results in Rule 2.6, shown in Box 2.3.

The power of a number does not have to be an integer. Thus, for example,

$$a^{0.5} * a^{0.5} = a^{(0.5+0.5)} = a^1 = a$$

or put another way the square root of $a = a^{0.5}$. Thus, the square root of the number 10 can be expressed either as $\sqrt{10}$ or as $10^{0.5}$. Rule 2.5 above applies to all values of m and n, not just integers.

This simple rule is the basis of a system of multiplication and division known as *logarithms*. The logarithm of any number n is the *power* of a fixed number—called the *base* of the system—that gives the same quantitative value as n. Whereas any base can be used, the most common system of logarithms uses the decimal number 10 as the base. Logarithms to the base 10 are known as *common logarithms* and written as "log" or $\log_{10}$. In Table 2.2 it can, for example, be seen that the logarithm of the decimal number 1.60 is 0.2041200 and of 8.00 is 0.9030900. We write this

TABLE 2.2
Example of Seven-Figure Common Logarithms

No.	Log	No.	Log	No.	Log
1.00	0.0000000	4.00	0.6020600	7.00	0.8450980
1.10	0.0413927	4.10	0.6127839	7.10	0.8512583
1.20	0.0791812	4.20	0.6232493	7.20	0.8573325
1.30	0.1139434	4.30	0.6334685	7.30	0.8633229
1.40	0.1461280	4.40	0.6434527	7.40	0.8692317
1.50	0.1760913	4.50	0.6532125	7.50	0.8750613
1.60	0.2041200	4.60	0.6627578	7.60	0.8808136
1.70	0.2304489	4.70	0.6720979	7.70	0.8864907
1.80	0.2552725	4.80	0.6812412	7.80	0.8920946
1.90	0.2787536	4.90	0.6901961	7.90	0.8976271
2.00	0.3010300	5.00	0.6989700	8.00	0.9030900
2.10	0.3222193	5.10	0.7075702	8.10	0.9084850
2.20	0.3424227	5.20	0.7160033	8.20	0.9138139
2.30	0.3617278	5.30	0.7242759	8.30	0.9190781
2.40	0.3802112	5.40	0.7323938	8.40	0.9242793
2.50	0.3979400	5.50	0.7403627	8.50	0.9294189
2.60	0.4149733	5.60	0.7481880	8.60	0.9344985
2.70	0.4313638	5.70	0.7558749	8.70	0.9395193
2.80	0.4471580	5.80	0.7634280	8.80	0.9444827
2.90	0.4623980	5.90	0.7708520	8.90	0.9493900
3.00	0.4771213	6.00	0.7781513	9.00	0.9542425
3.10	0.4913617	6.10	0.7853298	9.10	0.9590414
3.20	0.5051500	6.20	0.7923917	9.20	0.9637878
3.30	0.5185139	6.30	0.7993405	9.30	0.9684829
3.40	0.5314789	6.40	0.8061800	9.40	0.9731279
3.50	0.5440680	6.50	0.8129134	9.50	0.9777236
3.60	0.5563025	6.60	0.8195439	9.60	0.9822712
3.70	0.5682017	6.70	0.8260748	9.70	0.9867717
3.80	0.5797836	6.80	0.8325089	9.80	0.9912261
3.90	0.5910646	6.90	0.8388491	9.90	0.9956352

as $\log_{10}(1.60) = 0.2041200$, which means that $10^{0.2041200} = 1.60$. Likewise, $8.00 = 10^{0.9030900}$. Note also that $\log_{10}(5) = 0.6989700$ which when added to $\log_{10}(1.60)$ gives $0.2041200 + 0.6989700 = 0.9030900$ which is $\log_{10}(8)$, just as $5 * 1.6 = 8$.

For many scientific purposes, a different base is used, namely, the number "e" (also called the *Euler number* after the 18th century Swiss-born mathematician Leonhard Euler) that has a decimal value that is approximately equal to 2.7182818285.

Logarithms to the base "e" are known as *Natural* or *Naperian* logarithms after the 16th century scientist and mathematician John Napier. They are written as "ln" or as $\log_e$. The quantity "e" occurs in a variety of mathematical formulae relating to natural phenomena (hence, the name *natural logarithms*). Its value can be calculated from the formula

"e" = 1 + 1/1! + 1/2! + 1/3! + 1/4! + ... , and so on, where

2! (spoken as *two factorial*) = 1 * 2 = 2,

3! = 1 * 2 * 3 = 6,

4! = 1 * 2 * 3 * 4 = 24, and so forth.

In general, *n* factorial (where *n* is an integer) =

$$n! = 1 * 2 * 3* ...... *(n-2) * (n-1) * n$$

We will consider aspects of natural logarithms in Chapter 6. For now, we will focus on common logarithms or "logs." These use the base number 10.

Numbers between 1 and 10 have a logarithmic value between 0 and 1; between 10 and 100 the logarithmic value will be between 1 and 2, and so forth. The value for other numbers can either be calculated, for instance, using a pocket calculator, or obtained from special tables. These will show, for example, that the value for the number 5 is given as 0.6989700. Thus, the logarithm of $5 = \log_{10}(5) = 0.6989700$. Put another way, $5 = 10^{0.6989700}$.

Since $50 = 10 * 5 = 10^1 * 10^{0.6989700}$, then $50 = 10^{1.6989700}$. Thus, the log of

$$50 = \log_{10}(50) = 1.6989700$$

Similarly, the log of

$$500 = \log_{10}(500) = 2.6989700, \text{ and so on.}$$

As an example of the use of logarithms, consider $50^2 = 2500$. Since $50 = 10^{1.6989700}$ and the logarithm of $50 = 1.6989700$, then

$$50 * 50 = 10^{1.6989700} * 10^{1.6989700} = 10^{1.6989700+1.6989700}$$

$$= 10^{3.3979400} \text{ by the law of indices.}$$

Log 50 + log 50 = 1.6989700 + 1.6989700 = 3.3979400. This is the logarithm of the number 2500 (see the logarithm of 2.5 in Table 2.2). Thus, adding logarithms gives the same answer as multiplying the original numbers. This gives rise to Rule 2.7 in Box 2.4.

As another example, the logarithm of the square root of 50 = half of log 50 = 0.849485. This represents the number $10^{0.849485}$, which is the logarithm of the decimal number 7.071068. Thus, the square root of 50 = 7.071 to three decimal places.

BOX 2.4   RULES FOR LOGARITHMS

RULE 2.7

Adding two logarithms means that we are multiplying together two numbers. Subtracting two logarithms means that we are dividing.

Note: To use logarithms, all numbers must be treated as positive. There is no such thing as the logarithm of a negative number.

RULE 2.8

When using logarithms (common or natural),

(a) The log of $a * b = \log (a * b) = \log a + \log b$
(b) The log of $a/b = \log (a/b) = \log a - \log b$
(c) The log of $a^n = \log (a^n) = n * \log (a)$

Note: $\log(a \pm b)$ is not the same as $\log(a) \pm \log(b)$. Adding (or subtracting) two logarithms means multiplying (or dividing) the two numbers.

As a further example, $\log 100 = 2$ since $100 = 10^2$. Also, $\log 50 = 1.6989700$. To divide two numbers we subtract their logarithms so that:

$$\log (100/50) = \log 100 - \log 50 = (2 - 1.698700) = 0.3010300$$

This represents $10^{0.3010300}$ and is the logarithm of the number "2."

For numbers less than 1 and greater than 0 the power will be less than zero (see Box 2.5). For instance, $0.5 = 5 * 10^{-1}$ while $0.05 = 5 * 10^{-2}$. Hence, $\log (0.5) = -1 + 0.6989700$. This can be expressed in one of three ways: either

$$\log (0.5) = -0.3010300$$

or, we can separate the substantive number from the decimal point and write

$$\log (0.5) = \bar{1}.6989700 \text{ (read as "bar one plus } 0.6989700\text{"}$$

$$\text{where "bar one" or "}\bar{1}\text{" means "minus 1").}$$

Alternatively, we can add ten and call it 9.6989700 knowing that for most calculations in geomatics, we are not dealing with numbers as large as 10^{10}. So

$$\log (0.5) = 9.6989700$$

Thus, $\log (0.05)$ can appear either as -1.3010300 or else bar 2 plus 0.6989700 ($= \bar{2}.698970$) or as 8.698970.

The advantage of the second and third systems is that all logarithms are positive numbers to the right of the decimal point and all multiplications can be done by addition. On the other hand, most pocket calculators use the negative number approach.

BOX 2.5 LOGARITHMS TO THE BASE 10

$10^0 = 1$ $\log (1) = 0$

$10^1 = 10$ $\log (10) = 1$

$10^2 = 100$ $\log (100) = 2$

$10^3 = 1000$ $\log (1000) = 3$

$10^4 = 10000$ $\log (10{,}000) = 4$

$10^5 = 100000$ $\log (100{,}000) = 5$, and so on.

Likewise,

$10^{-1} = 0.1$ $\log (0.1) = -1$

$10^{-2} = 0.01$ $\log (0.01) = -2$

$10^{-3} = 0.001$ $\log (0.001) = -3$

$10^{-4} = 0.0001$ $\log (0.0001) = -4$

$10^{-5} = 0.00001$ $\log (0.00001) = -5$, and so on.

There is no real value for $\log (0)$—it would be minus infinity.

Note that a negative logarithm represents a positive number between 0 and 1. There can be no such thing as the logarithm of a negative number since plus and minus relate to multiplication and division and the location of the resulting decimal point. The basic rules for operating logarithms are given in Box 2.4.

Some calculations become cumbersome when using logarithms. In Example 2.5 we apply logarithms to calculating a distance between two points. The process is tedious, especially when pocket calculators are at hand, because it means we have to keep switching between logarithms and their inverse (known as *antilogarithms*). Their use is, of course, also dependent on access to tables of logarithms, known as *log tables*.

EXAMPLE 2.5: CALCULATING DISTANCES USING LOGARITHMS

Consider $\sqrt{(1243.18^2 + 656.91^2)}$. To square 1243.18, we must multiply it by itself.

Log (1243.18) = 3.0945340 (using seven-figure logarithms)

Hence, the log of $1243.18^2 = 2 * \log (1243.18) = 6.1890680$. This represents the logarithm of the number 1545496. Similarly, log $(656.91^2) = 2 * 2.8175059 = 5.6350118$. This represents the logarithm of the number 431530.8. Using seven-figure logarithms, we have

$$(1243.18^2 + 656.91^2) = (1545496 + 431531) = 1977027$$

Now,

$$\text{Log } (1977027) = 6.2960126$$

To take the square root, we halve the logarithm

$$\text{Half log } (1977027) = 3.1480063$$

To two decimal places, this is the logarithm of 1406.07. Hence,

$$\sqrt{(1243.18^2 + 656.91^2)} = 1406.07.$$

One consequence of Rule 2.8(c) in Box 2.4 is that

$$\log_{10}(e^y) = y * \log_{10}e.$$

Also,

$$\log_e(e^y) = y \log_e(e) \text{ or we can write this as } \ln(e^y) = y \ln(e)$$

But,

$$\log_e(e) = \ln(e) = 1 \text{ just as } \log_{10}(10) = 1$$

Hence,

$$\ln(e^y) = \log_e(e^y) = y \log_e(e) = y$$

As a result,

$$\log_{10}(e^y) = \log_e(e^y) * \log_{10}(e)$$

Replacing e^y by a,

$$\log_{10}(a) = \log_e(a) * \log_{10}(e)$$

EXAMPLE 2.6: CALCULATING THE VALUE OF LOGARITHMS

In Chapter 6, Section 6.3, we derive a formula known as *Taylor's expansion*, which allows us to calculate the value of natural logarithms. It takes the form

$$\ln(1 + z) = \log_e(1 + z) = z - z^2/2 + z^3/3 - z^4/4 + z^5/5 -$$

Using the modulus for common logarithms, we obtain

$$\log_{10}(1 + z) = 0.434294482 * (z - z^2/2 + z^3/3 - z^4/4 + z^5/5 -...)$$

If $z = 0.2$, then considering the first eight terms:

$$\log_{10}(1.2) = 0.434294482 * (0.2 - 0.02 + 0.0026667 -$$

$$0.0004 + 0.000064 - 0.0000107 + 0.0000018 - 0.0000003....) = 0.0791812$$

This confirms the value that we quoted in Table 2.2. Other logarithms can be calculated in a similar manner.

"e" has the value 2.718... and, hence, we can calculate the value of $\log_{10}(e)$. This number is known as the *modulus* of common logarithms.

$$\log_{10}(e) = 0.4342944819$$

Thus,

$$\log_{10}(a) = 0.434294482 * \log_e(a) = 0.434294482 \ln(a)$$

This applies for any positive value of a. Thus, we can convert common logarithms into natural logarithms and vice versa. We show an application for this in Example 2.6.

Logarithms are most effective when there are several multiplications or divisions or when dealing with certain statistical distributions. They also appear in integral calculus, as we will see in Chapter 6.

SUMMARY

Antilogarithm: The inverse of a *logarithm*. If, using four-figure logarithms, log (2) = 0.3010, then the "antilog" of 0.3010 is the number 2.

Binary system: A system of counting using only two numbers, 0 and 1.

Bit: Binary digits 0 and 1.

Byte: A number made up of 8 *bits*, hence in decimal terms a number from 0 to 255.

Common logarithm: The *power* to which the number 10 must be raised to equal a given number. For instance, the common logarithm for the number 100 is 2 since $100 = 10^2$. This is written as $\log_{10}(100) = 2$.

Complex number: A number containing a real and an imaginary part in the form $(x + iy)$ where $i = \sqrt{(-1)}$. The part x is *real*, y is real, but iy is an *imaginary number*.

Decimal places: The number of digits following a decimal point.

Decimal system: A method of counting in quantities of ten as distinct from, for example, *binary*.

Denominator: The lower part of a *fraction*. It is the denomination or category of that fraction, for instance, the number "33" in the fraction 6/33.

Exponent: A number or expression, usually written as a superscript that indicates the *power* to which a number or quantity has been raised, for instance, the number 1000 is ten cubed or 10 * 10 * 10 or 10 to the power of 3 or 10^3, where the 3 is the exponent or *index*.

Factor: Sometimes known as a divisor or submultiple, any quantity that divides exactly into a larger quantity—for instance—1, 2, 3, 4, and 6 are all factors of the number 12.

Factorial: A function that is equal to the first n numbers multiplied together. For example, factorial 5 (written as 5!) = 5 * 4 * 3 * 2 * 1 = 120.

Fraction: The ratio between two integers, for example, 6/33.

Highest common factor: The largest quantity that is a factor of both the *denominator* and *numerator* of a *fraction*, for example, 3 in the fraction 6/33.

Imaginary number: A *complex number* expressed in terms of the square root of a negative number, usually the square root of $-1 = \sqrt{(-1)} = i$.

Index: Used here as the *power* of a number.

Irrational number: A number that cannot be expressed as a ratio of two integers.

Iteration: The process of repeating a mathematical operation, where each step is applied to the answer of the previous operation.

Logarithm: The *power* to which some base number must be raised to equal a given number.

Lowest common denominator: The smallest number into which all the *denominators* can be divided—for instance 6/33 and 4/55 have a lowest common denominator of 165 (which is divisible by both 33 and by 55). Also known as the least common denominator.

Natural logarithms: The *power* to which the number "e" must be raised to equal a given number. "e" is the number 1 + 1/1! + 1/2! + 1/3! + 1/4! + ... where 4! means four *factorial* or 4 * 3 * 2 * 1. "e" is approximately equal to 2.7182818285.

Numerator: The top number in a *fraction* indicating how many of a certain *denominator*, for instance, the number "6" in the fraction 6/33.

Perfect square: An integer that is the square of another integer.

Power of a number: The number of times by which a number is multiplied by itself.

Prime number: An integer that has no factors other than itself and 1.

Pythagoras: The theorem that states that in any right-angled triangle the square on the hypotenuse (the length of the diagonal line) is equal to the sum of the squares on the other two sides.

Rational number: A number that can be expressed as the ratio of two integers.

Real number: A number that can take any value, not just integer, and is not an *imaginary number*.

Rounding up and down: Deciding whether 0.5 should be 1 or 0, for instance, the decimal number 8.1465 can be rounded up to 8.147 or down to 8.146.

Significant figures: The number of figures, starting from the left, after which we can write zeros for all the difference it will make to the accuracy of the answer.

Square root: A number which, when multiplied by itself, gives a given quantity. Usually written with the symbol $\sqrt{}$. Thus, 4 is the square root of 16 or 4 = $\sqrt{(16)}$ = 160.5 since 4 * 4 = 16.

3 Algebra: Treating Numbers as Symbols

3.1 THE THEOREM OF PYTHAGORAS

We can apply the ideas of arithmetic to unknown numbers. Algebra is the body of mathematical knowledge that deals with symbols. It deals with constants (such as the number e = 2.718..... mentioned in Chapter 2) and variables, such as (x, y) that can be the coordinates of a point with x and y taking any value. Each value of x and y is a number that can be added, subtracted, multiplied, or divided, and, hence, algebra may be regarded as generalized arithmetic. Algebra is about treating and manipulating numbers as symbols.

The proof of Pythagoras's theorem given in Box 3.1 uses a number of operations that were discussed in Chapter 2, namely,

1. $a^m * a^n = a^{(m+n)}$
2. $a^m/a^n = a^m * a^{-n} = a^{(m-n)}$
3. $(a^m)^n = a^{(m*n)}$

The proof also introduces the use of brackets so that

$$(a + b) * (c + d) = a * (c + d) + b * (c + d)$$

or

$$(a + b) * (c + d) = (a + b) * c + (a + b) * d$$

This in both cases is equal to

$$(a * c) + (a * d) + (b * c) + (b * d)$$

It also uses indices so that

$$(x + y)^1 * (x + y)^1 = (x + y)^{(1+1)}$$

$$= (x + y)^2 = x^2 + y^2 + 2xy$$

It also makes the point that we can often dispense with the "multiply" and "divide" symbols by writing the above as

$$(a + b)(c + d) = ac + ad + bc + bd$$

BOX 3.1   ONE PROOF OF PYTHAGORAS'S THEOREM

Consider a simple right-angled triangle with sides of length x, y, and z (the hypotenuse) as in Figure 3.1a. Copy the triangle three times and rotate it each time to form Figure 3.1b. This builds a square on the hypotenuse or long side.

The triangle in Figure 3.1a is half a rectangle with sides x and y. Hence, its area is half of x times y or ½ xy. The area of the outer square in Figure 3.1b is $(x + y) * (x + y) = (x + y)^2$. We can write this as $x * (x + y) + y * (x + y)$ or as $x * x + x * y + y * x + y * y$ or as $x^2 + y^2 + 2xy$ using the axioms and rules laid down in Chapter 2. The area of the outer square ABCD therefore $= x^2 + y^2 + 2xy$. It also equals four triangles plus the inner square whose sides are of length z. 4 times "1/2 xy" ($= 2xy$) plus the inner square area of $z * z$ gives a total area $= 2xy + z^2$.

Thus, $2xy + z^2 = x^2 + y^2 + 2xy$ or, subtracting $2xy$ from both sides of the equation,

$$z^2 = x^2 + y^2$$

The square on the hypotenuse is equal to the sum of the squares on the other two sides.

Using indices, we can also write

$$\frac{(a+b)}{(c+d)} = (a + b) * (c + d)^{-1}$$

using the -1 index to mean "divide by" or "one over."

Pythagoras deals with the sum of two squares. An equation, in which the highest power of any term is 2, is known as a *quadratic*. A particular quadratic form that is

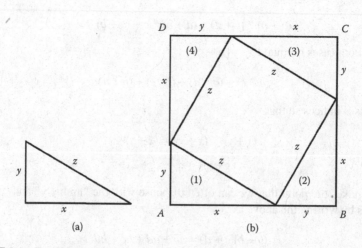

FIGURE 3.1   Rotating a triangle to prove Pythagoras's theorem.

often important is known as the *difference of two squares*. The expression $(x^2 - y^2)$ can be *factorized* (broken into factors) into the form $(x - y)(x + y)$. Multiplying these two factors together yields

$$(x - y)(x + y) = x^2 + xy - xy - y^2 = (x^2 - y^2)$$

We will have cause to use this on a number of occasions.

3.2   THE EQUATIONS FOR INTERSECTING LINES

The rules listed in Chapter 2 give powerful opportunities to solve problems. Consider a rectangular coordinate system with two lines A to B and C to D that intersect at P (as in Figure 3.2).

The *origin* of the coordinate system is at point O and the coordinates of

$$A = (x_A, y_A); B = (x_B, y_B); C = (x_C, y_C); \quad \text{and} \quad D = (x_D, y_D)$$

Let the line AB cut the x-axis at point Q $(d, 0)$. d is said to be the *intercept* on the x-axis. (Note that superscripts are used as indices, subscripts such as the $_A$ in x_A are used as identifiers.)

Consider first a straight line through the origin (Figure 3.3a). If the coordinates of any point on this line are (x, y), then the ratio between y and x (or y/x) will always be the same. This is the slope of the line—let us call it m. We then have $y = mx$ for all points on the line, with m being a constant number and x and y being variables.

If the line slopes backward (so that as y increases, so x decreases, and vice versa) then m will be a negative amount (Figure 3.3b).

Lines do not necessarily have to pass through the origin but they can always be made to do so by changing the position of the origin. Thus, in Figure 3.2, if the origin were at point Q where the line AB crosses the x-axis (which is the line where $y = 0$) then AB will pass through this *false origin*. To achieve this, all x values must be reduced by the amount $OQ = d$.

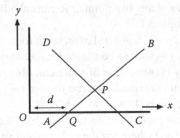

FIGURE 3.2   Intersecting lines.

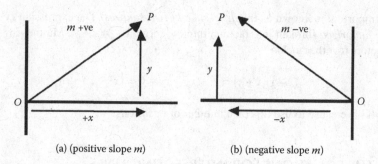

(a) (positive slope m)   (b) (negative slope m)

FIGURE 3.3   The slope of a line.

Thus, we can express the equation of the line AB using the original coordinates in the form $y = m(x-d)$ or more generally $y = mx + c$ where $c = -m * d$. m and c are constants for the line. When we know m and c, we can calculate the value of y given x or vice versa. The line $y = mx + c$ crosses the vertical y-axis where $x = 0$ and $y = c$. It crosses the x-axis where $y = 0$. At this point $x = -c/m$.

We can use the coordinate values to calculate m and c. Since the line AB passes through (x_A, y_A) and (x_B, y_B), these values must satisfy the conditions that

$$y_A = m_1 x_A + c_1$$

$$y_B = m_1 x_B + c_1$$

We can subtract the lower numbers from the upper so that

$$y_A - y_B = (m_1 x_A + c_1) - (m_1 x_B + c_1)$$

$$= m_1 x_A + c_1 - m_1 x_B - c_1 = m_1 x_A - m_1 x_B$$

$$= m_1(x_A - x_B)$$

Hence,

$$m_1 = (y_A - y_B)/(x_A - x_B) \quad \text{and} \quad c_1 = y_A - m_1 x_A = y_B - m_1 x_B$$

Thus, given the coordinates of any two points, we can calculate the parameters of the line that joins them (Example 3.1).

The relationship $y = mx + c$ is a general expression for a straight line in rectangular Cartesian coordinates. y is said to be the *dependent* variable and x the *independent*, meaning that for any chosen value of x, we can obtain a unique value of y. In fact, we could have written the equation in the form $x = ny + d$ (where $n = 1/m$ and $d = -c/m$). x would then be dependent on y.

In Chapter 9, we discuss parametric forms, which in the case of a straight line would be $x = at + c$, $y = bt + d$ where t is the independent variable or parameter and both x and y are dependent variables. a, b, c, and d are constants.

EXAMPLE 3.1: THE LINE JOINING TWO POINTS

Consider two points AB with coordinates

$$A\ (1234.56, 2345.67) \quad \text{and} \quad B\ (1296.32, 2417.38)$$

Then,

$$m_1 = (y_A - y_B)/(x_A - x_B)$$

$$= (2345.67 - 2417.38)/(1234.56 - 1296.32)$$

$$= (-71.71)/(-61.76) = +1.161108$$

$$c_1 = y_A - m_1\,x_A = 2345.67 - 1.161108 * 1234.56$$

or

$$c_1 = y_B - m_1\,x_B = 2417.38 - 1.161108 * 1296.32$$

In both cases, this gives $c_1 = 912.21$ (giving an independent check). Hence, the line joining A to B is $y = 1.161108x + 912.21$.

When $t = (-d/b)$ then $y = 0$ and we have the intercept of the line on the x-axis given by $x = c - ad/b$. When $t = (-c/a)$ then we have the intercept of the line on the y-axis, namely, $y = d - bc/d$. The slope of the line will be a/b.

We can now calculate the point of intersection of two straight lines, as shown in Example 3.2. Returning to Figure 3.2, we can describe the two lines,

$$\text{the first } (AB) \text{ as } y = m_1 x + c_1$$

$$\text{the second } (CD) \text{ as } y = m_2 x + c_2$$

The lines intersect where $(m_1 x + c_1) = (m_2 x + c_2)$ or

$$x = (c_2 - c_1)/(m_1 - m_2)$$

Here,

$$y = m_1(c_2 - c_1)/(m_1 - m_2) + c_1 = m_2(c_2 - c_1)/(m_1 - m_2) + c_2$$

$$= (m_1 c_2 - m_2 c_1)/(m_1 - m_2)$$

In Figure 3.4, let the line AB be $y = mx + c$. Consider the line $y = mx + c + 1$. For this line, for any value of x, the value of y is precisely 1 unit above the value in the line AB. This means that this new line is parallel to AB but 1 unit above it. More generally, every line $y = mx + d$, where d has any fixed value, will be

EXAMPLE 3.2: THE POINT OF INTERSECTION OF TWO LINES

Following on from the example in Example 3.1, for two points CD with

$$C\ (1300.24,\ 2351.77) \quad \text{and} \quad D\ (1212.45,\ 2431.78)$$

we have

$$m_2 = (-80.01)/(87.79) = -0.9113794 \quad \text{and} \quad c_2 = 3536.78$$

Or for the line CD

$$y = -0.9113794\ x + 3536.78$$

The point of intersection (P in Figure 3.2) between the lines AB and CD must satisfy both these conditions. Then,

$$\text{For } AB: y = 1.161108\ x + 912.21$$

$$\text{For } CD: y = -0.9113794\ x + 3536.78$$

Hence, at point P:

$$1.161108 * x + 912.21 = -0.9113794 * x + 3536.78$$

Adding $0.9113794 * x$ to both sides of the equation and taking 912.21 away from both sides gives:

$$1.161108 * x + 0.9113794 * x + 912.21 - 912.21$$

$$= -0.9113794 * x + 0.9113794 * x + 3536.78 - 912.21$$

which results in

$$2.072487 * x = 2624.57$$

Dividing both sides by 2.072487

$$x = 1266.39$$

Hence,

$$y = 1.161108 * 1266.39 + 912.21$$

or

$$y = 2382.62$$

The coordinates of P are therefore (1266.39, 2382.62). Thus, using the principles of arithmetic given at the start of Chapter 2, we have found the coordinates of the point of intersection of two lines.

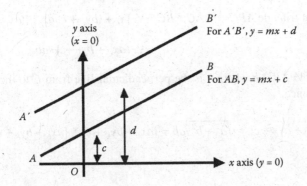

FIGURE 3.4   Parallel lines.

parallel to *AB*. *m* is a measure of the slope or angle and has the same value for all parallel lines.

Furthermore, in Figure 3.5 let the line through *AB* have the equation

$$ax + by + c = 0$$

Consider any point *C* that does not lie on that line and let *C* be (x_C, y_C). At point *A*, the *y* coordinate is y_C, hence, the *x* coordinate is $x_A = -(by_C + c)/a$.

At point *B*, the *x* coordinate is x_C, hence, the *y* coordinate is $y_B = -(ax_C + c)/b$.

By Pythagoras,

$$AB^2 = BC^2 + AC^2$$
$$= \{-(ax_C + c)/b - y_C\}^2 + \{x_C + (by_C + c)/a\}^2$$
$$= \{ax_C + by_C + c\}^2 * \{1/a^2 + 1/b^2\}$$

Now $\{1/a^2 + 1/b^2\} = \{a^2 + b^2\}/\{a^2b^2\}$. Hence,

$$AB = \{ax_C + by_C + c\} * \{\sqrt{a^2 + b^2} /ab\}$$

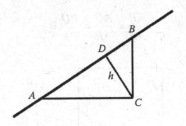

FIGURE 3.5   Distance offset from a line.

The area of triangle $ABC = \tfrac{1}{2} AC * BC = \tfrac{1}{2} \{x_C + (by_C + c)/a\} * \{(ax_C + c)/b + y_B\}$

$$= \tfrac{1}{2} \{ax_C + by_C + c\}^2/ab$$

It also equals $\tfrac{1}{2} AB * h$ where h is the perpendicular line from C to the line AB in Figure 3.5. Hence,

$$h * \{ax_C + by_C + c\} * \sqrt{a^2 + b^2}/ab = \{ax_C + by_C + c\} * \{ax_C + by_B + c\}/ab$$

or

$$h = (ax_C + by_C + c)/\sqrt{a^2 + b^2}$$

This gives a simple formula for calculating how far a known point is away from a known line. The value is, of course, zero if point C is actually on the line AB.

A further example of an application for the equation $y = mx + c$ arises when drawing lines on a map that cross the map sheet edge (at P in Figure 3.6) or any square window drawn around an area—a process known as *clipping*. The lines that cross the boundary of the area being displayed must be clipped so that all sections of the original data that lie outside the area under consideration can be discarded.

One way to compute the sections of a line to be included is given in Example 3.3. There is, however, an alternative approach to the truncation of the line AB that does not require the calculation of the equation $y = mx + c$. It is based entirely on the idea of *scale*. Consider the two triangles APQ and ABR in Figure 3.7.

The two triangles APQ and ABR in Figure 3.7 are exactly the same shape. They only differ in size and are said to be similar. *Similar triangles* do not necessarily have one angle a right angle; the key point is that they have corresponding angles of the same size. In fact, they just differ in terms of scale. This means that all linear distances are in the same proportion or ratio. If the *scale factor* between them is some number s then:

$$AP = s * AB; \; AQ = s * AR \quad \text{and} \quad PQ = s * BR.$$

Alternatively,

$$\frac{AP}{AB} = \frac{AQ}{AR} = \frac{PQ}{BR} = s$$

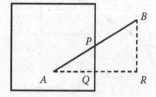

FIGURE 3.6   Clipping to a window.

EXAMPLE 3.3: INTERSECTION AT A MAP SHEET EDGE (1)

Referring to Figure 3.6 and using the coordinates of A and B as before with A (1234.56, 2345.67) and B (1296.32, 2417.38). In Example 3.2 we showed that

$$y = 1.161108 * x + 912.21$$

Now, if the edge of the map sheet is on a grid line QP whose x value is $x_P = 1250$, then at that point P

$$y_P = 1.161108 * 1250 + 912.21$$

$$= 2363.60$$

A computer-driven graphical plotter could be made to move from A (1234.56, 2345.67) to P (1250.00, 2363.60). It would then stop plotting at the map sheet edge, clipping the line AB at P.

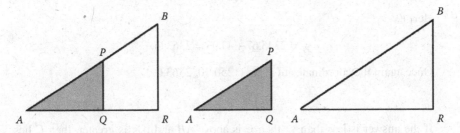

FIGURE 3.7 Similar triangles.

This will hold for similar triangles even when the triangle does not contain a right angle. In the case of Figure 3.6, we merely have to scale down BR to PQ to get the y value of P, as shown in Example 3.4.

3.3 POINTS IN POLYGONS

If we express the equation of the line through the two points A (x_A, y_A) and B (x_B, y_B) in the form

$$(y - y_A) = \frac{(y_B - y_A)}{(x_B - x_A)} * (x - x_A)$$

then it is easy to test whether any point is above or below the line AB. To do so, take any point C (x_C, y_C) and calculate

$$y = y_A + \frac{(y_B - y_A)}{(x_B - x_A)} * (x_C - x_A)$$

EXAMPLE 3.4: INTERSECTION AT A MAP SHEET EDGE (2)

Using the same numbers as in Example 3.3, then if R has the same x value as B and y value as A, then the distance

$$AR = x_B - x_A = 1296.32 - 1234.56 = 61.76$$

$$BR = y_B - y_A = 2417.38 - 2345.67 = 71.71$$

$$AQ = x_P - x_A = 1250.00 - 1234.56 = 15.44$$

Hence,

$$AQ/AR = 15.44/61.76 = 0.25 = s$$

$$QP = s * BR = 0.25 * 71.71 = 17.93$$

or

$$PQ = y_P - y_A = y_P - 2345.67 = 17.93$$

Hence,

$$y_P = 2345.67 + 17.93 = 2363.60$$

Once again the coordinates of P are (1250.00, 2363.60).

If the answer is less than y_C then C is above AB and if it is greater, then C lies below the line from A to B. Simple tests like this are useful when determining if two points are on the same or opposite sides of a line. Such tests arise in hidden line and surface removal, as discussed in Chapter 10.

This also provides one way to test whether a point P (x_P, y_P) lies within or outside a polygon. Consider the polygon $ABCDEF$ in Figure 3.8 where the coordinates of A are (x_A, y_A), of B are (x_B, y_B), and so forth. First, one should check whether x_P lies between the maximum and minimum values of x for the whole polygon, which in

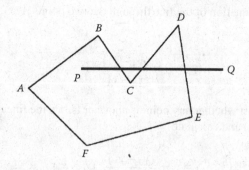

FIGURE 3.8   Point-in-polygon.

this case means that $x_A \leq x_P \leq x_E$ and similarly $y_F \leq y_P \leq y_D$ in order to reduce the amount of computation. (The notation $a \leq b$ means that a is less than or equal to b.) Testing whether something is greater or less than a specified value is a relatively trivial computation.

Next consider the section of the horizontal line through P that has x values greater than x_P. Let this be PQ where Q is such that $x_Q > x_E$ and $y_Q = y_P$. The number of occasions that this line crosses a side of the polygon must be counted.

In Figure 3.8 a series of checks must be carried out on each of the lines AB, BC, CD, and so forth, that form the polygon. In the case of AB, y_P (the y value for the horizontal line from P to Q) is greater than y_A and less than y_B so the line PQ will, if extended, intersect AB. From the coordinates of A and B we can work out the point on the line AB where the y value $= y_P$; the x value will be seen to be less than x_P and it can be ignored.

For the lines BC, CD, and DE their intersection with PQ has an x value greater than x_P and therefore they will count. Each of the lines EF and FA are wholly below P and can be ignored.

At the end, if the number of lines that are crossed to the right of P is an even number, then P will be outside the polygon; if the number is odd (as in this case), point P will be inside the polygon. Care must be taken if the line PQ goes exactly through one of the corner points as this can upset the count; this can, however, be avoided by a minor adjustment in the coordinates of P; for example, if the vertices are given to two decimal points (0.01) then by treating P to three decimal points (rounding it to 0.005), the problem will not arise.

3.4 THE EQUATION FOR A PLANE

We can extend the equation of a line to describe a plane. Consider the equation

$$z = mx + ny + c$$

For every fixed value of y (for example, $y = d$),

$$z = mx + (nd + c) = mx + c'$$

where $c' = nd + c$ and is a constant. This is the equation for a straight line in the plane $y = d$. Similarly, for every fixed value of x we have a set of straight lines. If z is regarded as the axis in the third dimension then $z = mx + ny + c$ must represent a *plane* surface (Figure 3.9).

3.5 FURTHER ALGEBRAIC EQUATIONS

Algebraic operations often entail rearranging equations, as illustrated in Box 3.2. Provided that we carry out the same operation on both sides of the equation by adding, subtracting, multiplying, or dividing by the same amount, we will not alter the basic relationship. This is a very important principle that helps solve many

---

BOX 3.2   THE INTERSECTION OF TWO PLANES

When two planes intersect we have equations of the form

$$z = m_1 x + n_1 y + c_1 \quad \text{and} \quad z = m_2 x + n_2 y + c_2$$

Subtracting one equation from the other:

$$z - z = m_1 x + n_1 y + c_1 - (m_2 x + n_2 y + c_2)$$

or

$$0 = m_1 x + n_1 y + c_1 - m_2 x - n_2 y - c_2$$

Adding $n_2 y$ to both sides of the equation:

$$n_2 y = m_1 x + n_1 y + c_1 - m_2 x - n_2 y - c_2 + n_2 y$$

Subtracting $n_1 y$ from each side of the equation:

$$n_2 y - n_1 y = m_1 x + n_1 y + c_1 - m_2 x - n_2 y - c_2 + n_2 y - n_1 y$$
$$= m_1 x + c_1 - m_2 x - c_2$$

or

$$(n_2 - n_1)y = (m_1 - m_2)x + (c_1 - c_2)$$

Dividing both sides by $(n_2 - n_1)$

$$y = \left\{ \frac{m_1 - m_2}{n_2 - n_1} \right\} x + \left\{ \frac{c_1 - c_2}{n_2 - n_1} \right\}$$

Replacing $((m_1 - m_2)/(n_2 - n_1))$ by m and $+((c_1 - c_2)/(n_2 - n_1))$ by c. The result is

$$y = mx + c$$

This has already been shown to be the equation of a straight line. Thus, two planes intersect in a straight line.

---

mathematical equations. The only thing that we cannot do is divide both sides by zero for although $2 * 0 = 1 * 0$, it does not follow that $2 = 1$.

All of the above relationships are examples of equations. An *equation* is a formula that asserts that two expressions have the same value. Thus, $y = mx + c$ is an *identical equation* or *identity* that is true for all values of the variables that lie along the line

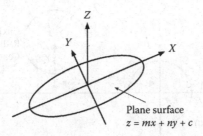

FIGURE 3.9   A plane surface.

that is determined by the constants m and c. A *conditional equation* is one that is only true for certain values of the variables. For instance, $x^2 - x = 2$ is a conditional equation that is only true when $x = 2$ or $x = -1$.

Two particular forms of equations commonly occur in geomatics—simultaneous equations and quadratic equations. A *simultaneous equation* is one of at least a pair of equations that must be satisfied simultaneously. An example was given in Example 3.2 where there were two straight lines that intersected. At the point of intersection P, the values of x and y had to satisfy both the equation for the line AB and the equation for the line CD simultaneously. If there are two unknowns (here x and y), then there must be at least two independent equations in order that their points of intersection can be found and the equations solved. For three unknowns (x, y, z), then there must be at least three independent equations each of which must be satisfied. (In Chapter 13, we consider what happens when there are more than the minimum number of equations.)

A *quadratic* is a function in which the variables may be "raised to the power of two" or be of "second degree"—for example—x^2, y^2, or xy (but not xy^2 which is of the third degree). $x^2 + y^2 = r^2$ is an example of a quadratic expression, one interpretation of which is that r is the radius of a circle with center $(0, 0)$ and (x, y) is any point on the circle.

The relationship $x^2 + y^2 = r^2$ only exists in real terms when both x and y are equal to or greater than $-r$ and less than or equal to $+r$. This may be written in the form

$$-r \le x \le + r$$

and

$$-r \le y \le + r$$

or

$$|x| \le r$$

and

$$|y| \le r$$

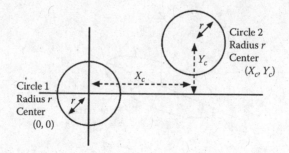

FIGURE 3.10 The equation of a circle.

where |x| and |y| mean the positive values of x and y. |x| and |y| are called the *modulus* of x and *modulus* of y or "mod x" and "mod y" or the "absolute values" of x and y.

For a circle of radius r but center at point X_c, Y_c (Circle 2 in Figure 3.10), then the general equation would take the form

$$(x{-}X_c)^2 + (y - Y_c)^2 = r^2$$

or

$$x^2 - 2X_cx + X_c^2 + y^2 - 2Y_cy + Y_c^2 = r^2$$

or

$$x^2 + y^2 + ax + by + c = 0$$

where

a, b, and c are constants with $a = -2X_c$; $b = -2Y_c$; and $c = X_c^2 + Y_c^2 - r^2$.

The term *quadratic* is commonly associated with *quadratic equation*. A quadratic equation takes the form

$$ax^2 + bx + c = 0$$

or

$$ax^2 + bx^1 + cx^0 = 0$$

In this form, the power of x takes the integer values 0, 1, and 2. Because it is of second degree, there are two possible solutions. Mathematics solves such equations in a variety of ways. For example, the equations can be rearranged by dividing every term by a (assuming a is not zero). Hence,

$$x^2 + (b/a)\,x + (c/a) = 0$$

or

$$x^2 + (b/a)\,x = -(c/a)$$

Next, we add a number n so that $x^2 + (b/a)\, x + n$ becomes a *perfect square*, that is, it is some value raised to the power of 2. Let this value be $(x + p)$ so that we have to find p such that $(x + p)^2 = x^2 + (b/a)\, x + n$.

We do not of course know n at this stage, but we do know that $(x + p)^2 = x^2 + 2px + p^2$. Hence, we need to choose p so that $2p = (b/a)$ and $n = p^2$. This means that $p = (b/2a)$ and hence, $n = p^2 = b^2/4a^2$. This gives a perfect square

$$(x + b/2a)^2 \quad \text{or} \quad x^2 + (b/a)x + b^2/4a^2$$

If we write our original equation $ax^2 + bx + c = 0$ as $x^2 + (b/a)\, x = -(c/a)$, then add $b^2/4a^2$ to both sides, we obtain:

$$x^2 + (b/a)\, x + b^2/4a^2 = b^2/4a^2 - (c/a)$$

or

$$(x + b/2a)^2 = b^2/4a^2 - (c/a) = b^2/4a^2 - (4ac/4a^2) = (b^2 - 4ac)/4a^2$$

Taking the square roots $(x + b/2a) = \pm\, \{\surd(b^2 - 4ac)\}/2a$

Hence,

$$x = \{-b \pm \surd(b^2 - 4ac)\}/2a$$

("x" equals minus "b" plus or minus the square root of "b" squared minus four "a" "c" all over two "a".)

Example 3.5 illustrates how this works in practice. It is known as "solving a quadratic." Note that if $b^2 < 4ac$ then $(b^2 - 4ac)$ is negative and since when using real numbers there is no such thing as the square root of a negative number, there is no solution to the problem since there is no real value of x that satisfies $ax^2 + bx + c = 0$.

When solving simultaneous equations a similar approach is used in which we alter both sides of an equation (Example 3.6). Consider two lines with equations:

$$ax + by + c = 0 \quad \text{and} \quad dx + ey + f = 0$$

Multiply both sides of the first equation by d and both sides of the second equation by a. This gives:

$$adx + bdy + cd = 0$$

and

$$adx + aey + af = 0$$

Subtracting one from the other:

$$adx + bdy - adx - aey + cd - af = 0$$

EXAMPLE 3.5: SOLVING A QUADRATIC

As shown in the main text if

$$ax^2 + bx + c = 0$$

Then,

$$x = \{-b \pm \sqrt{(b^2 - 4ac)}\}/2a$$

Consider $3x^2 + 4x + 1 = 0$. Here,

$$a = 3, b = 4, c = 1$$

Hence,

$$x = \{-4 \pm \sqrt{(16 - 12)}\}/6$$

So,

$$x = (-4 + 2)/6 = -1/3$$

or

$$x = (-4 - 2)/6 = -1$$

The two solutions to $3x^2 + 4x + 1 = 0$ are $x = -1/3$ or $x = -1$.

EXAMPLE 3.6: INTERSECTING LINES
USING SIMULTANEOUS EQUATIONS

As shown in the main text, if there are two equations

$$ax + by + c = 0 \quad \text{and} \quad dx + ey + f = 0$$

Then,

$$\frac{x}{(bf - ce)} = \frac{-y}{(af - cd)} = \frac{1}{(ae - bd)}$$

Consider the two lines $3x + 4y - 10 = 0$ and $5x - 2y - 8 = 0$. These will intersect where

$$\frac{x}{-52} = \frac{-y}{26} = \frac{1}{-26}. \text{ Hence, } x = 2 \quad \text{and} \quad y = 1$$

The values $x = 2$ and $y = 1$ satisfy both equations simultaneously. Hence, they must represent the point where the two lines intersect.

or

$$y(bd - ae) = (af - cd)$$

or

$$y = -(af - cd)/(ae - bd)$$

Similarly,

$$x = (bf - ce)/(ae - bd)$$

or

$$\frac{x}{(bf - ce)} = \frac{-y}{(af - cd)} = \frac{1}{(ae - bd)}$$

(In Chapter 7, we will rediscover these relationships in the context of determinants.)

3.6 FUNCTIONS AND GRAPHS

A *function* is a relationship in which, given the value of one or more variables, the overall value of the function can be calculated. For example, $y = f(x)$ can be read as "y is a function of x" with y as the value of the function and x the *argument*. The function can take any mathematical form such as

$$y = ax + b \text{ (which is a \textit{line})}$$

$$y = ax^2 + bx + c \text{ (a \textit{quadratic} or second-degree curve)}$$

$$y = ax^3 + bx^2 + cx + d \text{ (a \textit{cubic} or third-degree curve), and so on.}$$

A particular group of functions are called *polynomials*. A polynomial is essentially an expression that contains two or more terms such as

$$x^3 + 3x^2y + 3xy^2 + y^3 + x^2 + 3y^2 + xy + x + y + 4$$

More specifically, the term is used to describe a relationship such as:

$$y = a + bx + cx^2 + dx^3 + ex^4 + fx^5 + \ldots\ldots\ldots$$

where $a, b, c, d, e, f,$ and so on, have fixed values.

There may be a finite number of terms, for example, where

$$y = (1 + x)^4 = 1 + 4x + 6x^2 + 4x^3 + x^4$$

Alternatively, there may be an infinite number of terms in which case it will either generate an infinite number or else converge onto a specific value. The term *convergence* means that however many terms are added, the function will simply grow closer and closer to a specific value. Thus,

$$e = 1 + 1/1! + 1/2! + 1/3! + \dots$$

has an infinite number of terms but it never gets bigger than a certain value while the series of terms $1 + 2 + 3 + \dots$ just gets bigger and bigger and is not convergent.

A *series* is a sequence of terms that may be finite or infinite. An example of a series is

$$S = a_0 + a_1 + a_2 + a_3 + \dots + a_{(n-1)} + a_n$$

where n is a positive integer. It is often written as

$$\Sigma a_i \quad \text{or} \quad \sum_{i=0}^{i=n} a_i$$

where the symbol Σ (the Greek letter "sigma") means "the sum of." In practice, an infinite series will only have a finite sum if the series (a_0, $a_0 + a_1$, $a_0 + a_1 + a_2$, $a_0 + a_1 + a_2 + a_3$,.... etc.) converges.

Functions may involve one dependent and one independent variable or may have several independent variables such as: $z = f(x, y)$, meaning that z is a function of x and y where x and y are both independent variables. If z is dependent on x and y so that we have only one value of z for given values of x and y, then we have a surface. If z is linearly dependent on x and y (i.e., $z = f(x,y) = ax^1 + by^1 + c$) then, as we have seen above, this is a plane surface. If $z = ax^2 + by^2 + cx + dy + e$, then we have a second degree or quadratic surface. If $y = f(x)$ so that y is a function of x, then there is a unique value for y for every value of x. If the relationship between y and x is such that it is also possible to determine x uniquely given the value of y, then the relationship is said to have an *inverse function*, written as f^{-1}. Thus, if $y = f(x) = ax + b$, then $x = f^{-1}(y) = (y-b)/a$. This is often not possible; for instance, if $y = x^2$ then $x = +\sqrt{y}$ or $x = -\sqrt{y}$ and there are two possible relationships. Hence, there is no inverse function. In general, functions are mathematical relationships between two (or more) variables that may be in the form of one-to-one or one-to-many or many-to-many.

We have seen that the equation for a circle centered at the origin is $x^2 + y^2 = r^2$ where r = radius. For any value of x that represents a point on the circle there are two possible values of y (one positive and one negative); similarly, for any value of y there are two values of x and thus there is a one-to-two relationship between x and y and between y and x.

Relationships between two variables are often better expressed graphically than numerically since many people find a visual image easier to understand than a mathematical equation. A *graph* is a drawing showing the relationship between certain sets of quantities by means of points or lines plotted with respect to a set of coordinate axes.

TABLE 3.1

Values of y and x for $y = 0.5x^2$

x	0	±0.2	±0.4	±0.6	±0.8	±1	±1.2	±1.4	±1.6	±1.8	±2
y	0	0.02	0.08	0.18	0.32	0.5	0.72	.98	1.28	1.62	2

As an example of a graph, consider the function $y = 0.5\,x^2$. The values of points on the graph can be calculated if we choose a series of values for x and then compute y (see Table 3.1). Every pair of (x, y) coordinates can then be plotted. Intermediate points can be calculated as necessary.

When plotted, the coordinates form the curve as shown in Figure 3.11. In general, when drawing a graph, a series of points are plotted and then are joined by straight lines. A circle may be drawn on a piece of paper using a compass, resulting in a nice smooth shape, but in most cases this is not possible and as described in Chapter 1, curves are either plotted in the form of dots (the raster method) or as a series of short vectors with the precise location of points on the curve being calculated from the algebraic formulae. In practical terms, not all points can be plotted as they may extend far beyond the limits of any piece of paper.

In Figure 3.11, if $x = 100$ then $y = 5000$ and as a result, the axes would have to be scaled down to accommodate such a range.

Figure 3.12 shows the curve for $y = 1/x$. When $x = 1$ then $y = 1$; when $x = 0.5$ then $y = 2$; when $x = 0.1$ then $y = 10$; when $x = 0.001$ then $y = 1000$, and so on. As x gets smaller and smaller and approaches zero from the positive side, y goes off to plus infinity $(+\infty)$. As x approaches zero from the negative side, y goes off to minus infinity $(-\infty)$.

The value of y when $x = 0$ can only be expressed mathematically as $y = \infty$. It cannot be shown on any graph. It also has the strange property that it flips from $+\infty$ to $-\infty$ or vice versa as the value of x passes through zero.

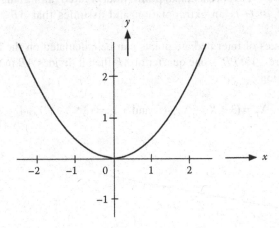

FIGURE 3.11   Graph of the function $y = 0.5x^2$.

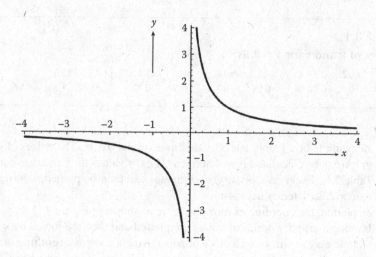

FIGURE 3.12   Graph of $y = 1/x$.

3.7   INTERPOLATING INTERMEDIATE VALUES

Both the graphical and numerical approaches can be used to determine values
in between or beyond those that have been measured. The process of determining intermediate values is known as *interpolation*. Taking a guess at what happens beyond what is already known is called *extrapolation*; for instance, we
know approximately the number of people who were in the world at the end of
each decade over the last century and by plotting the numbers on a graph, it is
possible to predict how many people will be in the world in AD 2050 or even AD
2100 assuming that the present trends continue. For now, we will only consider
simple linear interpolation and delay until later any discussion on more complex
forms.

In Figure 3.13, P represents a point interpolated along the straight line
between A and B; Q is an extrapolation and assumes that AB continues in a
straight line.

The coordinates of intermediate points can be calculated on the basis of ratios.
Thus, if in Figure 3.13, PB is one quarter of AB then it divides AB in the ratio 3 to 1
so, by interpolation:

$$X_P = (3 * X_B + X_A)/4 \quad \text{and} \quad Y_P = (3 * Y_B + Y_A)/4$$

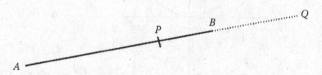

FIGURE 3.13   Linear interpolation and extrapolation.

**BOX 3.3   INTERPOLATION—THE COORDINATES
OF THE CENTROID OF A TRIANGLE**

In Figure 3.14, let D be the midpoint of BC and let G divide AD in the ratio 2:1 (so that $GD = 1/3$rd of AD). Let the coordinates be $A\,(x_A, y_A)$, $B\,(x_B, y_B)$, and C (x_C, y_C). By proportion the coordinates of D are

$$(x_D, y_D) = \left\{ \frac{x_B + x_C}{2}, \frac{y_B + y_C}{2} \right\}$$

The coordinates of $G = \left\{ \dfrac{2x_D + x_A}{3}, \dfrac{2y_D + y_A}{3} \right\}$

$$= \left\{ \frac{x_A + x_B + x_C}{3}, \frac{y_A + y_B + y_C}{3} \right\}$$

Note: Since this is symmetrical it would be the same for BE where E is the midpoint of AC and CF where F is the midpoint of AB. The lines joining the vertices of a triangle with the midpoints of the opposite sides pass through one point G known as the centroid. If the triangle were made of uniform density, then the centroid would be its center of mass or gravity.

If Q is the point on the line AB such that $BQ = (1/2)\,BA$, then B divides AQ in the ratio 2 to 1 so $B = (2Q + A)/3$ or $Q = (3B - A)/2$. Thus,

$$X_Q = (3 * X_B - X_A)/2 \quad \text{and} \quad Y_Q = (3 * Y_B - Y_A)/2$$

We can use these types of relationships to show that in any triangle ABC the lines joining each apex to the midpoint of the opposite side all pass through one point (see Box 3.3 and Figure 3.14).

The process of reducing graphs to straight-line sections helps in interpolation, making the calculation of intermediate values relatively simple. An illustration of such a calculation was given previously in Example 3.3, which showed the cutting off of a line

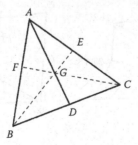

FIGURE 3.14   The midpoints of the sides of a triangle.

FIGURE 3.15   Interpolation of heights down a slope.

at a map sheet edge or window boundary. Given two points, each one being on opposite sides of the required value, the coordinates of the intermediate point can be determined. On the other hand, if as in Figure 3.15, A is the top of a slope and B the bottom, linear interpolation will overestimate lower heights and underestimate the higher ones.

Intermediate values are often interpolated on the assumption that change is linear. In particular, this can happen when constructing lines of equal value such as contour lines. Neighboring points of known value (such as spot heights) are joined and the positions where a contour line crosses the lines of the resulting triangles are then interpolated. These positions in turn are joined to form the contour. The procedure can be applied to any third-dimensional value such as air pressure or temperature at a point; but here we will consider heights.

In Figure 3.16a, the five spot heights have been joined by straight lines to form triangles. In Figure 3.16b, the values equivalent to 25 meters have been linearly interpolated along each line. Thus, between the 27.5 m and 21.2 m spot heights the level drops by 6.3 m. To drop to 25 m from 27.5 means to drop 2.5 m out of the total of

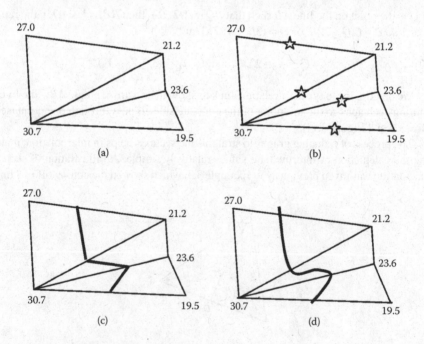

FIGURE 3.16   Interpolating contours between spot heights.

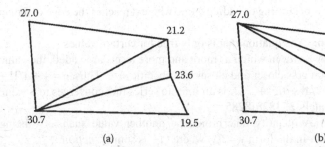

FIGURE 3.17   Alternative triangulation networks.

6.3 m. The contour crosses the line at a distance of (2.5/6.3) of its length. If the end points are (x_A, y_A) and (x_B, y_B), then the coordinates of the interpolated point will be

$$\left\{ \frac{3.8x_A + 2.5x_B}{6.3}, \frac{3.8y_A + 2.5y_B}{6.3} \right\}$$

The interpolated points can then be joined either (Figure 3.16c) by straight lines or (Figure 3.16d) by a curve drawn freehand or by using a computer algorithm as discussed in Chapter 9.

The assumption behind the interpolation is that the lines chosen are of uniform slope. In fact, other triangles could have been formed as in Figure 3.17b. Each will give a different contour line. Which is right?

The answer is that neither will give a perfect answer, which is why contours are generally regarded as only being reliable within a band representing about a third of the contour interval (for instance, within just over three meters for a 10-meter contour interval). The main problem is with the assumption that height changes evenly between spot heights. In some cases the surveyor may have made sure that points are chosen to fulfill this criterion but, in general, slopes are not uniform and a linear interpolation is just a convenient guess.

In Chapter 4, we will describe a technique that uses what is known as the *Theissen* or *Thiessen polygon* to select good-shaped triangles. For now, it should just be noted that the accuracy of interpolation techniques depends on (i) the accuracy of the original data; (ii) assumptions behind the interpolation; and (iii) the guesswork that enters into smoothing the contour line into an acceptable curve. We will explore ways in which a smooth curve can be fitted through a series of points in Chapter 9.

SUMMARY

Argument: An element to which a mathematical operation or function is to be applied.

Centroid: The point in a triangle where the lines joining each point of the triangle to the midpoint of the opposite side intersect (such lines being known as medians). For the triangle *ABC*, its coordinates are

$$\{(x_A + x_B + x_C)/3, (y_A + y_B + y_C)/3\}$$

Clipping: The process of starting or ending a line where it reaches the edge of a map square.

Conditional equation: An equation that is only true for certain values.

Convergent series: A *series* in which as more and more terms are added, the value of the function gets closer and closer to a specific size. Thus, $e = 1 + 1/1! + 1/2! + 1/3! + 1/4! + 1/5! + \ldots\ldots$ is an infinite series that converges to a value of approximately 2.7182818285.

Dependent variable: A quantity determined by another value such as y being dependent on x in the form $y = f(x)$ where "f" is some *function*.

Difference of two squares: An expression such as $(a^2 - b^2)$ that can be factorized into the form $(a + b)(a - b)$.

Equation: A formula that asserts that two expressions have the same value.

Equation of a circle: With center (X_C, Y_C) and radius r then $(x - X_C)^2 + (y - Y_C)^2 = r^2$.

Equation of a line: In plane rectangular Cartesian coordinates, $y = mx + c$ where m is the slope and c a constant representing the *intercept* of the line on the y-axis where $x = 0$. For a line passing through A and B then

$$m = (y_A - y_B)/(x_A - x_B).$$

Equation of a plane: In three-dimensional rectangular Cartesian coordinates, an equation of the form $z = mx + ny + c$ where m, n, and c are constants.

Extrapolation: Estimating the value of a function beyond the limits already known.

Function: A relationship in which, given the value of one or more variables, then the overall value of the function can be calculated. For instance, the function $f(x) = ax^3 + bx^2 + cx + d$ is a cubic or third-degree curve for which the value $y = f(x)$ can be calculated for any given value of x.

Graph: A visual demonstration of a function in the form of points and lines.

Identical equation or identity: A relationship that is true for all values that form part of the expression, for example, all values of x and y that lie along the line $y = mx + c$.

Independent variable: A variable quantity whose value determines that of a *dependent variable*.

Intercept: Where a line cuts an axis.

Interpolation: Estimating the value of a function within limits already known.

Inverse function: If y is a *function* of x, that is, $y = f(x)$, then the inverse function written as f^{-1} is such that $x = f^{-1}(y)$. Many functions do not have an inverse.

Perfect square: An expression that is the product of two identical components, for example, $x^2 + (b/a)x + b^2/4a^2 = (x + b/2a)^2$ is a perfect square.

Point in polygon (PiP): An algorithm to determine whether a point lies inside or outside a polygon.

Polynomial: An expression that is the sum of a number of terms, each consisting of a constant and one or more variables raised to a nonnegative integer power, such as $a + bx + cx^2 + cx^3 + dx^4$.

Quadratic: An equation of the second degree, that is, one where no term is raised to a higher power than 2, such as $ax^2 + bx + c = 0$.

Scale: The ratio between the lengths on one object with those of another, also known as the *scale factor*. The term is also used to mean a measuring instrument to determine quantities such as lengths.

Scale factor: The multiplying factor applied to one measurement to obtain another.

Series: The sum of a finite or infinite number of terms, expressible in the form Σx – (sigma x).

Similar triangles: Triangles that are the same shape and differ only in scale.

Simultaneous equations: Two or more equations that must be satisfied at the same time. For example, when finding the intersection of two lines, the equations for each line must be satisfied simultaneously.

Thiessen polygon: (Also sometimes spelled *Theissen*) With a series of given points scattered around an area, a Thiessen polygon surrounds each given point in such a way that every other point within the polygon is nearer to that point than any other given point.

4 The Geometry of Common Shapes

4.1 TRIANGLES AND CIRCLES

Geometry is the study of constructible shapes. In this chapter we will review some of the shapes that occur in geomatics and GIS that occupy ordinary two- or three-dimensional "Euclidian" space. Euclid was a Greek mathematician of the 3rd century BC who worked out a series of axioms or postulations concerning points, lines, angles, surfaces, and volumes. From these, he derived 465 theorems.

His basic axioms included such statements as that for any two distinct points there is only one straight line that passes through them and if three distinct points are not on a straight line then there is only one plane that will pass through them. Euclid identified 10 axioms but subsequently a further one was added, namely that only one straight line can be drawn parallel to a given line through any point not on that line.

In Euclidian space, the shortest distance between two points is a straight line. Two lines that are either parallel or intersect form a plane—in fact parallel lines may be said to intersect at a point at infinity, an important consideration when drawing images of three-dimensional (3D) objects in perspective on a plane (2D) surface, as discussed in Chapter 10.

The triangle is the simplest shape that is made up of straight lines. In fact, all 2D shapes can be regarded as being made up from a series of triangles, just as every curve can be thought of as a series of short straight lines. Although this can give rise to a number of errors, for example, when calculating an area enclosed by a curved line, the approximation can be adequate for many practical purposes.

Triangles come in all sorts of shapes and sizes but the basic fact is that the angles of a plane triangle add up to half of a complete turn or 180°; the angles of a spherical triangle, which is one drawn on the surface of a sphere, add up to more than 180°. For the present, we will only consider plane triangles.

In the triangle ABC in Figure 4.1a, at A face C then turn clockwise to B; at B turn clockwise to face backward at C; at C turn clockwise to face A; you will have turned through the three angles ABC and half a complete turn. It is sometimes convenient to measure this in units of radians where π radians ("pi" radians) = 180°. Measures in radians are particularly important when using differential and integral calculus, as discussed in Chapter 6. For now, all angles will be considered as being in degrees, each degree being divided into 60 minutes (60′) and each minute containing sixty seconds (60″).

It is conventional for the angles of the triangle formed by points A, B, and C to be designated by capital letters while the sides opposite the angles have lowercase letters a, b, and c. In Figure 4.1b, the internal angles 2 and 3 are both less than 90° and are *acute* while the internal angle A is greater than 90° and is *obtuse*.

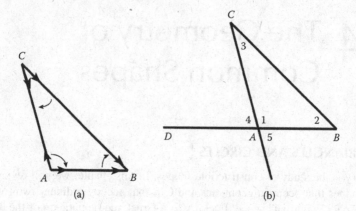

FIGURE 4.1 The angles of a triangle.

Note that since the angles $\angle 1 + \angle 2 + \angle 3 = 180°$ and also the angle marked "4" is such that $\angle 4 + \angle 1 = 180°$; hence, the external angle of the triangle is equal to the sum of the two internal opposite angles. The symbol "$\angle$" means "angle."

This also applies to the angle marked 5 in Figure 4.1b. Thus, $\angle 4 = \angle 2 + \angle 3 = \angle 5$. In fact, whenever two lines intersect the opposite angles are equal (Rule 4.2 in Box 4.1).

BOX 4.1   ANGLES OF A TRIANGLE AND IN A CIRCLE

RULE 4.1

The external angles of a triangle obtained by extending each of the sides equal the sum of their interior opposite angles.

RULE 4.2

When two straight lines intersect, opposite angles are equal.

RULE 4.3

The angle subtended by an arc at the center of a circle equals twice the angle subtended at the circumference.

RULE 4.4

The angles subtended by an arc in the same segment of the circumference of a circle are all equal.

RULE 4.5

The opposite angles of a quadrilateral bounded by a circle add up to 180°.

RULE 4.6

The angle subtended by a diameter at the circumference of a circle equals 90°.

RULE 4.7

The perpendicular bisector of a chord passes through the center of the circle.

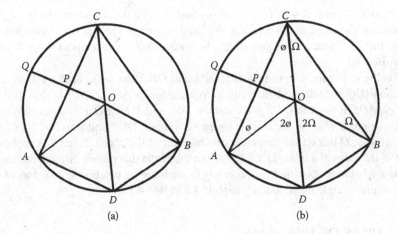

FIGURE 4.2 Angles subtended by arcs.

Triangles with unequal sides are said to be *scalene* while those that have two angles equal are called *isosceles*. Isosceles triangles not only have two angles equal, but also two sides are of the same length. If all three angles are equal, then all three sides will be of equal length and the triangle is said to be *equilateral*. The angles of an equilateral triangle are all equal to 60°.

In Figure 4.2a, a circle, known as the *circumscribing circle*, has been drawn around the triangle *ABC*. Its center is *O*, hence *OC* = *OA* as both lines are radii of the circumscribing circle. Now draw the line *OQ* perpendicular to *AC*, cutting *AC* at *P*.

∠*ACO* = ∠*CAO* since the triangle *AOC* is isosceles because *OA* = *OC* = the radius of the circumscribing circle. ∠*APO* = ∠*CPO* = 90° by construction. The two triangles *APO* and *CPO* share a common side (*OP*) and *OC* = *OA* then they must be of exactly the same size and hence point *P* is the midpoint of *AC* and *AP* = *CP*. Triangles that are exactly the same size and shape are said to be *congruent*. Two triangles that have the same angles but are of different size are said to be *similar*. Similar triangles differ merely in scale.

In Figure 4.2a,b, the line *CO* has been extended to *D* so that *CD* is a diameter of the circle. Any straight line that crosses a circle is called a *chord* (e.g., *AC* and *CB*) while the curved section that a chord cuts off is an *arc* (for example, the curved sections *AQC* or *ADB*). In Figure 4.2a, *OA* = *OB* = *OC* = the radius of the circle. Hence, triangles *AOC* and *BOC* are isosceles and ∠*OAC* = ∠*OCA*. Call this angle ø ("phi"). Hence, ∠*AOD* = 2ø (as per Rule 4.1 in Box 4.1). Also, ∠*OCB* = ∠*OBC* = Ω ("omega") and hence ∠*BOD* = 2Ω. Thus, ∠*ACB* = ø + Ω while ∠*AOB* = 2(ø + Ω). This is true wherever *C* is along the section of the circle *AQCB*.

In Figure 4.2b, the external angle *AOB* = 360 − (2ø + 2Ω) will (by Rule 4.3 in Box 4.1) be twice the angle *ADB*, which therefore equals 180 − (ø + Ω). Thus, ∠*ACB* + ∠*ADB* = 180° so long as *C* and *D* are in opposite segments (a *segment* being a portion of a circle bounded by an arc and a chord; any chord divides a circle into two segments). It follows, too, that angles subtended in the same segment are all equal (Rule 4.4 in Box 4.1). It also follows that as in Figure 4.2 where the line CD is a diameter, the angle at the center = 180° and hence the angle at the circumference would be 90°. Thus, $\angle DAC$ and $\angle DBC$ both equal 90°. This leads to Rules 4.5 and 4.6 in Box 4.1.

Finally, in Figure 4.2a in triangles OPA and OPC, the angle at $P = 90°$ by construction (OQ was defined as being perpendicular to AC). $\angle OAP = \angle OCP$ since triangle AOC is isosceles, OP is a common side, and $AO = CO$ as radii of the circle. Hence, triangle OAP is the mirror image of triangle OCP and thus $AP = PC$. Put another way, O lies on the perpendicular bisector of the chord. Since this applies to each of the sides of a triangle it follows that the circle that passes through each apex of a triangle has as its center the point where the three perpendicular bisectors of the sides of the triangle meet, leading to Rule 4.7 in Box 4.1.

4.2 AREAS OF TRIANGLES

We will return to aspects of the geometry of the circle shortly but for now, consider the triangle ABC in Figure 4.3. The triangle is half a rectangle whose area is its base times its height. Therefore, the area of a triangle = (1/2) * base * height = $0.5hc$. It also equals $\sqrt{s(s-a)(s-b)(s-c)}$ where s is half its perimeter as demonstrated in Box 4.2 and Example 4.1.

We can extend the calculations of the area of a triangle to the *trapezium*, which is a four-sided figure with two sides parallel as shown in Figure 4.4.

In Figure 4.4, the trapezium is made up from two triangles (ABP and QCD) and a rectangle $PBCQ$. Let $BP = CQ = h$, $AP = t$, $PQ = u$, and $QD = v$. Then,

$$\text{area} = (½) \, t * h + u * h + (½) \, v * h = (½) \, (t + u + v) * h + (½) \, u * h$$

$$= (½) \, (AD + BC) \, h$$

= half the sum of the parallel sides times the distance between them (Box 4.3).

We can extend this simple relationship to the coordinates of A, B, and C in Figure 4.5 where the area of triangle ABC = trapeziums $\{A'ACC' + C'CBB' - A'ABB'\}$.

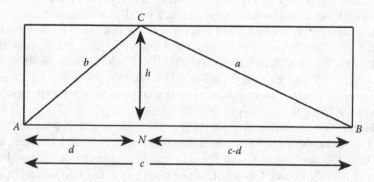

FIGURE 4.3 The area of a triangle.

BOX 4.2   THE AREA OF A TRIANGLE
FROM ITS SEMIPERIMETER

The triangle ABC in Figure 4.3 may be divided by the perpendicular CN into two areas with $AN = d$ and $NB = (c-d)$. The area of triangle $ABC = (1/2)$ base * height $= (1/2)hc$. Using Pythagoras for triangle ANC $b^2 = h^2 + d^2$ or $h^2 = b^2 - d^2$. Also, in triangle BNC $a^2 = h^2 + (c - d)^2 = h^2 + c^2 - 2cd + d^2$

$$\text{Or } a^2 = h^2 + d^2 + c^2 - 2cd$$

Since $b^2 = h^2 + d^2$, this means $a^2 = b^2 + c^2 - 2cd$. Rearranging,

$$2cd = b^2 + c^2 - a^2$$

or

$$d = (b^2 + c^2 - a^2)/2c$$

Hence,

$$h^2 = b^2 - d^2 = (b - d)(b + d) \text{ (the difference of two squares)}$$

$$= (b - (b^2 + c^2 - a^2)/2c)(b + (b^2 + c^2 - a^2)/2c)$$

$$= \{(2bc - b^2 - c^2 + a^2)/2c\} \; \{(2bc + b^2 + c^2 - a^2)/2c\}$$

$$= \{a^2 - (b - c)^2\} \; \{(b + c)^2 - a^2\}/4c^2$$

These again are the differences of two squares. Hence,

$$h^2 = \{(a - b + c)(a + b - c)\} \; \{(b + c - a)(a + b + c)\}/4c^2, \text{ or}$$

$$4c^2h^2 = (a + b + c - 2b) * (a + b + c - 2c) * (a + b + c - 2a) * (a + b + c)$$

$$= (2s - 2b)(2s - 2c)(2s - 2a)(2s) \text{ where } 2s = a + b + c$$

$$= 16 \, s(s - a)(s - b)(s - c)$$

Or, dividing both sides by 16,

$$(1/4) \, c^2h^2 = s(s - a)(s - b)(s - c)$$

Taking the square roots,

$$(1/2)ch = \text{the area of } ABC = \sqrt{s(s - a)(s - b)(s - c)}$$

where s is the semiperimeter of the triangle.

Thus, the area of a triangle $= \sqrt{s(s-a)(s-b)(s-c)}$ where s is half its perimeter.

EXAMPLE 4.1: THE AREA OF A TRIANGLE

Consider the triangle A, B, C in Figure 4.3. Assume its coordinates are

$$A\ (1, 10),\ B\ (9, 10),\ C\ (4, 14)$$

$$AB = c = 8;\ d = 3, \quad \text{and} \quad (c - d) = 5$$

The height $h = 4$; hence its area = ½ base * height = 16.
By Pythagoras,

$$BC = a = \sqrt{(h^2 + (c - d)^2)} = \sqrt{(41)} = 6.40312$$

Likewise,

$$AC = b = \sqrt{(h^2 + d^2)} = \sqrt{(25)} = 5$$

Hence, the semiperimeter = s = (1/2) * (8 + 5 + 6.40312) = 9.70156. Using the formula for the area, which is

$$\sqrt{\{s(s - a)(s - b)(s - c)\}}$$

The area = $\sqrt{(9.70156 * 3.29844 * 4.70156 * 1.70156)}$ = $\sqrt{(256)}$ = 16 as before.

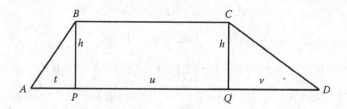

FIGURE 4.4 The area of a trapezium.

BOX 4.3   THE AREA OF A TRIANGLE AND A TRAPEZIUM

RULE 4.8

The area of a triangle is "half base times height." It also equals $\sqrt{s(s - a)(s - b)(s - c)}$ where a, b, c are the lengths of the sides and s is the semiperimeter = (½)$(a + b + c)$.

RULE 4.9

The area of a trapezium equals half the sum of the length of the parallel sides times the distance between them.

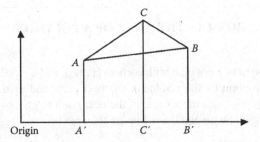

FIGURE 4.5 The area of a triangle by coordinates.

For the trapezium $A'ACC'$, $A'A = y$ coordinate of A, $C'C = y$ coordinate of C while the distance between the two sides $= x_C - x_A$. The area $= \frac{1}{2}\{(y_A + y_C)(x_C - x_A)\}$.

Likewise, for trapezium $C'CBB'$, the area $= \frac{1}{2}\{(y_C + y_B)(x_B - x_C)\}$ and for $A'ABB'$ the area $= \frac{1}{2}\{(y_A + y_B)(x_B - x_A)\}$. Combining the above and simplifying, the area of the triangle ABC

$$= \frac{1}{2}\{x_A y_B + x_B y_C + x_C y_A - y_A x_B - y_B x_C - y_C x_A\}$$

In fact, since every polygon can be subdivided into a series of triangles this sequence can be extended. Thus, for a six-sided polygon with points A, B, C, D, E, and F (see Example 4.2) the area would be:

$$\frac{1}{2}\{x_A y_B + x_B y_C + x_C y_D + x_D y_E + x_E y_F + x_F y_A -$$
$$y_A x_B - y_B x_C - y_C x_D - y_D x_E - y_E x_F - y_F x_A\}$$

For an eight-sided polygon A, B, C, D, E, F, G, and H, then the area would be (Box 4.4)

$$\text{area} = \frac{1}{2}\{x_A y_B + x_B y_C + x_C y_D + x_D y_E + x_E y_F + x_F y_{G} + x_G y_H + x_H y_A$$

$$- y_A x_B - y_B x_C - y_C x_D - y_D x_E - y_E x_F - y_F x_G - y_G x_H - y_H x_A\}$$

and so on.

EXAMPLE 4.2: CALCULATING THE AREA OF A POLYGON

Consider the six-sided polygon whose corners are A, B, C, D, E, F with A (1, 11), B (6, 13), C (10, 18), D (14, 12), E (12, 2), F (6, 7). Using the formula in Box 4.4, twice the area
$= (1 * 13 + 6 * 18 + 10 * 12 + 14 * 2 + 12 * 7 + 6 * 11 - 11 * 6 - 13 * 10 - 18 * 14 - 12 * 12 - 2 * 6 - 7 * 1)$
$= (13 + 108 + 120 + 28 + 84 + 66 - 66 - 130 - 252 - 144 - 12 - 7) = -192$

Hence, area $= 96$

Note: The minus sign would not arise if the area were computed in the reverse order.

BOX 4.4   THE AREA OF A POLYGON

RULE 4.10

To obtain the area of a polygon with corners (x_A, y_A), and so forth, multiply the x value of each point by the y value of the next point and then subtract the y value of each point times the x value of the next point to give twice the area.

As an example, for an eight-sided figure the area is

$$\tfrac{1}{2}\{x_A y_B + x_B y_C + x_C y_D + x_D y_E + x_E y_F + x_F y_{G+} x_G y_H + x_H y_A$$

$$- y_A x_B - y_B x_C - y_C x_D - y_D x_E - y_E x_F - y_F x_G - y_G x_H - y_H x_A\}$$

4.3   CENTERS OF A TRIANGLE

In Chapter 3, Box 3.3, we showed that the lines joining the corners of a triangle to the middle of their opposite sides pass through a point G known as the *centroid*. On the basis that the area of a triangle is half its base times its height, then it follows that for the left-hand area in Figure 4.6, the area of ACF = area of FCB since $AF = FB$ by construction. In effect, the triangle ABC is balanced about the line CF; similarly, for the lines from A and B through G. Thus, G is the point of balance or center of gravity of the triangle. Its coordinates were shown in Box 3.3 to be $\{1/3(x_A + x_B + x_C), 1/3(y_A + y_B + y_C)\}$.

In Figure 4.7, AD is constructed so as to be perpendicular to BC, and BE to be perpendicular to AC. The two lines intersect at O and line CO meets AB at H. We will now show that CH is perpendicular to AB.

Since $\angle AEB = 90° = \angle ADB$ then a circle with AB as diameter passes through E and D (Rule 4.5), hence $\angle EDO = \angle EBA$ (angles in the same segment—Rule 4.4) = $90 - A$. Since $\angle ODC = 90°$ by construction, $\angle EDC = A$. But E and D also lie on a circle with diameter OC since the angles subtended by a diameter are $90°$. Hence, $\angle EDC = \angle EOC = A$. In triangle EOC, $\angle OEC = 90°$, hence $\angle OCE = 90 - A$. Now in triangle CAH, $\angle CAH = A$ and $\angle ACH = \angle OCE = 90 - A$, hence $\angle AHC = 90°$. Thus,

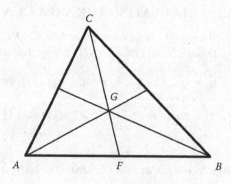

FIGURE 4.6   The centroid.

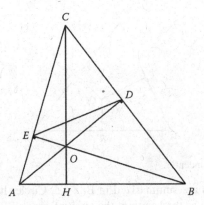

FIGURE 4.7   The orthocenter.

CH is perpendicular to AB and the perpendiculars from each point of a triangle to its opposite side intersect at one point, known as the *orthocenter*.

In Figure 4.8, let the line CI be the bisector of the angle ACB and AI bisect CAB. Let P, Q, and R be the feet of the perpendiculars from I to the corresponding sides. Then in triangles IPC and IQC, $\angle ICP = \angle ICQ$ since IC bisects PCQ, and $\angle IPC = 90° = \angle IQC$. Thus, the two triangles have the same angles and a common side (CI) and must therefore be congruent. That means that $IP = IQ$. Similarly, triangles API and ARI are congruent, so $IP = IR$. IB is a common side between triangles IRB and IQB, and since the angles at Q and R are right angles, then by Pythagoras, QB must be the same size as RB. Thus, the two triangles IRB and IQB are congruent so that IB is the bisector of angle ABC.

The point where the bisectors of all three angles meet is known as the *incenter*. A circle can be drawn around this point that touches each of the three sides of the triangle (Figure 4.10a).

Finally, the perpendicular bisectors of each side also meet in a point known as the *circumcenter* (CC) as we saw in Figure 4.2. In Figure 4.9, P' is the midpoint of AC, Q' is the midpoint of CB, and R' is the midpoint of BA. CC is the point from which a circle can be drawn through all three vertices A, B, C.

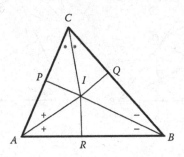

FIGURE 4.8   The incenter.

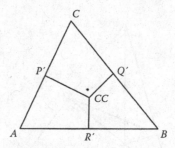

FIGURE 4.9 The circumcenter.

The various centers are summarized in Box 4.5. Circles have a number of well-known properties, such as the length of the circumference = $2\pi r$ where r = length of the radius. The area = πr^2. A line that just touches the circle (e.g., AB at point R in Figure 4.10a) is called a *tangent*. Any other line than a tangent will either not touch the circle at all or else will intersect it at two points. A tangent and the radius to the point where it touches the circle are at 90° and this radius is said to be *normal* to the tangent.

BOX 4.5   CENTERS OF A TRIANGLE

1. The perpendiculars from the vertices to the opposite sides all pass through the orthocenter.
2. The bisectors of each angle pass through the incenter.
3. The perpendicular bisectors of each side pass through the circumcenter.
4. The lines joining each vertex to the midpoint of the opposite side all pass through the centroid.

Note: Though not proven here, the three points marking the orthocenter, centroid, and circumcenter all lie on a straight line, known as the *Euler line*. The incenter does not, however, lie on this line.

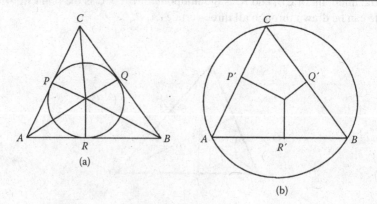

(a)

(b)

FIGURE 4.10 Inscribed and circumscribed circles.

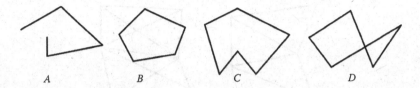

FIGURE 4.11   Straight-line figures.

4.4   POLYGONS

A *polygon* is a closed figure bounded by three or more straight lines that do not cross; it has the same number of vertices as there are sides. The polygon is said to be concave if any of its interior angles are greater than 180°, otherwise it is said to be convex.

In Figure 4.11, *A* is not a polygon; *B* is a convex polygon; *C* is a concave polygon, and *D* is essentially two polygons. The best-known examples of polygons are triangles, quadrilaterals (especially squares and rectangles), pentagons (five-sided), and hexagons (six-sided figures). A contour line that closes back on itself and is constructed out of straight-line sections would also form a polygon.

Any polygon can be broken down into a series of triangles though even with a rectangle these are not unique (see Figure 4.12). The three sides of the first triangle create internal angles that add up to 180°. A further point adds a further triangle and a further 180° to bring the total sum of the internal angles to 360°. A five-sided polygon will have its internal angles adding up to 540°, and so forth. Since the whole polygon can be thought of as a series of adjacent triangles, the sum of its interior angles will be $= (n - 2) * 180°$ where $n =$ the number of sides (Box 4.6).

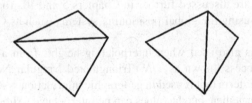

FIGURE 4.12   Two ways to divide a quadrilateral.

BOX 4.6   THE ANGLES OF A POLYGON

RULE 4.11

The sum of the interior angles of a polygon $= (n - 2) * 180°$ where $n =$ the number of sides.

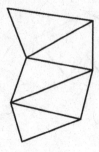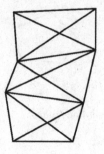

FIGURE 4.13 Triangulation networks.

Given any set of points there are many ways in which they can be joined to form a set of triangles in a process known as *triangulation* (Figure 4.13). Until recently, the whole basis of map-making was constructed around a set of points between which the angles had been observed. The result was a network of polygons and triangles, some overlapping.

By measuring all three angles in each triangle the shape of the mesh could be rigidly determined and there would be independent checks on the accuracy of the measurements since for each triangle the measured angles had, after suitable adjustment, to add up to 180°. The overall scale was determined by measuring at least one of the sides of the network.

Although the first triangulation networks were observed during the 18th century, it was during the latter part of the 19th and most of the 20th centuries that many areas of the globe became covered by such networks. The fact that the surface of the Earth is curved meant that none of the networks of triangles were flat and hence some corrections had to be applied to the measured angles and distances in order to calculate the coordinates of the triangulation or "trig" stations. The relationships between angles and distances on a curved surface and their corresponding values on the flat are discussed further in Chapters 5 and 10. Today, positions are often determined using the global positioning systems (such as GPS) discussed in Chapter 14.

Triangulation is also used when interpolating heights from a series of irregular points in a process known as *TIN* (Triangulated Irregular Network). TIN has become the generic term that is used in geographic information systems to describe a way in which intermediate height values can be interpolated. Although height is the most commonly used third dimension in geomatics, the technique can be extended to other characteristics at a point, such as air pressure or the strength of the magnetic field. The process involves the conversion of a set of irregularly distributed points of known third-dimensional value ("spot heights") into a network of triangles within which further height values can be interpolated, as discussed in Chapter 3 (see Figures 3.16 and 3.17).

Consider a series of points *A*, *B*, *C*, *D*, *E* in Figure 4.14a, for which the eastings, northings, and height or (*x*, *y*, *z*) values are known. The points can be joined by straight lines and the intermediate heights interpolated. The question here is "which points should be connected?" A favored method is known as the *Delaunay*

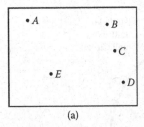
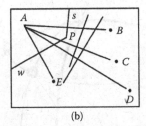
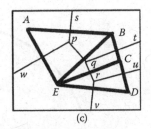

(a)   (b)   (c)

FIGURE 4.14   Thiessen polygons and the Delaunay triangles.

triangulation. It is related to the creation of a series of polygons known variously as *Dirichlet* or *Voronoi tessellations* or *Thiessen polygons.*

In Figure 4.14a, we have a set of points of known height. We can plot these on a plane surface and then draw lines from A to all neighboring points and construct the perpendicular bisector of each of the lines AB, AC, AD, and AE. Two of these (WP and PS in Figure 4.14b) together with the bounding area mark out a polygon in which all the points are nearer to A than to any of the other points B, C, D, and E. That is because all the perpendicular bisectors separate out areas one side of which is nearer A and the other is nearer to the second generating point.

This process can be extended to all the other points, creating a series of polygons within which all parts of the area are closer to the generating point than to any other. For example in Figure 4.14c, the lines of the bounding rectangle together with W to P to Q to R to V define a polygon around E that is the area within which all points lie closer to E than to A, B, C, D, and so forth.

Such polygons are known as *Thiessen polygons* and are of particular interest in certain types of statistical mapping. The lines connecting nearby points in the final creation of these polygons form a series of triangles known as the *Delaunay triangles.* They are deemed to be the "best shaped" triangles, and are the nearest to being equilateral that can be constructed from the original points. These triangles are shown in Figure 4.14c by the darker lines.

4.5   THE SPHERE AND THE ELLIPSE

Let us now return to the geometry of the circle since it is important in two areas of geomatics—the study of positioning fixing and in the representation of the shape of the Earth, which is approximately spherical.

A plane intersects a sphere in a circle; if the plane passes through the center of the sphere, as in the case of the meridians of longitude on the Earth, the line on the surface is said to be a *great circle.* Other circles such as parallels of latitude (except the special case of the equator) are called *small circles.*

The geometry of any figure on the surface of a sphere differs from that on a plane because triangles become spherical triangles and their three angles no longer add up to 180°. Thus, in Figure 4.15 the angle $NG'P = 90° =$ angle NPG'. The angle $G'NP$ at the North Pole is the difference in longitude between the meridian or great circle through Greenwich and the meridian through P; this means that the three angles of

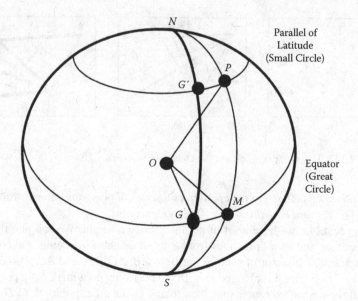

FIGURE 4.15   Great and small circles.

the triangle *G'NP* are in excess of 180°. When the sides of the triangle are defined by great circles the amount by which the sum of the three angles exceeds 180° is called the *spherical excess*. Computations involving spherical triangles will be considered in Chapter 5 on trigonometry.

A sphere is only an approximation to the shape of the earth; in reality an *ellipsoid* (obtained by rotating an ellipse about its shorter or "minor" axis) gives a better representation. An *ellipse* (Figure 4.16) is a closed figure shaped like a compressed circle that is symmetrical about two axes known as the *major axis* and the *minor axis*. The circle that is formed using the major axis as diameter is known as the *auxiliary circle*. The radius of this circle is known as the semimajor axis and is normally referred to as *a* while the length of the semiminor axis is *b*. In the geometry of the ellipse compared with the auxiliary circle, the distances in the *x*-direction remain the

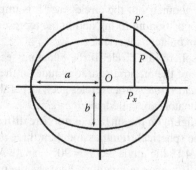

FIGURE 4.16   Ellipse and auxiliary circle.

same but the distances in the y-direction are compressed or scaled down so that P' on the circle becomes P on the ellipse.

4.6 SECTIONS OF A CONE

The ellipse and circle are two of four special second-order curves known as *conic sections*, the others being the parabola and the hyperbola (see below). They take the form of

$$ax^2 + by^2 + cxy + dx + ey + f = 0$$

where a, b, c, d, e, and f are constants. For a circle, $a = b$ and $c = 0$. Geometrically, they can be derived from a line known as the *directrix* shown as the line CD in Figure 4.17. F_1 and F_2 are two points on the major axis known as the *foci*, each *focus* being the same distance either side of the center O on the major axis. Q is the foot of the perpendicular from P onto the line CD so that PQ is parallel to the major axis.

The ellipse has the following characteristics:

1. For all conics, the ratio between the distances F_2P and PQ is a constant that is known as the *eccentricity e* (not to be confused with the Euler number "e" that equals 2.718...). For an ellipse, the value of e is such that $0 < e < 1$.
2. If the origin O has coordinates $(0, 0)$ then for the ellipse F_1 is located at point $(-ae, 0)$ and F_2 at $(+ae, 0)$ where a is the length of the semimajor axis and e is the eccentricity.
3. The directrix is located at a distance (a/e) from the center of the ellipse. It thus cuts the major axis at point $(a/e, 0)$.
4. The sum of the distances $F_1P + PF_2$ is a constant for the ellipse. (This is why it is possible to draw an ellipse by placing a pin at each of the foci, a loop of string of length $F_1P + PF_2 + F_2F_1$ around the pins; then, by holding a pencil

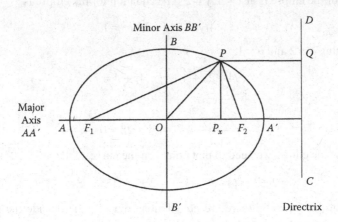

FIGURE 4.17 An ellipse and its directrix.

at P and keeping the string taught, the pencil can be moved around the pins to draw the ellipse).

5. $ae = \sqrt{(a^2 - b^2)}$ or $e = \sqrt{(1 - b^2/a^2)}$.
6. The area of an ellipse is πab.
7. The length of the circumference of an ellipse is a complicated thing to calculate.
8. If P is a point on the ellipse with coordinates (x, y) and center $(0, 0)$, then $x^2/a^2 + y^2/b^2 = 1$.

These relationships will be explained more fully in Chapter 9 after we have discussed trigonometric functions. The circle can be regarded as a special case of an ellipse in which $a = b = r$, the radius of the circle and $x^2/r^2 + y^2/r^2 = 1$. The two foci are coincident at the center of the circle. The eccentricity $e = 0$ and the directrix is located a distance $(a/0)$ from the center, which means that it is an infinite distance away.

For the ellipse $0 < e < 1$. When $e = 1$, we have $FP = PQ$ (see Figure 4.19) and instead of being gently rounded, the ellipse becomes infinite in size. The curve is known as a *parabola*. It has only one focus and its directrix is symmetrically placed on the other side of the origin, so that if F is at $(a, 0)$ the directrix passes through $(-a, 0)$. A simple example of a parabola is the function $y^2 = mx$ where m is a constant.

EXAMPLE 4.3: THE EQUATION OF AN ELLIPSE

In many graphics packages, ellipses are constructed from their enclosing rectangle. Consider the rectangle whose sides stretch from $x = 2$ to $x = 6$ and $y = 1$ to $y = 3$ (see Figure 4.18).

For an ellipse to fit within this rectangle, the major axis must have $2a =$ width of box or $a = 2$; the minor axis must have $2b =$ height of box or $b = 1$. The center of the ellipse is at $x = 4$, $y = 2$. The equation for the ellipse is

$$(x - 4)^2/a^2 + (y - 2)^2/b^2 = 1$$

or, putting $a = 2$ and $b = 1$:

$$x^2/4 - 2x + 4 + y^2 - 4y + 4 = 1$$

or

$$x^2 - 8x + 4y^2 - 16y + 28 = 0$$

To draw the ellipse we need to plot points on the curve

$$y = 2 \pm \tfrac{1}{2} \sqrt{\{4 - (x - 4)^2\}} = 2 \pm \tfrac{1}{2} \sqrt{\{8x - x^2 - 12\}}$$

Note: In this equation there are no solutions if x or y lie outside the range, namely, $2 \leq x \leq 6$ or $1 \leq y \leq 3$.

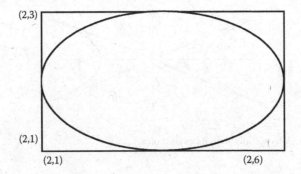

(2,3)

(2,1)

(2,1)   (2,6)

FIGURE 4.18 An enclosed ellipse.

The parabola has only one focus and the tangent to the curve at point P exactly bisects the lines FP and QP. This means that the line PR in Figure 4.19 is parallel to the axis and also at the same angle to the tangent at P as is the line PF. Hence, if a light source were placed at F and the inside of the parabolic surface were perfectly reflecting, then a beam of light from F would emerge parallel to the major axis. Conversely, a parabolic mirror can focus a parallel beam of light or other form of radiation, for instance from the sun, onto a single point (the focus, F).

If the eccentricity e has a value greater than 1, then the shape of the resulting curve is a *hyperbola*. It has an equation of the form $x^2/a^2 - y^2/b^2 = 1$. It consists of two branches formed by a pair of intersecting lines known as the *asymptotes* (AB and CD in Figure 4.20) to which it approaches more and more closely but only reaching them at infinity. The slope of these asymptotes is given by the equations

$$y = +(b/a)x$$

and

$$y = -(b/a)x$$

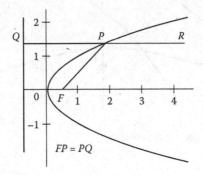

FIGURE 4.19 A parabola.

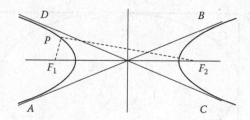

FIGURE 4.20   A hyperbola with its asymptotes.

With the hyperbola there are two foci (F_1 and F_2) but whereas with the ellipse as shown in Figure 4.17, the sum of the two distances $PF_1 + PF_2$ is a constant, in the case of the hyperbola the difference between the two distances is fixed. Thus, ($PF_1 - PF_2$) is constant in Figure 4.20. This has allowed navigation systems, especially at sea, to track a hyperbolic course using the difference in time for signals sent from two known points using what is called the *Doppler effect*. With transmitters at three known points there will be two intersecting sets of hyperbolae, allowing for the position of a vessel to be determined. In practice the two sets of hyperbolae may indicate four possible points but if the initial position is known the location of the vessel can be tracked. In Figure 4.21, the foci for the two hyperbolae are FF_1 and FF_2 where the transmitters would be located. Three possible locations are shown (A, B, and C). If there are more than three transmitters then the ambiguity can be resolved.

The circle, ellipse, parabola, and hyperbola all have solid equivalents (the sphere, the ellipsoid, the paraboloid, and hyperboloid). They are all quadratic forms and

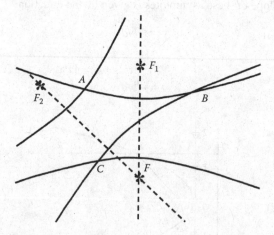

FIGURE 4.21   Two intersecting hyperbolae.

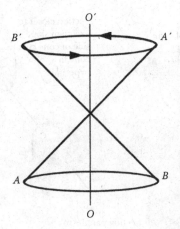

FIGURE 4.22 A double cone.

in their plane (i.e., 2D) form they are cross-sections of a three-dimensional cone. Consider two intersecting lines (*AA'* and *BB'* in Figure 4.22). These form a plane. Draw one of the lines of bisection (*OO'*) and then in three dimensions rotate the original lines around this bisector to generate a double cone (one cone stuck upside down on top of the other). Cut this solid along a plane perpendicular to the axis of rotation and you have a circle; cut it at an angle less than the slope of the sides of the cone and the cross-section will be an ellipse (Figure 4.23). If the section of the double cone is cut parallel to the edges of the cone then if the cone is infinitely large then the section will be infinitely large and be a parabola. If, however, the section is greater, it will cut the double cone twice (top and bottom) in a hyperbola (Figure 4.24).

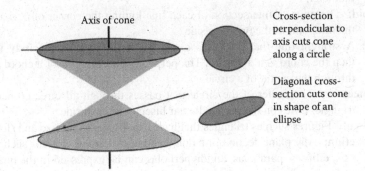

FIGURE 4.23 Sections of a cone—circle and ellipse.

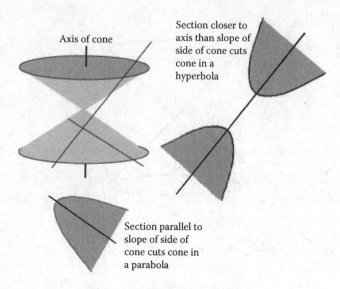

FIGURE 4.24   Sections of a cone—parabola and hyperbola.

SUMMARY

Acute angle: An angle in a triangle that is less than a *right angle*.

Arc: A section of the circumference of a circle. Angles subtended by an arc to any point opposite and in the same segment of the circumference of a circle are equal and half the size of the angle subtended at the center.

Area of triangle: ½ Base * Height or $\sqrt{s(s-a)(s-b)(s-c)}$ where s is the *semi-perimeter* or for the triangle ABC, or ½ $(x_A y_B + x_B y_C + x_C y_A - y_A x_B - y_B x_C - y_C x_A)$. The latter formula can be extended to give the area of a polygon.

Auxiliary circle: The circle surrounding an ellipse based on the *major axis* as its diameter.

Centroid: The point of intersection of each line joining the corner of a triangle to the midpoint of the opposite side.

Chord: A straight line that divides a circle. If the division is precisely in half then the chord is a diameter. The perpendicular bisector of a chord passes through the center of a circle.

Circumcenter: The center of the circle that passes through all three corners of a triangle. It lies on the perpendicular bisector of each side.

Congruent: Figures such as triangles that have exactly the same size and shape.

Conic section: Any plane section of a double cone. All conic sections such as circles, ellipses, parabolas, and hyperbolas can be expressed in the quadratic form $ax^2 + by^2 + cxy + dx + ey + f = 0$.

Degree: The measure of an angle in which there are 360 divisions in a full turn. Each division (written as 1°) can be subdivided into 60 minutes (60'), which in turn can be subdivided into 60 seconds (60"). The angles of a plane triangle always add up to 180°. On a curved surface they add up to more than 180°.

Delaunay triangulation: The formation of a network of triangles whose sides are perpendicular to the sides of a set of *Thiessen polygons*.

Directrix: A line perpendicular to the major axis of an *ellipse, parabola,* or *hyperbola* that is used to generate that conic section.

Dirichlet tessellation: *See* Thiessen polygon.

Eccentricity: The ratio between the distance from a focus of a conic section to a point on the curve and the distance from that point to the *directrix*. With an ellipse the eccentricity is between 0 and 1, with a parabola it is 1, and with a hyperbola it is greater than 1.

Ellipse: A *conic section* that is a closed figure that with suitable axes can be expressed in the form $x^2/a^2 + y^2/b^2 = 1$.

Ellipsoid: A geometrical surface or solid formed by rotating an ellipse about one of its axes.

Equilateral triangle: A triangle in which all the sides are equal and each angle = $60°$.

Focus: A fixed point on the major axis of an ellipse, hyperbola, or parabola that with the *directrix* and *eccentricity* determines the shape of the *conic section*.

Grad: Also known as a *gon*. It is an angular measure in which 100 grads (or gons) equals one-quarter turn or $90°$.

Hyperbola: A *conic section* that extends to infinity and with suitable axes can be expressed in the form $x^2/a^2 - y^2/b^2 = 1$.

Incenter: The point that is equidistant from each side of a triangle so that a circle can be drawn touching each side. It lies on the bisector of each angle.

Isosceles: A triangle in which two sides (and hence, two angles) are equal.

Major axis: The longest axis or line through the center of an *ellipse*.

Minor axis: The shortest axis or line through the center of an *ellipse*.

Normal: A line that is perpendicular to another line or curve.

Obtuse angle: An angle in a triangle that is greater than a *right angle*.

Orthocenter: The point where the perpendiculars from each corner of a triangle to the opposite side intersect.

Parabola: A *conic section* that extends to infinity and with suitable axes can be expressed in the form $y^2 = 4ax$.

Polygon: A closed figure bounded by three or more sides. The sum of the interior angles of a polygon with n sides add up to $(n-2) * 180°$.

Radian: An angular measurement in which there are 2π radians in a full turn where 2π is also the ratio between the circumference of a circle and its radius. $\pi \approx 3.14159265$.

Right angle: An angle that is precisely a quarter turn or $90°$ or $\pi/2$ *radians*.

Scalene: A triangle in which all sides are of unequal length.

Segment: The portion of a circle bounded by an *arc* and a *chord*.

Semiperimeter: Half the sum of the length of the sides of a triangle.

Spherical excess: The amount by which the sum of the three angles of a triangle on a sphere exceeds $180°$.

Tangent: A line that just touches a curve. Also used in trigonometry (see Chapter 5) to describe the ratio between the side opposite and the side adjacent in a right-angled triangle.

Thiessen polygon: A polygon around a point such that any point within that polygon is nearer to its center than to any other point. Also known as *Dirichlet* or *Voronoi tessellations*. Often spelled as *Theissen*.

TIN (Triangulated Irregular Network): A network of irregularly shaped triangles.

Trapezium: A four-sided figure with two sides parallel.   •

Triangulation: The connection of a series of points to form a network of triangles.

Voronoi tessellation: *See* Thiessen polygon.

5 Plane and Spherical Trigonometry

5.1 BASIC TRIGONOMETRIC FUNCTIONS

Trigonometry is concerned with ratios between the sides and angles of triangles. Although at its simplest it is concerned with right-angled triangles on a plane surface, its applications extend to many areas of geomatics and geographic information systems (GIS) including calculations on curved surfaces such as that of the Earth. Trigonometry is used extensively in surveying and navigation.

In Chapter 3, Figure 3.7 showed similar triangles that have a common shape but differ in scale. In Figure 5.1, the triangles ABC, $AB'C'$, and $AB''C''$ are all the same shape (they have the same angles) but they differ in size or scale. In these triangles the ratio $\frac{BC}{AB} = \frac{B'C'}{AB'} = \frac{B''C''}{AB''}$ = "side opposite over side adjacent."

Given the fixed angle A, then in any right-angled triangle, the ratio BC/AB is constant. This is called the *tangent* of angle A or tan A. Similarly, the ratio BC/AC is constant = "side opposite over hypotenuse." This is called the *sine* of angle A or sin A. Likewise, the ratio AB/AC is a constant = "side adjacent over hypotenuse." It is called the *cosine* of angle A or cos A. (See Box 5.1 and Example 5.1.)

The ratios sin, cos, and tan are not independent. Given the theorem by Pythagoras

$$AB^2 + BC^2 = AC^2$$

Dividing both sides by AC^2 we obtain $(AB/AC)^2 + (BC/AC)^2 = 1$, or $(\sin A)^2 + (\cos A)^2 = 1$. Also,

$$(\sin A / \cos A) = \left(\frac{AB}{AC}\right) \bigg/ \left(\frac{BC}{AC}\right) = \left(\frac{AB}{AC}\right) = \tan A$$

Hence,

$$\tan = \frac{\sin}{\cos}$$

It is sometimes helpful to deal with the reciprocal ratios AC/AB and AC/BC. These (and the reciprocal of tan A) have special names. 1 divided by sine is called *cosecant* or *cosec*; 1 divided by cosine is the *secant* or *sec*; and 1 divided by tangent = *cotangent* or *cot*.

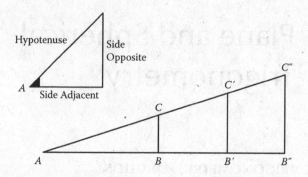

FIGURE 5.1 Similar right-angled triangles.

By convention, just as $x * x = x^2$, so $\sin A * \sin A = (\sin A)^2$ and is written as $\sin^2 A$. Likewise, $\cos A * \cos A = \cos^2 A$, and so forth. However, although $1/x = x^{-1}$, $1/\sin A$ is *not* written as $\sin^{-1} A$ since this has a reserved meaning, which is "the angle whose value is sine A." This, in turn, is called *arc-sine A*. $\mathrm{Cos}^{-1} A$ means "the angle whose value is cosine A." It is called *arccos*. Similarly, $\tan^{-1} = arctan$, $\mathrm{cosec}^{-1} = arccosec$, $\sec^{-1} = arcsec$, and $\cot^{-1} = arccot$.

In Figure 5.1, ABC is a right-angled triangle and the angle at C (the angle ACB) has as its cosine the ratio BC/AC. Thus, $\cos C$ is the same as $\sin A$. This only happens when the angle $C = 90° - A$. For all angles A, $\sin A = \cos (90° - A)$; $\cos A = \sin (90° - A)$; hence, $\tan A = \cot (90 - A)$.

We can always convert non-right-angled triangles into ones with right angles as shown in Figure 5.2. Even though the triangle ABC has no right angle, there is still a value for "$\sin A$" since this is dependent on the size of the angle A, not the shape or size of the triangle. We can create a right-angled triangle (Figure 5.2) by drawing

---

BOX 5.1   BASIC RELATIONSHIPS

Sine = sin = side opposite/hypotenuse

Cosine = cos = side adjacent/hypotenuse

Tangent = tan = side opposite/side adjacent

Cosecant = cosec = 1/sin

Secant = sec = 1/cos

Cotangent = cot = 1/tan

$\sin A/\cos A = \tan A$

$(\sin A)^2 + (\cos A)^2 = 1$

$\sin A = \cos (90 - A)$; $\cos A = \sin (90 - A)$

EXAMPLE 5.1: FUNCTIONS OF ANGLE A

Sin A, cos A, and tan A are functions of the angle A. Their values can be calculated using special formulae (see Box 6.3 in Chapter 6) or be obtained from tables, pocket calculators, and so on.

As an example for $A = 35°$ from tables with seven decimal places

$$\sin 35° = 0.5735764$$

$$\cos 35° = 0.8191520$$

$$\tan 35° = 0.7002075$$

Note: $(\sin 35) * (\sin 35) = (\sin 35)^2 = \sin^2 35 = 0.3289899$.

$$(\cos 35) * (\cos 35) = (\cos 35)^2 = \cos^2 35 = 0.6710101$$

Thus,

$$\sin^2(35) + \cos^2(35) = 1$$

Likewise,

$$\sin 35/\cos 35 = 0.5735764/0.8191520 = 0.7002075 = \tan 35$$

the perpendicular line from C onto the horizontal line AB so that angle $ANC = 90°$. We represent the sides of the triangle as lowercase letters (with $BC = a$, $CA = b$, and $AB = c$, and the height $CN = h$, known as the *altitude*) and the angles as uppercase letters A, B, C.

In Figure 5.2,

$$\sin A = CN/AC = h/b$$

or

$$h = b \sin A$$

Also,

$$\sin B = CN/BC = h/a$$

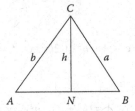

FIGURE 5.2 The height or altitude of a triangle.

or

$$h = a \sin B$$

Thus,

$$b \sin A = a \sin B$$

or

$$\frac{\sin A}{a} = \frac{\sin B}{b}$$

By constructing the perpendicular from B we can also show that $(\sin A)/a = (\sin C)/c$. Thus, for any triangle ABC with sides a, b, c

$$\frac{\sin A}{a} = \frac{\sin B}{b} = \frac{\sin C}{c}$$

This is known as the *sine formula* for triangle ABC.

Note also that the area of the triangle $= (1/2)$ base * height $= (1/2) ch$

$$= (1/2) bc \sin A = (1/2) ca \sin B = (1/2) ab \sin C$$

Further, in Figure 5.2, $\cos A = AN/AC$. Thus, $AN = AC \cos A = b \cos A$.
Using Pythagoras,

$$AN^2 + CN^2 = AC^2 \quad \text{or} \quad CN^2 = AC^2 - AN^2 = b^2 - b^2\cos^2A$$

Likewise, in triangle BNC where $NB = (c - AN) = (c - b \cos A)$. Hence, we have

$$CN^2 = BC^2 - BN^2 = a^2 - (c - b \cos A)^2 = a^2 - \{c^2 - 2bc \cos A + (b \cos A)^2\}$$
$$= a^2 - c^2 + 2bc \cos A - b^2 \cos^2A$$

Also,

$$CN^2 = b^2 - b^2 \cos^2A \text{ (as shown just above)}$$

Hence,

$$a^2 - c^2 + 2bc \cos A = b^2 \text{ or } 2bc \cos A = b^2 + c^2 - a^2$$

Thus,

$$\cos A = (b^2 + c^2 - a^2)/2bc$$

**BOX 5.2   SINE AND COSINE FORMULAE
FOR ANY PLANE TRIANGLE**

1. $(\sin A)/a = (\sin B)/b = (\sin C)/c$
2. $\cos A = (b^2 + c^2 - a^2)/2bc$

   $\cos B = (c^2 + a^2 - b^2)/2ca$

   $\cos C = (a^2 + b^2 - c^2)/2ab$

3. The area of the triangle $= (1/2)\, bc \sin A = (1/2)\, ca \sin B = (1/2)\, ab \sin C$

Similarly,

$$\cos B = (c^2 + a^2 - b^2)/2ca$$

and

$$\cos C = (a^2 + b^2 - c^2)/2ab$$

These relationships are known as the *cosine formulae*. Thus, if we know the lengths of the sides of a triangle, we can calculate the size of each of the angles using the formulae in Box 5.2.

5.2   OBTUSE ANGLES

So far we have assumed that the angles A, B, and C are less than or equal to a right angle. If we think of the case where ABC has B as a right angle, then the larger that A becomes, the smaller must be the angle C.

As A approaches a 90° angle, C approaches zero, while BC and AC become parallel and of similar length (Figure 5.3). In the limit $\sin 90° = 1$ while $\cos 90° = 0$. Since $\tan 90° = \sin 90°/\cos 90° = 1/0$ and we cannot divide by zero, it means that as A approaches 90°, $\tan A$ gets very much bigger until it becomes infinitely large. But what happens for angles over 90°? The sine formula and the cosine formula apply to triangles where there is a fixed relationship between the angles A, B, and C. The functions sin, cos, and

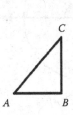

FIGURE 5.3   Toward a right angle.

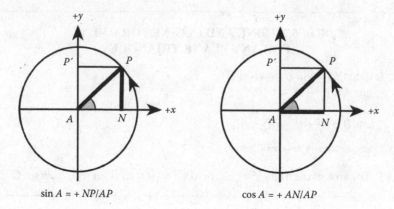

$$\sin A = + NP/AP \qquad\qquad \cos A = + AN/AP$$

FIGURE 5.4   A circle with unit radius.

tan give values that relate to the size of the angle, not to the shape or size of a triangle. We can look at them in a different way. Consider a circle of unit radius (Figure 5.4) and center A and a point P in the quadrant or quarter of the circle between the $+x$ and $+y$ axes.

If $\sin A = NP/AP$ and $AP = 1$ (unit radius), then the length of $NP = \sin A$. AP' is the projection of P onto the $+y$ axis and also $= \sin A$. P' is on the $+y$ side of the origin. Likewise, the length $AN = \cos A$. This is the projection of P onto the $+x$ axis.

As P moves anticlockwise around the circle, the angle that AP makes with the $+x$ axis increases until at $90°$ the projection of AP onto the $+y$ axis reaches the point where $NP = AP' = 1$ and $AN = 0$. As the angle increases further the length of NP grows shorter. It still has the value of $\sin A$. On the other hand, AN has moved from 1 down to 0 and is now on the negative side of the axis. Comparing Figures 5.4 and 5.5 with P in Figure 5.5, the mirror image of what it was in Figure 5.4, then the lengths AP' (the projection of P onto the $+y$ axis) is exactly the same but AN, the projection of P on the $+x$ axis, has in the second case become negative.

For angles where $90° \leq A \leq 180°$

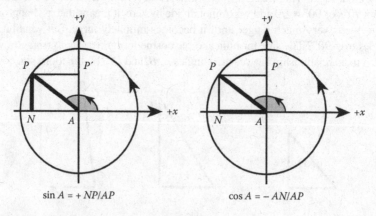

$$\sin A = + NP/AP \qquad\qquad \cos A = - AN/AP$$

FIGURE 5.5   Angles in the second quadrant.

$$\sin A = +\sin (180 - A)$$

$$\cos A = -\cos (180 - A)$$

Hence, since tan = sin/cos

$$\tan A = -\tan (180 - A)$$

A similar argument applies to angles greater than 180° and less than 270° in what is called the *third quadrant*. In Figure 5.6a, both AN and NP are negative

$$\sin A = -\sin (A - 180)$$

$$\cos A = -\cos (A - 180)$$

$$\tan A = +\tan (A - 180)$$

And for the fourth quadrant (Figure 5.6b) where AN has again become positive but NP is still negative

$$\sin A = -\sin (360 - A)$$

$$\cos A = +\cos (360 - A)$$

$$\tan A = -\tan (360 - A)$$

In particular, $(360 - A)$ is a complete turn less A $(= -A)$. Thus,

$$\sin (-A) = -\sin A$$

$$\cos (-A) = +\cos A$$

$$\tan (-A) = -\tan A$$

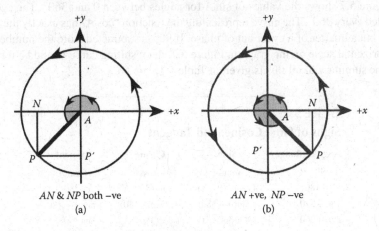

AN & NP both −ve	AN +ve, NP −ve
(a)	(b)

FIGURE 5.6   Angles in the third and fourth quadrants.

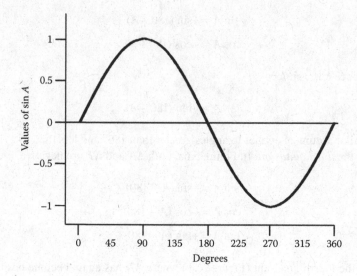

FIGURE 5.7   The cycle of values of sin A.

Also, since sin A = cos $(90 - A)$, then

$$\sin (90 + A) = \cos \{90 - (90 + A)\} = \cos (-A) = +\cos A$$

And since cos (A) = sin $(90 - A)$, then

$$\cos (90 + A) = \sin \{90 - (90 + A)\} = \sin (-A) = -\sin A$$

Hence,

$$\tan (90 + A) = -\cos A$$

Figure 5.7 shows the values of sin A for angles between 0 and 360°. The cycle is repeated every 360°. The curve representing the function "cos A" has exactly the same shape but a quarter of a cycle out of phase. If 90° is subtracted from the numbers on the horizontal scale on the graph in Figure 5.7, the resulting curve would be for "cos A." The summary of all this is given in Table 5.1.

TABLE 5.1

Signs of Sine, Cosine, and Tangent

Angular Range	Sine	Cosine	Tangent
0 – 90	$+ \sin A$	$+ \cos A$	$+ \tan A$
90 – 180	$+ \sin(180 - A)$	$- \cos(180 - A)$	$- \tan(180 - A)$
180 – 270	$- \sin(A - 180)$	$- \cos(A - 180)$	$+ \tan(A - 180)$
270 – 360	$- \sin(360 - A)$	$+ \cos(360 - A)$	$- \tan(360 - A)$

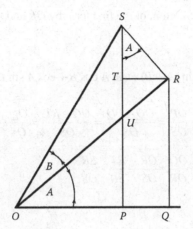

FIGURE 5.8 Combining adjacent angles.

5.3 COMBINED ANGLES

In later chapters we will use this information but for now, there are two more relationships that need to be established. Consider two adjacent angles A and B in Figure 5.8 defined by the lines OS, OR, and OQ. From point S (anywhere along the line OS—scale does not matter) draw the perpendicular from S onto OQ at P cutting OR at U. Also, choose R as the point that gives angle $ORS = 90°$ and draw the other horizontal or perpendicular lines RT and RQ. None of this alters the angles A or B.

$$\angle OUP = 90 - A; \ \angle OUS = 180 - (90 - A) = 90 + A$$

Hence,

$$\angle SUR = 90 - A$$

Also,

$$\angle SUR = \angle OUP \text{ (Rule 4.2 in Chapter 4).}$$

Since

$$\angle SRU = 90 \text{ by construction, } \angle USR = A$$

$$\sin B = \frac{RS}{OS}; \cos B = \frac{OR}{OS}; \sin A = \frac{QR}{OR}; \quad \text{and} \quad \cos A = \frac{QR}{OR}$$

$$\sin(A + B) = \frac{PS}{OS} = \frac{PT + TS}{OS} = \frac{QR + TS}{OS} \text{ (since } PT = QR)$$

$$\sin(A + B) + \frac{QR}{OS} + \frac{TS}{OS} = \frac{QR}{OR} * \frac{OR}{OS} + \frac{TS}{SR} * \frac{SR}{OS}$$

(by dividing the top and bottom of the first term by OR and the second term by SR).

Hence,

$$\sin (A + B) = \sin A \cos B + \cos A \sin B$$

Similarly, $\cos (A + B) = \dfrac{OP}{OS} = \dfrac{OQ - PQ}{OS} = \dfrac{OQ - RT}{OS} = \dfrac{OQ}{OS} - \dfrac{RT}{OS}$

$$= \dfrac{OQ}{OR} * \dfrac{OR}{OS} - \dfrac{RT}{SR} * \dfrac{SR}{OS}$$

Hence,

$$\cos (A + B) = \cos A \cos B - \sin A \sin B$$

The results are summarized in Box 5.3.

We have already shown in Box 5.2 that $\cos A = (b^2 + c^2 - a^2)/2bc$ and in Box 5.3 that $\cos A = 2 \cos^2 (\frac{1}{2}A) - 1$. Hence,

$$2 \cos^2(\tfrac{1}{2}A) = (b^2 + c^2 - a^2)/2bc + 1 = (b^2 + c^2 + 2bc - a^2)/2bc$$

BOX 5.3   COMBINED ANGLES

$$\sin (A + B) = \sin A \cos B + \cos A \sin B$$

$$\cos (A + B) = \cos A \cos B - \sin A \sin B$$

$$\sin 2A = \sin (A + A) = 2 \sin A \cos A$$

Hence,

$$\sin A = 2 \sin(\tfrac{1}{2}A) \cos(\tfrac{1}{2}A)$$

$\cos 2A = \cos (A + A) = \cos^2 A - \sin^2 A;$ and since $\cos^2 A + \sin^2 A = 1$, then

$$\cos 2A = 1 - 2\sin^2 A = 2\cos^2 A - 1$$

Hence,

$$\cos A = \cos^2(\tfrac{1}{2}A) - \sin^2 (\tfrac{1}{2}A) = 1 - 2\sin^2(\tfrac{1}{2}A) = 2\cos^2(\tfrac{1}{2}A) - 1$$

Since $\sin (-B) = -\sin B$ and $\cos (-B) = +\cos B$, then

$$\sin (A - B) = \sin A \cos B - \cos A \sin B$$

$$\cos (A - B) = \cos A \cos B + \sin A \sin B$$

Or

$$\cos^2(\tfrac{1}{2}A) = (b^2 + c^2 + 2bc - a^2)/4bc = \{(b+c)^2 - a^2\}/4bc$$

Since the numerator is the difference of two squares

$$\cos^2(\tfrac{1}{2}A) = (b + c + a)(b + c - a)/4bc = (a + b + c)(a + b + c - 2a)/4bc$$

$$= \{(a + b + c)/2\}\{(a + b + c)/2 - a\}/bc = s(s - a)/bc$$

where s is the semiperimeter $= \frac{1}{2}(a + b + c)$. Thus,

$$\cos(\tfrac{1}{2}A) = \sqrt{s(s-a)/bc}$$

Similarly,

$$\sin^2(\tfrac{1}{2}A) = (s - b)(s - c)/bc,$$ hence the results shown in Box 5.4.

5.4   BEARINGS AND DISTANCES

The formulae given in Box 5.3 are necessary and sufficient for the solution of a wide range of problems in surveying and mapping. Before considering some examples we need to clarify the difference between a mathematician's approach and a surveyor's. Traditionally in mathematics, angles are measured from the horizontal

---

**BOX 5.4   TRIGONOMETRICAL FUNCTIONS
AND THE SEMIPERIMETER**

$$\sin(\tfrac{1}{2}A) = \sqrt{(s-b)(s-c)/bc}$$

$$\cos(\tfrac{1}{2}A) = \sqrt{s(s-a)/bc}$$

$$\tan(\tfrac{1}{2}A) = \sqrt{(s-b)(s-c)/s(s-a)}$$

$$\sin(\tfrac{1}{2}B) = \sqrt{(s-c)(s-a)/ca}$$

$$\cos(\tfrac{1}{2}B) = \sqrt{s(s-b)/ca}$$

$$\tan(\tfrac{1}{2}B) = \sqrt{(s-c)(s-a)/s(s-b)}$$

$$\sin(\tfrac{1}{2}C) = \sqrt{(s-a)(s-b)/ab}$$

$$\cos(\tfrac{1}{2}C) = \sqrt{s(s-c)/ab}$$

$$\tan(\tfrac{1}{2}C) = \sqrt{(s-a)(s-b)/s(s-c)}$$

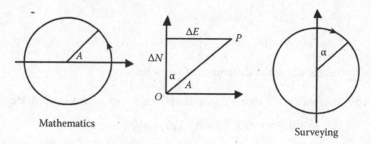

FIGURE 5.9 Angle and bearing measurements.

anticlockwise. Surveyors and navigators start with the north and measure angles and bearings clockwise (Figure 5.9).

When dealing with bearings, the Greek alphabet is often used. Bearings are angles and conform to all the rules that affect the angles A, B, and C that we have been discussing. Thus, in the center triangle of Figure 5.9 the value $\tan \alpha = \Delta E / \Delta N$ while $\tan A = \Delta N / \Delta E$. In what follows we will often use the Greek alphabet for both bearings and angles measured clockwise.

The accuracy of the process of calculating bearings and distances from coordinates depends upon the number of significant figures that are used throughout the computation, as illustrated in Examples 5.2 and 5.3.

Taking the relationships a stage further, consider Figure 5.10 where point Q forms a triangle with OP. Q is to the right of the line from O to P and the measured angles are α and β. Let the coordinates of O be (E_O, N_O), P (E_P, N_P), and Q (E_Q, N_Q).

If the bearing from O to P is ϕ, then the bearing $OQ = \phi + \alpha$ and the bearing P to $O = 180 + \phi$ so that the bearing $PQ = 180 + \phi - \beta$. $\angle Q = 180 - (\alpha + \beta)$. Hence,

$$\sin Q = \sin\{180 - (\alpha + \beta)\} = \sin(\alpha + \beta) = \sin \alpha \cos \beta + \cos \alpha \sin \beta$$

EXAMPLE 5.2: BEARINGS AND DISTANCES FROM COORDINATES

If point O in Figure 5.10 is (2624.81 E, 3427.64 N) and P is (3056.61 E, 4058.18 N), then $\Delta E = 431.80$ and $\Delta N = 630.54$. Let the bearing $OP = \phi$

$$\Delta E / \Delta N = 0.6848098 \quad \tan^{-1}(0.6848098) = 34.40373 \text{ degrees}$$

$$\phi = 34.40373 \text{ degrees} = 34°\ 24'\ 13.4'' = \text{bearing from } O \text{ to } P$$

$$\sin \phi = \Delta E / OP, \text{ hence } OP = \Delta E \operatorname{cosec} \phi = 431.8 * 1.76985 = 764.22$$

$$\cos \phi = \Delta N / OP, \text{ hence } OP = \Delta N \sec \phi = 630.54 * 1.21201 = 764.22$$

This checks the calculation.

EXAMPLE 5.3: COORDINATES FROM BEARINGS AND DISTANCES

If O in Figure 5.10 has coordinates (2624.81 E, 3427.64 N) and the distance OP = 764.22 and bearing ø = 34° 24′ then

$$\Delta E = OP \sin ø = 764.22 * 0.5649670 = 431.76$$

$$\Delta N = OP \cos ø = 764.22 * 0.8251135 = 630.57$$

These figures differ slightly from the values given in Example 5.2. This is because the bearing should in fact be ø = 34° 24′ 13.4″. Using the more accurate value would give slightly different values for sin and cos, which would result in

$$\Delta E = 764.22 * 0.5650206 = 431.80$$

$$\Delta N = 764.22 * 0.8250766 = 630.54$$

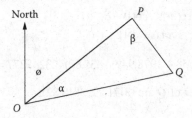

FIGURE 5.10   Fixing points from observed angles.

Using the sine formula, $OQ = OP \sin \beta / \sin Q$. As we have seen,

$$OP = (E_P - E_O)/\sin ø = (N_P - N_O)/\cos ø$$

$$E_Q = E_O + OQ \sin (ø + \alpha) = E_O + OQ \sin ø \cos \alpha + OQ \cos ø \sin \alpha$$

$$= E_O + \{OP \sin \beta \sin ø \cos \alpha + OP \sin \beta \cos ø \sin \alpha\}/[\sin \alpha \cos \beta + \cos \alpha \sin \beta]$$

By substituting for OP we obtain the relations shown in Box 5.5.

BOX 5.5   COORDINATES FROM OBSERVED ANGLES

If point Q is to the right of the line from O to P as in Figure 5.10, then

$$E_Q = \frac{E_P \cot \alpha + E_O \cot \beta + N_P - N_O}{\cot \alpha + \cot \beta}$$

$$N_Q = \frac{N_P \cot \alpha + N_O \cot \beta - E_P - E_O}{\cot \alpha + \cot \beta}$$

EXAMPLE 5.4: COMPUTING A POINT FROM TWO OBSERVED ANGLES

Using the previous data from Example 5.2 and in Figure 5.10:
Let O have coordinates $(2624.81\ E, 3427.64\ N)$ and $P\ (3056.61\ E, 4058.18\ N)$

$$\Delta E = 431.80$$

$$\Delta N = 630.54$$

Let $\alpha = 20°$ and $\beta = 60°$. Then

cot $\alpha = 2.7474774$; cot $\beta = 0.5773503$; cot α + cot $\beta = 3.3248277$

$$E_Q = \frac{E_P \cot\alpha + E_O \cot\beta + N_P - N_O}{\cot\alpha + \cot\beta}$$

$$= (8397.967 + 1515.435 + 630.54)/3.3248277 = 3171.27$$

$$N_Q = \frac{N_P \cot\alpha + N_O \cot\beta - E_P + E_O}{\cot\alpha + \cot\beta}$$

$$= (1149.758 + 1978.949 - 431.80)/3.3248277 = 3818.82$$

Hence the coordinates of Q are (3171.27, 3818.82).

If Q were on the other side of the line OP then the two last terms in the upper line of both expressions in Box 5.5 would be of the opposite sign and references to P and O would be reversed.

As an example of the use of trigonometric functions in geomatics, consider the case of traversing, which is a technique commonly used in land surveying. It entails measuring a series of angles and distances.

The traverse would start at a known point (A in Figure 5.11) and end at another known point (B) in order to confirm the accuracy of the work. The bearings and distances from A to P to Q to B would be derived from observations so that the coordinates of the two new points P and Q could be calculated. The measurement of the distances AP, PQ, and QB would be relatively straightforward, especially with modern electronic distance measuring devices.

The measurement of bearings is more problematic since the bearing of AP is an angle relative to North; but in what direction is North? The solution is to have another

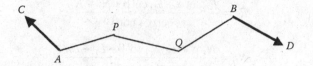

FIGURE 5.11 A traverse.

EXAMPLE 5.5: COMPUTING A TRAVERSE

Station	Distance	Observed Angle	Bearing	Sin Bearing	Cos Bearing	Δ East	Δ North	East	North
C								5663.28	13794.22
	497.68		141 21 02	0.6245538	−0.7809818	310.83	−388.68		
A		108 17 08						5974.11	13405.54
	498.93		69 38 10	0.9375015	0.3479813	467.75	173.62		
P		228 33 27						6441.86	13579.16
	318.95		118 11 37	0.8813561	−0.4724525	281.11	−150.69		
Q		108 44 11						6722.97	13428.47
	399.55		46 55 48	0.7305199	0.6828914	291.88	272.85		
B		246 54 02						7014.85	13701.32
	750.38		113 49 50	0.9147443	−0.4040332	686.41	−303.18		
D								7701.26	13398.14

known point (here C), calculate the bearing of AC, measure the angle between AC and AP clockwise and hence derive the bearing of AP. We can then derive the bearing PQ by measuring the angle at P.

Because all measurements and calculations are prone to error, either through human mistakes or the accumulation of small inaccuracies, it is necessary to verify the positions by some independent means. This is achieved by measuring the angle at Q, the distance QB, and a check angle at B to another known point (D).

In the calculation shown in Example 5.5, the given values are in bold, there are three observed distances and four observed angles from which the coordinates of the two new points have been derived. Note that values have been used that do not require adjustment; normally in a traverse there will be a need to apply corrections to ensure exact mathematical consistency. The theory of errors is explored from a statistical perspective in later chapters.

5.5   ANGLES ON A SPHERE

In practice, the world is not flat and we must extend some of the ideas above to what happens on a curved surface. For now, we will assume for practical purposes that the world is a sphere. As explained in Chapter 4, on a sphere any plane that passes through the center of the sphere cuts the surface along a line that is called a *great circle*. Thus, assuming that the Earth is a sphere, the equator and all meridians of longitude are great circles. The shortest distance between any two points on the surface of a sphere follows the line of a great circle.

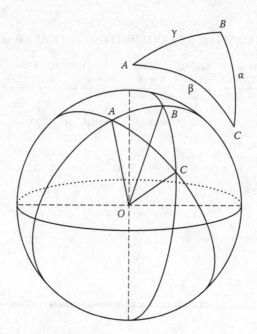

FIGURE 5.12   The spherical triangle.

A triangle on the surface of a sphere that has each of its arcs as part of a great circle is called a *spherical triangle*. Thus, if O is the center of the sphere in Figure 5.12, then the plane OAB cuts the surface along the arc AB, which is part of a great circle; similarly, for BC and CA. The length of the arc AB is the radius of the sphere multiplied by the angle AOB in radians. This is usually expressed in letters from the Greek alphabet—for example, π radians for 180°. The angle on the surface (for instance, BAC) is written in normal text—here as A. For a spherical triangle ABC the angles would then be A, B, and C while the side opposite A (i.e., BC) would be α (alpha), CA would be β (beta), and AB would be γ (gamma).

Consider a sphere of radius R (Figure 5.13). Let P be the foot of the perpendicular from A onto the plane OBC; let the line PS be perpendicular to OB and PQ be perpendicular to OC. Triangles APS and APQ both have right angles at P since AP is perpendicular to the plane that contains all the points $OQCBS$.

Furthermore, since the points A, S, and P form a plane that is at right angles to the line OB it follows that the line AS (not drawn in Figure 5.13) must be at right angles to the line OB or that ASO is a right-angled triangle with the angle at $S = 90°$.

It also follows that the plane ASP is parallel to the plane that touches the sphere at B, known as the tangent plane at B since both are at right angles to the line OB, which is a radius of the sphere. The angle ASP must therefore be the same as the angle ABC at the surface.

Thus, $\angle ASP = B$. Finally, the $\angle AOS = \angle AOB = \gamma$. Similarly, $\angle APQ = 90° = \angle AQO$. Also, $\angle AQP = C$ and $\angle AOQ = \beta$. The relationships are shown in Figure 5.14.

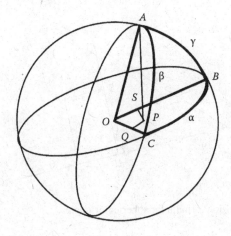

FIGURE 5.13 Spherical angles.

All this may seem difficult to visualize until we separate out the triangles. Then we have

$$\text{From Figure 5.14c} \quad AQ = OA \sin \beta = R \sin \beta$$

$$\text{From Figure 5.14d} \quad AS = OA \sin \gamma = R \sin \gamma$$

Combining these with Figure 5.14a,b, we have

$$\text{From Figure 5.14a} \quad AP = AQ \sin C = R \sin \beta \sin C$$

$$\text{From Figure 5.14b} \quad AP = AS \sin B = R \sin \gamma \sin B$$

Hence,

$$\sin \beta \sin C = \sin \gamma \sin B \text{ or } \sin B / \sin \beta = \sin C / \sin \gamma$$

And similarly,

$$= \sin A / \sin \alpha$$

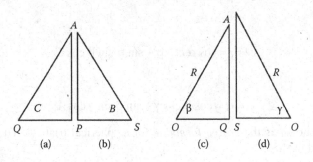

FIGURE 5.14 The sine formula for spherical triangles.

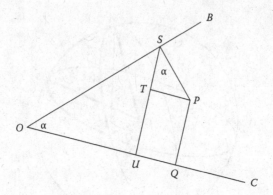

FIGURE 5.15   The cosine formula for spherical triangles.

Thus,

$$\frac{\sin A}{\sin \alpha} = \frac{\sin B}{\sin \beta} = \frac{\sin C}{\sin \gamma}$$

This is known as the *sine formula* for a spherical triangle.

Taking this a stage further, in the plane OBC as shown in Figure 5.13, draw SU perpendicular to OC and PT perpendicular to SU. This is shown in Figure 5.15 where $\angle BOC = \alpha$ and hence, $\angle OSU = 90° - \alpha$. Since PS is perpendicular to OB, then $\angle PST = \alpha$.

From Figure 5.14, $AS = R \sin \gamma$ and $SP = R \sin \gamma \cos B$; also $OQ = R \cos \beta$ and $OS = R \cos \gamma$. From Figure 5.15, $TP = SP \sin \alpha = R \sin \alpha \sin \gamma \cos B$. Also, $TP = UQ$ (since $PTUQ$ is a rectangle) $= OQ - OU = R \cos \beta - OS \cos \alpha$
$$= R \cos \beta - R \cos \gamma \cos \alpha.$$

Hence,

$$R \sin \alpha \sin \gamma \cos B = R \cos \beta - R \cos \gamma \cos \alpha$$

Dividing through by R and rearranging

$$\cos \beta = \cos \gamma \cos \alpha + \sin \gamma \sin \alpha \cos B$$

Similarly,

$$\cos \gamma = \cos \alpha \cos \beta + \sin \alpha \sin \beta \cos C$$

and

$$\cos \alpha = \cos \beta \cos \gamma + \sin \beta \sin \gamma \cos A$$

These are known as the *cosine formulae* for a spherical triangle. The results are summarized in Box 5.6.

**BOX 5.6   SINE AND COSINE FORMULAE
FOR SPHERICAL TRIANGLES**

1. $\dfrac{\sin A}{\sin \alpha} = \dfrac{\sin B}{\sin \beta} = \dfrac{\sin C}{\sin \gamma}$

2. $\cos \alpha = \cos \beta \cos \gamma + \sin \beta \sin \gamma \cos A$

   $\cos \beta = \cos \gamma \cos \alpha + \sin \gamma \sin \alpha \cos B$

   $\cos \gamma = \cos \alpha \cos \beta + \sin \alpha \sin \beta \cos C$

As an example of the application for the sine and cosine formulae, consider two points A and B with the following latitude and longitude (see Figure 5.16):

$$A\ (40°\ N,\ 10°\ E)\ \text{and}\ B\ (55°\ N,\ 15°\ E)$$

The angle ANB at the North Pole = difference in longitude between A and B = 5°. The latitude of A is the angle that OA makes with the equator which means that the angle NOA = length of arc AN = 90° – latitude of A (known as the *colatitude* of A) = 50°.

Similarly, the length of arc NB = 35° (the colatitude of B). We therefore have a spherical triangle in which we know two sides (AN and BN) and an included angle (ANB). Hence, using the spherical sine and cosine formulae as shown in Box 5.6, we can calculate the bearing and distance of A to B, as shown in Example 5.6.

Spherical trigonometry underpins the conversion of coordinates on the surface of the Earth into coordinates that can be used on a flat surface such as a map. In Chapter 11, we examine map projections and some of the ways in which the data may be transformed.

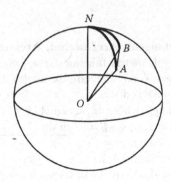

FIGURE 5.16   Colatitudes AN, BN.

EXAMPLE 5.6: BEARING AND DISTANCE
FROM LATITUDE AND LONGITUDE

In Figure 5.16, N is the North Pole. A ($40°$ N, $10°$ E) and B ($55°$ N, $15°$ E) differ in longitude by $5°$. Angle $AN = 50°$, $BN = 35°$, and $ANB = 5°$.
    From the cosine formula,

$$\cos (AB) = \cos 50 \cos 35 + \sin 50 \sin 35 \cos 5$$

$$= 0.5265408 + 0.4377130 = 0.9642538$$

Hence, the angle $AB = 15°\ 22'$ degrees or 0.2681836 radians. This is the angle AOB subtended by AB at the center.
    Assuming the radius of the Earth to be $6.4 * 10^6$ meters then the length of $AB = 0.2681836^r * 6.4 * 10^6 = 1716.4$ kilometers.
    Using the sine formula,

$$(\sin ANB)/(\sin AB) = (\sin NAB)/(\sin NB), \text{ or}$$

$$\sin NAB = \sin NB * \sin ANB/\sin AB = 0.1886573$$

Hence,

$$\text{angle } NAB = 10°\ 52' = \text{Bearing from } A \text{ to } B$$

Similarly,

$$\sin NBA = \sin NA * \sin ANB/\sin NB = 0.2519627$$

Hence,

$$\text{Angle } NBA = 165°\ 24'$$

$$\text{Bearing } BA = 194°\ 36'$$

Note that in general, unlike on a plane surface, the direction from A to B on a sphere does not normally differ by $180°$ from the direction from B to A.

SUMMARY

Arcsine, Arccosine, Arctangent, Arccosecant, Arcsecant, Arccotangent: The inverse of the trigonometric functions *sine*, *cosine*, and so on. Written as $\sin^{-1} ø$, $\cos^{-1} ø$, and so on. They are the angles that give rise to the values sine ø, cosine ø, and so on.

Area of triangle: In a triangle whose angles are A, B, C, and sides a, b, c, then
     area $= ½\ bc \sin A = ½\ ca \sin B = ½\ ab \sin C$
     If $s = ½(a + b + c)$ then area $= \sqrt{s(s-a)(s-b)(s-c)}$

Combined angles: For two angles A, B
     $\sin (A + B) = \sin A \cos B + \cos A \sin B$
     $\cos (A + B) = \cos A \cos B - \sin A \sin B$

$$\sin (A - B) = \sin A \cos B - \cos A \sin B$$
$$\cos (A - B) = \cos A \cos B + \sin A \sin B$$
$$\sin (2A) = 2 \sin A \cos A$$
$$\cos (2A) = 1 - 2 \sin^2 (A) = 2 \cos^2 (A) - 1$$
$$\sin (A) = 2 \sin \tfrac{1}{2}A \cos \tfrac{1}{2}A$$
$$\cos (A) = 1 - 2 \sin^2 (\tfrac{1}{2}A) = 2 \cos^2 (\tfrac{1}{2}A) - 1$$

If $s = \tfrac{1}{2}(a + b + c)$

$$\sin^2 (\tfrac{1}{2}A) = (s - b)(s - c)/bc$$
$$\cos^2 (\tfrac{1}{2}A) = s(s - a)/bc$$
$$\tan^2 (\tfrac{1}{2}A) = \{(s - b)(s - c)\}/\{s(s - a)\}$$

Coordinates from angles: If in triangle ABC, C is to the right of the line A to B and the coordinates of A are (x_A, y_A) and of B are (x_B, y_B) then point C is
$$x_C = \{x_B \cot A + x_A \cot B + y_B - y_A\}/\{\cot A + \cot B\}$$
$$y_C = \{y_B \cot A + y_A \cot B + x_B - x_A\}/\{\cot A + \cot B\}$$

Cosecant: The reciprocal of *Sine*. Written as cosec ø. cosec ø = 1/sin ø.

Cosine: The ratio in a right-angled triangle between the side adjacent and the hypotenuse. Written as cos ø. $\cos^2 ø = 1 - \sin^2 ø$.

In the triangle ABC, $\cos(\tfrac{1}{2}A) = \sqrt{s(s-a)/bc}$, and so forth.

Cosine formula: In a plane triangle whose angles are A, B, C, and sides a, b, c, and semiperimeter s, then
$$\cos A = (b^2 + c^2 - a^2)/2bc$$
$$\cos B = (c^2 + a^2 - b^2)/2ca$$
$$\cos C = (a^2 + b^2 - c^2)/2ab$$
$$\cos(\tfrac{1}{2}A) = \sqrt{s(s-a)/bc}$$
$$\cos(\tfrac{1}{2}B) = \sqrt{s(s-b)/ca}$$
$$\cos(\tfrac{1}{2}C) = \sqrt{s(s-c)/ab}$$

Cotangent: The reciprocal of *Tangent*. Written as cot ø.
$$\cot ø = 1/\tan ø$$

Obtuse angles: Angles greater than 90° and less than 180°. For an angle A:
If $90° \le A \le 180°$: $\sin A = +\sin (180 - A)$, $\cos A = -\cos (180 - A)$,
$\tan A = -\tan (180 - A)$
If $180° \le A \le 270°$: $\sin A = -\sin (A - 180)$, $\cos A = -\cos (A - 180)$, and
$\tan A = \tan (A - 180)$
If $270° \le A \le 360°$: $\sin A = -\sin(360 - A)$, $\cos A = +\cos (360 - A)$, and
$\tan A = -\tan (360 - A)$
Also, $\sin (90 + A) = \cos A$, $\cos (90 + A) = -\sin A$, and
$\tan (90 + A) = -\cot A$
And, $\sin (-A) = -\sin A$, $\cos (-A) = \cos A$, $\tan (-A) = -\tan A$

Secant: The reciprocal of *cosine*. Written as sec ø. sec ø = 1/cos ø.

Sine: The ratio in a right-angled triangle between the side opposite and the hypotenuse. Written as sin ø. $\sin^2 ø = 1 - \cos^2 ø$.

In the triangle ABC, $\sin (\tfrac{1}{2}A) = \sqrt{(s-b)(s-c)/bc}$, and so forth.

Sine formula: In a plane triangle whose angles are A, B, C, and sides a, b, c, and semiperimeter s, then
$$(\sin A)/a = (\sin B)/b = (\sin C)/c$$

$$\sin(\tfrac{1}{2}A) = \sqrt{(s-b)(s-c)/bc}$$
$$\sin(\tfrac{1}{2}B) = \sqrt{(s-c)(s-a)/ca}$$
$$\sin(\tfrac{1}{2}C) = \sqrt{(s-a)(s-b)/ab}$$

Spherical triangle: A triangle on the surface of a sphere formed by the arcs of three great circles. If the angles on the surface where these arcs intersect are A, B, C, and if the arcs subtend angles α, β, γ at the center of the sphere then

$\sin A/\sin \alpha = \sin B/\sin \beta = \sin C/\sin \gamma$

$\cos \alpha = \cos \beta \cos \gamma + \sin \beta \sin \gamma \cos A$

$\cos \beta = \cos \gamma \cos \alpha + \sin \gamma \sin \alpha \cos B$

$\cos \gamma = \cos \alpha \cos \beta + \sin \alpha \sin \beta \cos C$

Tangent of an angle: The ratio in a right-angled triangle between the side opposite and the side adjacent. Written as tan ø.

tan ø = sin ø/cos ø.

In the triangle ABC, $\tan(\tfrac{1}{2}A) = \sqrt{(s-b)(s-c)/s(s-a)}$, and so forth.

The word tangent is also used to describe a line that just touches a curve or a plane that just touches a solid.

Traverse: The locating of points by measuring a series of distances and associated bearings, derived from successive angular measurements.

6 Differential and Integral Calculus

6.1 DIFFERENTIATION

Calculus is a branch of mathematics the origins of which were the basis of an acrimonious dispute between two great mathematicians—Sir Isaac Newton and Gottfried Leibniz—both of whom claimed to be the discoverer. *Differential calculus* is concerned with the rate at which a function changes while *integral calculus* extends the idea that the sum of a finite number of separate values can be used to form a continuous value as, for instance, in determining the area under a curve.

First, consider a function such as the parabola plotted as *ABCD*, and so forth, in Figure 6.1. It can take the form $y = ax^2$ where a is some constant. If we travel from a point B with coordinates (x, y) on the curve to a point C, the direction in which we are traveling along the curve changes from BB' to CC'. These directions are called the *tangents* at B and C. They are related to the idea of the tangent of an angle, as discussed in Chapter 5, since the slope of the straight line from B to C is the ratio shown as $\delta y/\delta x$ in Figure 6.2 (δ being the Greek letter "delta" and is shown here in lowercase). $\delta y/\delta x$ is the tangent of the angle θ (the Greek letter "theta") of the slope of the straight line BC.

If B is (x, y) and the distance marked as "δx" is small (δ standing for "a small amount of") and if, on moving from B to C along the curve, the y value increases by an equivalent small amount δy, then the coordinates of C must be: $(x + \delta x, y + \delta y)$.

These must also satisfy the basic equation $y = ax^2$. Hence:

$$y + \delta y = a(x + \delta x)^2 = ax^2 + 2ax\delta x + a\delta x^2$$

Since for point B, $y = ax^2$, it follows that

$$\delta y = 2ax.\delta x + a\delta x^2 = (2ax + a\delta x)\,\delta x$$

Dividing both sides by δx,

$$\delta y/\delta x = 2ax + a\delta x$$

As δx gets smaller and smaller, so does δy and the ratio between the two, namely, $\delta y/\delta x$ tends to the value $= 2ax$. In the limit, we use the ordinary alphabet and say that $dy/dx = 2ax$. This process is known as *differentiation*. Thus, if we differentiate y with respect to x where $y = ax^2$, we obtain $dy/dx = 2ax$.

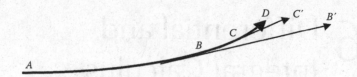

FIGURE 6.1 Tangents to a curve.

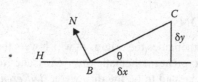

FIGURE 6.2 The slope and the normal to a curve.

Note that δx means "a little bit of x" but dy/dx means "in the limit the value of $\delta y/\delta x$ when both δy and δx have become very, very small."

If we have a function of the form $y = ax^2 + bx + c$, then

$$y + \delta y = a(x + \delta x)^2 + b(x + \delta x) + c = ax^2 + 2ax\delta x + a(\delta x)^2 + bx + b\delta x + c$$

or

$$\delta y = 2ax\delta x + a(\delta x)^2 + b\delta x$$

On dividing both sides by δx,

$$\delta y/\delta x = 2ax + b + a\delta x$$

In the limit as δx tends to zero,

$$dy/dx = 2ax + b$$

We can extend the argument to a cubic and find that if

$$y = ax^3 + bx^2 + cx + e$$

then

$$dy/dx = 3ax^2 + 2bx + c$$

And in general if $y = ax^n$, then $dy/dx = nax^{(n-1)}$.

If we have two functions u and v that are both functions of x and if $y = uv$, then

$$y + \delta y = (u + \delta u)(v + \delta v) = uv + u\delta v + v\delta u + \delta u.\delta v$$

Thus,

$$\delta y = u\delta v + v\delta u + \delta u.\delta v$$

hence

$$\delta y/\delta x = u\delta v/\delta x + v\delta u/\delta x + \delta u.\delta v/\delta x$$

The expression $\delta u.\delta v$ gets smaller twice as fast as either δu or δv separately; hence $+\delta u.\delta v/\delta x$ tends to zero. The result is that

If $y = uv$, then

$$dy/dx = u\,dv/dx + v\,du/dx$$

(Note: If $y = x^2 = x * x$, then $u = v = x$ and $dy/dx = x(dx/dx) + x(dx/dx) = 2x$, as we would expect.) This is summarized in Box 6.1.

"dy/dx" is the rate of change of the function y with respect to x or, looked at another way as in Figure 6.1, it is the slope of the curve at point (x, y). The line that is perpendicular to the tangent (BN in Figure 6.2) is known as the *normal* to the curve. The slope of $BN = \theta + 90°$. Also,

$$\tan (\theta + 90°) = \sin (\theta + 90°)/\cos (\theta + 90°)$$

$$= -\cos \theta/\sin \theta = -\cot \theta$$

$$= -(\delta x/\delta y) = -1/(\delta y/\delta x)$$

Thus, the tangent of the slope of the normal $= -1$ divided by the tangent of the slope of the curve.

In general if y is a function of x, we can write this as $y = f(x)$ where we use the expression "$f(x)$" to mean "the function (x)," which in turn means "the value of the function given the value x." When we differentiate the function $y = f(x)$, we obtain dy/dx which we can write as $f'(x)$. This can in turn be differentiated to give the *second derivative*.

BOX 6.1   BASIC DIFFERENTIALS

If $y = ax^n + bx^{(n-1)} + cx^{(n-2)} + \ldots.. px^2 + qx + r$, then

$$dy/dx = nax^{(n-1)} + (n-1)bx^{(n-2)} + (n-2)cx^{(n-3)} + \ldots.2px + q$$

If $y = uv$ where u and v are both functions of x, then

$$dy/dx = u\,dv/dx + v\,du/dx$$

This can be written either as d^2y/dx^2 or as $f''(x)$. Thus, if $y = f(x) = a\,x^n$, then

$$dy/dx = f'(x) = n\,a\,x^{(n-1)}$$

and

$$d^2y/dx^2 = f''(x) = n\,(n-1)\,a\,x^{(n-2)}$$

Thus, if

$$y = ax^3 + bx^2 + cx + e$$

then

$$dy/dx = f'(x) = 3ax^2 + 2bx + c$$

$$d^2y/dx^2 = f''(x) = 6ax + 2b$$

$$d^3y/dx^3 = f'''(x) = 6a$$

$$d^4y/dx^4 = f^{IV}(x) = 0$$

($f'''(x)$ is the *third derivative* while f^{IV} is the *fourth derivative,* which for a cubic is zero.)

If dy/dx represents the rate of change of a curve then d^2y/dx^2 represents the rate of change; put in another context, if y is a distance, dy/dx is the rate of change of the rate of change distance or the velocity or speed, and d^2y/dx^2 is the acceleration. d^3y/dx^3 would be the rate of change of the acceleration.

Consider the function

$$y = 1 + 9x - 6x^2 + x^3$$

$$dy/dx = 9 - 12x + 3x^2 = 3(1 - x)(3 - x)$$

$$d^2y/dx^2 = -12 + 6x$$

$$d^2y/dx^2 = 0 \text{ when } x = 2 \text{ and } y = 3$$

$$dy/dx = 0 \text{ when } x = 1 \text{ and } y = 5$$

$$\text{and when } x = 3 \text{ and } y = 1$$

In Table 6.1 we list some values of x and y for this function and have interpolated a curve between the points in Figure 6.3a. It will be seen from Figure 6.3a that there are two points where the slope is horizontal, that is, $dy/dx = 0$ {here (1,5) and (3,1)}. These points are known as the *maxima* or *minima* for the curve. A cubic has one maximum and one minimum point—although note that this does not mean a maximum value for all points on the curve, which can come from and go off to ± infinity.

TABLE 6.1
Data for $y = 1 + 9x - 6x^2 + x^3$

x	-3	-2	-1	-0.5	0	0.5	1	1.5	2	2.5	3	3.5	4	4.5	5	5.5	6
y	-107	-49	-15	-5.1	1	4.1	5	4.4	3	1.6	1	1.9	5	11.1	21	35.4	55

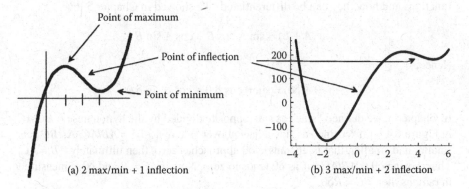

(a) 2 max/min + 1 inflection (b) 3 max/min + 2 inflection

FIGURE 6.3 (a) A cubic curve and (b) a quartic.

Maxima and minima are turning points where the curve having gone upward now turns downward or vice versa.

A point where $d^2y/dx^2 = 0$ is known as a *point of inflection* and a cubic curve has one of them. At a point of inflection the curve stops bending in one direction and starts bending in the other; in effect, the curve crosses its tangent at that point. In the cubic in Figure 6.3a, the curve starting from the bottom left turns clockwise until it reaches the point of inflection where $x = 2$ and $y = 3$ and then starts bending in an anticlockwise direction. An example based on a quartic equation is given in Example 6.1 and illustrated in Figure 6.3b.

EXAMPLE 6.1: POINTS OF INFLECTION

The quartic equation $y = x^4 - 8x^3 - 2x^2 + 120x + 6$ has three maxima/minima and two points of inflection.

$$dy/dx = 4x^3 - 24x^2 - 4x + 120 = 4(x + 2)(x - 3)(x - 5)$$

This is zero when $x = -2, +3,$ and $+5$. Hence, there are three places where the curve reaches a peak or a trough, namely, $(-2, -162), (3, 213), (5, 181)$

$$d^2y/dx^2 = 12x^2 - 48x - 4 = 4(3x^2 - 12x - 1)$$

Using the formula from Chapter 3 where the solution to the quadratic equation

$$ax^2 + bx + c = 0 \text{ was given as } x = \{-b \pm \sqrt{(b^2 - 4ac)}\}/2a, \text{ then}$$

$(3x^2 - 12x - 1)$ is zero when $x = \{12 \pm \sqrt{(144 + 23)}\}/6$ or $x \approx -0.08$ or 4.08.

In Chapter 8, we discuss the idea of a radius of curvature at a point on a curve; at a point of inflection the radius of curvature is infinitely large.

6.2 DIFFERENTIATING TRIGONOMETRIC FUNCTIONS

Before illustrating the use of differentiation, we need to consider the trigonometric functions and how they can be differentiated. We showed in Chapter 5 that

$$\sin (A + B) = \sin A \cos B + \cos A \sin B$$

Hence,

$$\sin (\theta + \delta\theta) = \sin \theta \cos \delta\theta + \cos \theta \sin \delta\theta$$

In Chapter 5, we defined "sine" as side opposite divided by the hypotenuse $= BC/AC$ in Figure 6.4 with "cosine" as side adjacent over side opposite $= AB/AC$. As BC gets smaller and smaller, that is, the angle $\delta\theta$ approaches zero, then ultimately $AB = AC$. Thus, cosine $\delta\theta = AB/AC = 1$ as $\delta\theta$ tends to zero. Now if $AC = R$ and $\delta\theta$ is measured in radians then $BC = R\delta\theta$.

As $\delta\theta$ gets smaller, BC/AC or sin $\delta\theta$ becomes $R\delta\theta/R = \delta\theta$. Hence, for small values of $\delta\theta$, sin $\delta\theta = \delta\theta$. If $y = \sin \theta$, then

$$y + \delta y = \sin (\theta + \delta\theta) = \sin \theta \cos \delta\theta + \cos \theta \sin \delta\theta$$

$$= \sin \theta + \delta\theta \cos \theta$$

Since $y = \sin \theta$, it means that $\delta y = \delta\theta \cos \theta$, or $\delta y/\delta\theta = \cos \theta$.

If we revert to the notation that $y = \sin x$, then $dy/dx = \cos x$.

We have also seen in Chapter 5 that $\cos (A + B) = \cos A \cos B - \sin A \sin B$. Hence, if $y = \cos \theta$ then

$$y + \delta y = \cos (\theta + \delta\theta) = \cos \theta \cos \delta\theta - \sin \theta \sin \delta\theta$$

$$= \cos \theta - \delta\theta \sin \theta$$

Therefore, $\delta y/\delta\theta = -\sin \theta$. Thus, if $y = \cos x$, then $dy/dx = -\sin x$.

The results are summarized in Box 6.2. See also Example 6.2.

Before considering the other trigonometric functions, we need to consider how to differentiate products. Consider $y = \sin^2 x = \sin x * \sin x$. Let $u = \sin x$, then $du/dx =$

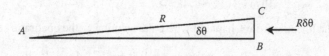

FIGURE 6.4 Small angles.

EXAMPLE 6.2: THE RATE OF CHANGE OF SIN X

We have shown that if $y = \sin x$, then $dy/dx = \cos x$. To illustrate this, let $x = 30°$ $= \pi/6$ radians $= 0.5235987756^r$

$$\sin (0.5235988^r) = 0.50000002113249$$

$$\sin (0.5235989^r) = 0.50000010773503$$

The difference in sines $= 0.00000008660254 =$ rate of change for sin (30°) for a difference in angle of 0.0000001 radians

$$\cos 30° = 0.8660254$$

Hence, the rate of change of sin (30°) = cos (30°).

$\cos x$. If $y = \sin^2 x$, then $y = u^2$, and $dy/du = 2u$. Remembering that dy/du means the limit of $\delta y/\delta u$ as δy and δu approach zero, then

$$\frac{\delta y}{\delta x} = \frac{\delta y * \delta u}{\delta x * \delta u} = \frac{\delta y * \delta u}{\delta u * \delta x} = \frac{\delta y}{\delta u} * \frac{\delta u}{\delta x}$$

In the limit,

$$(dy/dx) = (dy/du) * (du/dx) = 2u * \cos x = 2 \sin x \cos x$$

Thus, if

$$y = \sin^2 x$$

then

$$dy/dx = 2 \sin x \cos x$$

Similarly, if

$$y = \sin^3 x$$

then

$$dy/dx = 3 \sin^2 x \cos x, \text{ and so on.}$$

Hence, if

$$y = \sin^n x$$

then

$$dy/dx = n \sin^{(n-1)} x \cos x$$

We showed above that if $y = u * v$ where u and v are both functions of x, then

$$dy/dx = u\ dv/dx + v\ du/dx$$

We can use this to differentiate $\tan x$.
If

$$y = \tan x = (\sin x)/(\cos x)$$

then let

$$u = \sin x$$

for which

$$du/dx = \cos x$$

If

$$v = \sec x = 1/(\cos x) = (\cos x)^{-1}$$

then

$$dv/dx = -1 * (\cos x)^{-2} * (d(\cos x)/dx)$$

Hence,

$$dv/dx = -1 * (\cos x)^{-2} * (-\sin x) = \sin x/(\cos^2 x) = \tan x \sec x$$

(We can write this as $y = \sec x$ and $dy/dx = \sec x \tan x$.)

·If

$$y = \tan x = (\sin x)/(\cos x) = u * v$$

then

$$du/dx = \cos x \text{ and } dv/dx = \sec x \tan x$$

By substituting these values in "$\{dy/dx = u\ dv/dx + v\ du/dx\}$," we obtain for

$$y = \tan x$$

then

$$dy/dx = \sin x * \{\sin x/(\cos^2 x)\} + (1/\cos x) * \cos x$$

$$= \sin^2 x/\cos^2 x + 1 = (\sin^2 x + \cos^2 x)/\cos^2 x$$

$$= 1/\cos^2 x = \sec^2 x$$

(since $\sin^2 + \cos^2 = 1$). Hence, if

$$y = \tan x, \text{ then } dy/dx = \sec^2 x$$

Similar relationships exist for $\operatorname{cosec} x$ and $\cot x$. These are summarized in Box 6.2.

BOX 6.2   DIFFERENTIATING TRIGONOMETRICAL FUNCTIONS

If $y = \sin x$	then	$dy/dx = \cos x$
If $y = \cos x$		$dy/dx = -\sin x$
If $y = \tan x$		$dy/dx = \sec^2 x$
If $y = \operatorname{cosec} x$		$dy/dx = -\operatorname{cosec} x \cot x$
If $y = \sec x$		$dy/dx = \sec x \tan x$
If $y = \cot x$		$dy/dx = -\operatorname{cosec}^2 x$

6.3   POLYNOMIAL FUNCTIONS

Let us return to the general polynomial of the form:

$$y = f(x) = a_0 + a_1 x + a_2 x^2 + a_3 x^3 + a_4 x^4 + \cdots + a_n x^n$$

(Note as an aside that $a_0 = a_0 x^0$ while $a_1 x = a_1 x^1$ and that $f(0)$, the value of $f(x)$ when $x = 0$ is such that $f(0) = a_0$.)

$$dy/dx = f'(x) = a_1 + 2a_2 x + 3a_3 x^2 + 4a_4 x^3 + \cdots + na_n x^{(n-1)}$$

When x becomes zero then $f'(0) = a_1$ or $a_1 = f'(0)/1$. If we differentiate $f'(x)$, we obtain $f''(x)$ and so:

$$d^2y/dx^2 = f''(x) = 2a_2 + 2 * 3a_3 x + 3 * 4a_4 x^2 + \cdots + (n-1)na_n x^{(n-2)}$$

Thus,

$$f''(0) = 2a_2$$

or

$$a_2 = f''(0)/(1 * 2)$$

Again,

$$d^3y/dx^3 = f'''(x) = 1 * 2 * 3a_3 + 2 * 3 * 4a_4 x + \cdots + (n-2)(n-1)nx^{(n-3)}$$

Thus,

$$f'''(0) = 1 * 2 * 3 * a_3$$

or

$$a_3 = f'''(0)/(1 * 2 * 3)$$

If we keep on repeating this process, we will find that

$$a_n = f^n(0)/(1 * 2 * 3 * \cdots * n) = f^n(0)/n!$$

where $f^n(0)$ is the value of the nth derivative when $x = 0$. Hence, we can write our original equation as

$$y = f(0) + x^1 f'(0)/1! + x^2 f''(0)/2! + x^3 f'''(0)/3! + \cdots + x^n f^n(0)/n!$$

This is known as Maclaurin's theorem after the 18th century Scottish mathematician Colin Maclaurin. Its particular relevance relates to functions such as $f(x) = \sin x$ because then we have $f'(x) = \cos x$, $f''(x) = -\sin x$, $f'''(x) = -\cos x$, and $f^{iv}(x) = \sin x$.

Since $\sin(0) = 0$ and $\cos(0) = 1$, then by measuring x in radians and using Maclaurin's formula, we obtain:

$$\sin x = x - x^3/3! + x^5/5! - x^7/7! + \cdots$$

Likewise,

$$\cos x = 1 - x^2/2! + x^4/4! - x^6/6! + \cdots$$

Thus, we have a way of calculating $\sin x$ and $\cos x$.

If we define the exponential function $f(x) = e^x$ as the function for which its differential is itself, then $f'(x) = e^x = f''(x)$, and so on. Since $e^0 = 1$ (anything to the power of zero = 1), then

$$e^x = 1 + x^1/1! + x^2/2! + x^3/3! + x^4/4! + \cdots$$

$$e^1 = 1 + 1/1! + 1/2! + 1/3! + 1/4! + \cdots$$

$$= 2.7182818\ldots. \text{ As quoted earlier.}$$

These series are infinite in that they go on forever, but they converge fairly rapidly allowing for the values of functions such as sine, cosine, and e^x to be computed relatively simply (Box 6.3 and Example 6.3). It is not possible to express all functions in the form of a polynomial but many can be expressed in this way.

Note that in the expansion of e^x given above, we can replace e^x by e^{nx} where n is some constant. The general term in the expansion of e^{nx} will be $(nx)^r/r! = n^r x^r/r!$ Differentiating,

$$d(n^r x^r/r!)/dx = r\, n^r x^{r-1}/r! = n\{n^{r-1} x^{r-1}/(r-1)!\}$$

This applies to every term in the expansion of e^{nx}. Hence,

$$\text{the differential of } e^{nx} = n * e^{nx}$$

or

$$d(e^{nx})/dx = n * e^{nx}$$

BOX 6.3   MACLAURIN'S THEOREM

If $y = f(x)$, then

$$y = f(0) + xf'(0)/1! + x^2f''(0)/2! + x^3f'''(0)/3! + \cdots x^nf^n(0)/n!$$

In particular,

$$e^x = 1 + x^1/1! + x^2/2! + x^3/3! + x^4/4! + \cdots$$

And, where x is in radians,

$$\sin x = x - x^3/3! + x^5/5! - x^7/7! + \cdots$$

$$\cos x = 1 - x^2/2! + x^4/4! - x^6/6! + \cdots$$

Also (see Section 6.4), $\log_e(1 + x) = \ln(1 + x) = x - x^2/2 + x^3/3 - x^4/4 + \cdots$

In particular, the differential of $e^{-x} = d(e^{-x})/dx = -e^{-x}$.

Also, if $y = e^x$, then $\log_e y = \log_e(e^x) = x\log_e e = x$ (as was shown in Chapter 2). Differentiating with respect to y rather than x we obtain:

$$\frac{d(\log_e y)}{dy} = dx/dy$$

But we defined e^x so that $dy/dx = e^x = y$. Hence,

$$dx/dy = 1/y$$

EXAMPLE 6.3: SINE AND COSINE VALUES USING MACLAURIN'S THEOREM

First, let us evaluate sin 30 where $30° = \pi/6$ radians $= 0.5235987756$r. Using the formula from Box 6.3, namely, $\sin x = x - x^3/3! + x^5/5! - x^7/7! + \ldots$ then,

$$\sin 30 = 0.5235988 - 0.0239246 + 0.0003280 - 0.0000021 + \ldots$$

$$= 0.5$$

Using the formula for $\cos x = 1 - x^2/2! + x^4/4! - x^6/6! + \ldots$ then,

$$\cos 30 = 1 - 0.1370778 + 0.0031317 - 0.0000286 + 0.0000001 - \ldots$$

$$= 0.8660254$$

These values agree with the tables for sine and cosine.

Thus,

$$\frac{d(\log_e y)}{dy} = 1/y$$

In general,

$$\frac{d(\log_e x)}{dx} = 1/x$$

Also,

$$\frac{d(\log_e y)}{dx} = \frac{d(\log_e y)}{dy} * dy/dx = (1/y) * dy/dx$$

In the case of $y = e^{x^2}$, then $\log_e y = x^2$ and $(1/y)dy/dx = 2x$. Hence,

$$\text{if } y = e^{x^2}$$

then

$$dy/dx = 2yx = 2x\,e^{x^2}$$

If we extend all these ideas to three or more dimensions, we will have functions such as

$$z = f(x, y) = 3x^3 + 2x^2y + y^3 + 4x + 5y$$

If we assume that y is treated as a constant, then we can differentiate z with respect to x. Similarly, we can keep x constant and differentiate with respect to y. These are known as the *partial derivatives* and are shown with a curly delta ∂. Thus, in this example,

$$\partial z/\partial x = 9x^2 + 4xy + 4$$

and

$$\partial z/\partial y = 2x^2 + 3y^2 + 5$$

We will use partial derivatives in later chapters.

6.4 LINEARIZATION

The process of converting complex relationships into linear combinations of variables is sometimes referred to as *linearization*. Sir Isaac Newton, for example, developed a way to solve quite complex nonlinear equations in which he reduced the problem to one dimension and repeated the process iteratively until he had a satisfactory solution. In Example 2.4 in Chapter 2, we showed how to obtain a square root by iteration, although we did not explain why it worked.

Consider the general case of $f(x) = k$ where k is a constant and $f(x)$ is a function of x such as the polynomial $ax^3 + bx^2 + cx$.

EXAMPLE 6.4: NEWTON'S METHOD
FOR SOLVING POLYNOMIALS

In Table 6.1 we considered the function $y = f(x) = 1 + 9x - 6x^2 + x^3$ for which $dy/dx = f'(x) = 9 - 12x + 3x^2$. What is the value of x for which $y = 0$? From Table 6.1 an answer must lie between $x = -0.5$ and 0. Let us try

$$x_0 = -0.4 \text{ and use } x_{new} = x_{old} - \{f(x_{old}) - k\}/f'(x_{old})$$

where

$$k = 1 + 9x - 6x^2 + x^3 = 0$$

Our next tries should be:

$$x_1 = -0.4 - \{1 - 3.6 - 0.96 - 0.064\}/\{9 + 4.8 + 0.48\}$$
$$= -0.4 + 3.624/14.28 = -0.146$$

So now try $x = -0.146$

$$x_2 = -0.146 - \{1 - 1.314 - 0.128 - 0.003\}/\{9 + 1.752 + 0.064\}$$
$$= -0.146 + 0.445/10.816 = -0.105$$

So now try $x = -0.105$

$$x_3 = -0.105 - \{1 - 945 - 0.066 - 0.001\}/\{9 + 1.260 + 0.033\}$$
$$= -0.105 + 0.012/10.293 = -0.104$$

To three decimal places, the solution $= -0.104$

Solving the equations $ax^3 + bx^2 + cx = k$ can be complicated. In fact, *Newton's method* for solving equations (see Example 6.4) is particularly useful where the polynomial is of fifth or higher degree since there are no general algebraic methods for solving higher-order equations.

However, if we can take a guess at an approximate solution (in the case of a cubic there will of course be three possible answers, though some may involve imaginary numbers), then we can calculate

$$x_{new} = x_{old} - \left\{ \frac{f(x_{old} - k)}{f'(x_{old})} \right\} = x_{old} - \{f(x_{old} - k)\}/f'(x_{old})$$

where $f(x_{old})$ means the value of the function using the currently assumed value for x and f' is the first derivative evaluated using the same value.

We then repeat this process over and over until we have a sufficient number of significant figures for the answer to be acceptable. In fact, this is what happened in Example 2.4 where $f(x) = x^2 = k$ (with k being the number 27392834) and $f'(x) = 2x$.

We then set $x_{new} = [x_{old} - \{(x_{old})^2 - k\}/2x_{old}] = [(1/2)\{x_{old} + k/x_{old}\}]$. On the third iteration we found the value we were seeking. Example 6.4 illustrates the procedure for a cubic.

In Section 6.3, we introduced Maclaurin's theorem in which

$$y = f(0) + x^1 f'(0)/1! + x^2 f''(0)/2! + x^3 f'''(0)/3! + \cdots + x^n f^n(0)/n!$$

Another 18th century mathematician, Brook Taylor, has given his name to an extension of this theorem, which states that if $f(x)$ is a polynomial of the form $a_0 + a_1 x + a_2 x^2 + \cdots + a_n x^n$, then if δ is any number

$$f(x + \delta) = f(\delta) + x f'(\delta)/1! + x^2 f''(\delta)/2! + x^3 f'''(\delta)/3! + \cdots + x^n f^n(\delta)/n!$$

where f^n is the nth derivative of $f(x)$. The proof of this is given in Box 6.4. By putting $\delta = 0$, we have the same expression as in the Maclaurin theorem.

BOX 6.4   TAYLOR'S THEOREM

Given that

$$f(x) = a_0 + a_1 x + a_2 x^2 + \cdots + a_n x^n$$

we have

$$f'(x) = a_1 + 2a_2 x + \cdots + na_n x^{n-1}$$

$$f''(x) = 2a_2 + 3 * 2a_3 x + \cdots + n * (n-1)a_n x^{n-2}, \text{ and so on.}$$

Also, for

$$f(x + \delta) = a_0 + a_1(x + \delta) + a_2(x + \delta)^2 + \cdots + a_n(x + \delta)^n$$

then

$$f'(x + \delta) = a_1 + 2a_2(x + \delta) + \cdots + na_n(x + \delta)^{n-1}, \text{ and so on.}$$

$$f''(x + \delta) = 2a_2 + 6a_3(x + \delta) + \cdots + n * (n-1)a_n(x + \delta)^{n-2}, \text{ and so on, for all}$$
values of x. Hence, putting $x = 0$,

$$f(\delta) = a_0 + a_1(\delta) + a_2(\delta)^2 + \cdots + a_n(\delta)^n$$

$$f'(\delta) = a_1 + 2a_2(\delta) + \cdots + na_n(\delta)^{n-1}$$

$$f''(\delta) = 2a_2 + 6a_3(\delta) + \cdots + n * (n-1)a_n(\delta)^{n-2}, \text{ and so on.}$$

Combining all these, we obtain Taylor's theorem, which states that

$$f(x + \delta) = f(\delta) + x f'(\delta)/1! + x^2 f''(\delta)/2! + x^3 f'''(\delta)/3! + \cdots + x^n f^n(\delta)/n!$$

We can also express this as

$$f(x + \delta) = f(x) + \delta f'(x)/1! + \delta^2 f''(x)/2! + \delta^3 f'''(x)/3! + \cdots + \delta^n f^n(x)/n!$$

If, for instance, $y = f(x) = \log_e(x)$, then

$$f'(x) = (x)^{-1}; f''(x) = -(x)^{-2}; f'''(x) = +2(x)^{-3}; \ldots; f^n(x) = (-1)^{(n-1)}(n-1)! (x)^{(-n)}$$

If $\delta = 1$, then

$$\log_e(x+1) = f(1) + x f'(1)/1! + x^2 f''(1)/2! + x^3 f'''(1)/3! + \cdots + x^n f^n(1)/n! + \cdots$$

Since $\log_e(1) = 0$;

$$f^n(1) = (-1)^{(n-1)}(n-1)! (1)^{(-n)} \text{ and } x^n f^n(1)/n! = (-1)^{(n-1)} x^n/n$$

Hence,

$$\log_e(1+x) = \ln(1+x) = x - x^2/2 + x^3/3 - x^4/4 + \cdots$$

Given as shown in Box 6.4 that

$$f(x+\delta) = f(x) + \delta f'(x)/1! + \delta^2 f''(x)/2! + \delta^3 f'''(x)/3! + \cdots + \delta^n f^n(x)/n!$$

then if δ is small, we can ignore terms in δ^2 and rearrange this as

$$\delta = \left\{ \frac{f(x+\delta) - f(x)}{f'(x)} \right\} = \{f(x+\delta) - f(x)\}/f'(x)$$

leading to

$$x_{new} = (x+\delta) = x_{old} - \{f(x_{old}) - k\}/f'(x_{old})$$

where k is the value of $f(x)$ that we are seeking. This is what we call *Newton's method* for solving complex equations.

In fact, Taylor's expansion can be extended for functions other than polynomials, including functions of the form $z = f(x, y)$. This involves partial derivatives of the form $\frac{\partial^{(r+s)} f(x,y)}{\partial x^r \partial y^s}$, which means all the combinations of the partial derivatives up to $n = r + s$. As stated at the end of Section 6.3, a first-order partial derivative is the differential of an expression in terms of one variable, all other variables being treated as constants. We do not intend to explore higher-order partial derivatives because here, fortunately, we can use only the first derivatives if we have made a reasonable approximation; if not, we will just have to repeat the process several times. This gives us:

$$f\{x + \delta x, y + \delta y\} = f(x,y) + (\partial f/\partial x)\, \delta x + (\partial f/\partial y)\, \delta y + \text{some small terms}$$

where $\partial f/\partial x$ means the differential of $f(x, y)$ with respect to x, with y being treated as a constant.

We will not demonstrate the proof of this here, but simply note that if we can ignore the terms of higher order, then we are left with a linear relationship of the form

$$f_{new} = f_{old} + (\partial f_{old}/\partial x)\, \delta x + (\partial f_{old}/\partial y)\, \delta y$$

where f_{old} means the value that we use when inserting the initial or iterated values in the function $f(x, y)$ and its partial derivative with respect to x and the partial derivative with respect to y. We will work through an example of how this applies when we discuss least square adjustment in Chapter 14.

6.5   BASIC INTEGRATION

The process of differentiation can be put into reverse in what is known as *integration*. We have shown that if $y = x^n$, then $dy/dx = nx^{(n-1)}$. Consider the function $f(x) = c + x^{(n+1)}/(n + 1)$ where c is any constant number. If we differentiate this, we obtain $f'(x) = x^n$. The constant c that had been introduced disappears when we differentiate since it is in effect cx^0 and becomes 0.

If $y = x^n$, then the function $\{c + x^{(n+1)}/(n + 1)\}$ is said to be the *indefinite integral* of y. It is called *indefinite* since the number c is at this stage unknown. c is said to be the *constant of integration*.

The integral of y is often shown with an elongated letter s of the form $\int$. If $y = bx$, then:

the integral of y with respect to x, $= \int y\, dx = \int (bx)\, dx = c + bx^2/2$

Rather than keep writing "with respect to x," we use the notation from before and write "$\int y\, dx$" meaning "what happens to y when we integrate with respect to x."

If $y = \sin x$, then $\int y\, dx = c - \cos x$ while $\int \cos x\, dx = c + \sin x$. Examples of the results of integration are given in Box 6.5.

As with basic arithmetic

$$\int \{f(x) + g(x)\}dx = \int f(x)dx + \int g(x)dx$$

Also,

$$\int k f(x)\, dx = k \int f(x)\, dx \text{ where } k \text{ is a constant.}$$

We have seen that if

$$y = f(x) * g(x)$$

then

$$dy/dx = g(x) * df/dx + f(x) * dg/dx$$

Unfortunately, $\int \{f(x) * g(x)\}\, dx$ cannot be treated in such a neat way. It does not obey this rule.

Given that integration is the reverse of differentiation and given that the integral of x^n equals $\int x^n dx = c + \{1/(n+1)\}x^{n+1}$, we can apply the same principles if n is negative, with one exception. Thus, the integral of $x^{-2} = \int x^{-2}dx = c - x^{-1}$.

The one exception occurs when $n = -1$. We cannot integrate x^{-1} in this way since $\{1/(n + 1)\}$ would become 1/0 and be infinite. We have, however, seen that

$$\frac{d(\log_e x)}{dx} = 1/x = x^{-1}$$

BOX 6.5   SUMMARY OF INTEGRALS

$$\int x^n \, dx = c + \{1/(n+1)\} \, x^{n+1} \text{ provided } n \neq -1$$

$$\int x^{-1} \, dx = c + \log_e x$$

$$\int \sin x \, dx = c - \cos x$$

$$\int \sin 2x \, dx = c - (1/2)\cos 2x$$

$$\int \cos x \, dx = \sin x + c$$

$$\int e^{nx} = (1/n)e^{nx} + c$$

Also,

$$\int \tan x = \log_e(\sec x) + c$$

$$\int \cot x = \log_e(\sin x) + c$$

$$\int \sec x = \log_e(\sec x + \tan x) + c$$

{This is shown in Box 6.6 to equal $\log_e \tan (\pi/4 + x/2) + c$}

$$\int \operatorname{cosec} x = \log_e(\operatorname{cosec} x - \cot x) + c$$

{This is shown in Box 6.6 to equal $\log_e(\tan x/2) + c$}

Knowing this,

$$\int x^{-1} \, dx = c + \log_e x$$

Since the differential of $e^x = e^x$, it follows that $\int e^x dx = e^x + \text{constant}$. Also,

$$\int e^{nx} dx = (1/n)e^{nx} + \text{constant}.$$

For trigonometric functions, the easiest way to show that the integral of $\tan x = \int \tan x = \log_e(\sec x)$ is by differentiating the answer. Thus,

$$\text{if } v = \sec x$$

then

$$dv/dx = \sec x \tan x$$

Also,

$$d(\log_e v)/dv = 1/v$$

Hence,

$$d(\log_e(\sec x))/dx = \{d(\log_e v)/dv\} * \{(dv/dx)\} = (1/\sec x) * \sec x \tan x = \tan x$$

BOX 6.6   COMBINED FUNCTIONS

In Box 6.5 we quote the fact that (sec x + tan x) = tan ($\pi/4 + x/2$). To prove this we use the formulae for sin $(A + B)$ and cos $(A + B)$ in Chapter 5, Box 5.3,

$$\tan (\pi/4 + x/2) = \sin (\pi/4 + x/2)/\cos (\pi/4 + x/2)$$

$$= \{\sin \pi/4 \cos x/2 + \cos \pi/4 \sin x/2\}/\{\cos \pi/4 \cos x/2 - \sin \pi4 \sin x/2\}$$

$$= \{\cos x/2 + \sin x/2\}/\{\cos x/2 - \sin x/2\}$$

(since sin $\pi/4$ = cos $\pi/4 = 1/\sqrt{2}$).

Multiplying above and below by $\{\cos x/2 + \sin x/2\}$, then,

$$\tan (\pi/4 + x/2) = \{\cos^2 x/2 + \sin^2 x/2 + 2 \cos x/2 \sin x/2\}/\{\cos^2 x/2 -\sin^2 x/2\}$$

But cos $2A = \cos^2 A - \sin^2 A$, sin $2A = 2 \sin A \cos A$, and $\cos^2 A + \sin^2 A = 1$. Hence,

$$\tan (\pi/4 + x/2) = \{1 + \sin x\}/\cos x = \sec x + \tan x$$

Thus,

$$\int \sec x = \log_e(\sec x + \tan x) + c = \log_e\{\tan (\pi/4 + x/2)\} + c$$

Similarly,

$$\operatorname{cosec} x - \cot x = \{1 - \cos x\}/\sin x = \{2\sin^2 x/2\}/\{2 \sin x/2 \cos x/2\}$$

Hence,

$$\operatorname{cosec} x - \cot x = \tan x/2$$

Thus,

$$\int \operatorname{cosec} x = \log_e(\operatorname{cosec} x - \cot x) + c = \log_e(\tan x/2) + c$$

6.6   AREAS AND VOLUMES

Integration occurs in geomatics and GIS in a number of operations, one of which is in determining area. Consider two points on a curve (Figure 6.5), for instance,

$$P (x, y) \text{ and } P' (x + \delta x, y + \delta y)$$

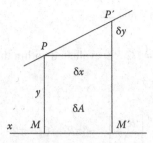

FIGURE 6.5   Area beneath a curve.

The area of the strip beneath the section of the curve $PP' = MPP'M'$ which we will call δA. Then, $\delta A = y\delta x + \frac{1}{2}\,\delta y\delta x$ = a rectangle plus a triangle. As in previous discussions, as δy and δx get small, so

$$\delta A/\delta x = y + \frac{1}{2}\,\delta y$$

In the limit,

$$dA/dx = y$$

We can express this in another way by saying that if we add all the little bits of δA together as we move along the curve, we will obtain the whole area under the curve, namely, $\Sigma\delta A$ where Σ (sigma) means "the sum of."

More particularly, as δx tends to zero we can change to the differential notation and in so doing change the summation symbol from Σ to $\int$ yielding

$$A = \int dA = \int y dx$$

Thus, $\int y dx$ is the area under a curve, between the curve and the x-axis.

If, for example, $y = x^2$ then the area under the curve $= c + x^3/3$. We still, of course, have the constant of integration that is unknown. If, however, we are interested in the area under the curve between the points, for instance, where $x = 1$ and $x = 2$, then it will be $\{(c + 2^3/3) - (c + 1^3/3)\}$ and the unknown c disappears, yielding 7/3.

By setting definite limits to the integration (say from $x = a$ to $x = b$), we obtain the *definite integral*. We write this as $\int_a^b y dx$ which means "the value of the integral $\int y dx$ when x equals b minus the value when x equals a."

If we set limits to the integration, then the result is sometimes written within squared brackets in the form $[I]_{x=a}^{x=b}$ where I is the indefinite integral that is evaluated at $x = b$ and $x = a$.

So,

$$\text{if } y = x^2$$

then

$$I = x^3/3 \text{ and } \left[\frac{x^3}{3}\right]_{x=a}^{x=b} = (b^3 - a^3)/3$$

Returning to Figure 6.4, the area of the very narrow triangle ABC is

$$\frac{1}{2}\text{ base } * \text{ height} = \frac{1}{2}\,R * R\delta\theta$$

If we integrate this between 0 and 2π,

$$\text{Area} = \int_0^{2\pi} 1/2\ R^2 d\theta = 1/2\ R^2\theta \text{ evaluated within the limits } \theta = 0 \text{ to } \theta = 2\pi$$

This is

$$\left[\frac{R^2\theta}{2}\right]_0^{2\pi} = [\pi R^2 - 0] = \pi R^2$$

This confirms the well-known formula for the area of a circle.

If $y = ax^2 + bx + c$ and is part of a parabola, then the area between y_1 at x_1 and y_2 at x_2 and the x-axis will be

$$[ax^3/3 + bx^2/2 + cx]_{x_1}^{x_2}$$

Area
$$= ax_2^3/3 + bx_2^2/2 + cx_2 - ax_1^3/3 - bx_1^2/2 - cx_1$$

$$= ax_2^3/3 - ax_1^3/3 + bx_2^2/2 - bx_1^2/2 + cx_2 - cx_1$$

Here, the lowest common denominator $= 6$. Also,

$$(x_2^3 - x_1^3) = (x_2 - x_1)(x_2^2 + x_2x_1 + x_1^2) \text{ and } (x_2^2 - x_1^2) = (x_2 - x_1)(x_2 + x_1)$$

Hence,

$$\text{Area} = (1/6)(x_2 - x_1)\{2ax_2^2 + 2ax_2x_1 + 2ax_1^2 + 3bx_2 + 3bx_1 + 6c\}$$

$$= (1/6)(x_2 - x_1)\{ax_2^2 + bx_2 + c + ax_1^2 + bx_1 + c$$

$$+ ax_2^2 + 2ax_2x_1 + ax_1^2 + 2bx_2 + 2bx_1 + 4c\}$$

$$= (1/6)(x_2 - x_1)\{y_2 + y_1 + a(x_2 + x_1)^2 + 2b(x_2 + x_1) + 4c\}$$

$$= (1/6)(x_2 - x_1)\{y_2 + y_1 + 4a\left(\frac{x_2 + x_1}{2}\right)^2 + 4b\left(\frac{x_2 + x_1}{2}\right) + 4c\}$$

$$= (1/6)(x_2 - x_1)\{y_2 + y_1 + 4y_m\}$$

where y_m is the value of y at the point midway between x_1 and x_2 where x has the value $(x_1 + x_2)/2$.

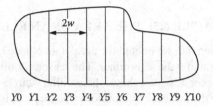

Y0 Y1 Y2 Y3 Y4 Y5 Y6 Y7 Y8 Y9 Y10

FIGURE 6.6   Area of an irregular shape.

This formula, known as *Simpson's rule* after the 18th century mathematician Thomas Simpson, provides a way in which to determine areas of irregular shapes drawn on a map. Thus, in Figure 6.6 we can divide an area up into a series of strips of equal width and measure the length of each straight line. If we assume that each strip is of width w then, if we treat each strip as a parallelogram, its area is

$$w * (Y_r + Y_{r+1})/2$$

For Figure 6.6,

$$\text{Area} = (w/2) * (Y_0 + 2Y_1 + 2Y_2 + 2Y_3 + 2Y_4 + 2Y_5 + 2Y_6 + 2Y_7 + 2Y_8 + 2Y_9 + Y_{10})$$

This assumes that the boundaries are all straight lines rather than curves. If we assume that each end of a strip is part of a parabola, we can use Simpson's rule (shown in Box 6.7). To do so, we would first need to make sure that there is an even number of strips and then treat each pair of strips. The area of the first pair, for example, is:

$$\text{Area of double strip} = (1/6) * 2w * (Y_0 + 4Y_1 + Y_2)$$

The area for the whole shape would then be

$$\text{Area} = (w/3)\{Y_0 + 4Y_1 + 2Y_2 + 4Y_3 + 2Y_4 + 4Y_5 + 2Y_6 + 4Y_7 + 2Y_8 + 4Y_9 + Y_{10}\}$$

This will usually give a better approximation as it assumes that the boundaries are curves, not straight lines. In Example 6.5 we apply the formula to an ellipse to show how reasonable even with only four strips, the area calculation can be.

BOX 6.7   AREA OF IRREGULAR SHAPES

For strips of width w and lengths Y_i ($i = 0$ to n) then:
By Trapezium rule:

$$\text{Area} = (w/2) * (Y_0 + 2Y_1 + 2Y_2 + 2Y_3 + 2Y_4 + 2Y_5 + \cdots + 2Y_{n-2} + 2Y_{n-1} + Y_n)$$

By Simpson's rule:

$$\text{Area} = (w/3)\{Y_0 + 4Y_1 + 2Y_2 + 4Y_3 + 2Y_4 + 4Y_5 + \cdots + 2Y_{n-2} + 4Y_{n-1} + Y_n\}$$

where n is an even number.

EXAMPLE 6.5: THE AREA OF AN ELLIPSE

In Chapter 4, Section 4.6, we quoted the area of an ellipse as "πab" where a is the semimajor axis and b the semiminor. The reason for this can be seen from Figure 4.16 in Chapter 4, which shows the auxiliary circle whose area is πa^2. Since in the ellipse the dimensions parallel to the major axis are unchanged but in the direction of the minor axis, the dimensions are all reduced by a factor b/a. Thus, the area of the ellipse is (b/a) times the area of its auxiliary circle, namely, $(b/a) * (\pi a^2)$ or πab.

Let us divide the ellipse into four strips of equal width parallel to the minor axis (Figure 6.7). If spaced at an interval along the major axis of $a/2$, then $y_0 = 0 = y_4$; $y_2 = 2b$; and $y_1 = b\sqrt{3} = y_3$ since the ellipse is defined by $x^2/a^2 + y^2/b^2 = 1$ and the x values at y_1 and y_3 are $\pm a/2$ giving half the lengths as $b(\sqrt{3})/2$.

Using Simpson's rule (Box 6.7), $w = a/2$.

$$\text{The area of the ellipse} = (a/6)(0 + 4 * b\sqrt{3} + 2 * 2b + 4 * b\sqrt{3} + 0)$$
$$= ab(4 + 8\sqrt{3})/6 \approx 3ab$$

This is not far short of the true value (πab or $3.14 * ab$), and we only divided the area into four strips.

The principles of integral calculus can also be used to determine volumes. The volume of a cylinder, whatever the shape of its base = base area * height provided the sides are parallel. Thus, the volume of the cylinder in Figure 6.8 = Ah.

For the cone (be it a pyramid or any other base shape with a base area A and height h) the area A_z of cross-sectional slice at height z above the base will be in proportion to the base A but reduced both in terms of length and breadth to an amount

$$\{(h - z)/h\}$$

Its area will be

$$A_z = \{(h - z)/h\} * \{(h - z)/h\} * A = \{A/h^2\}\{h^2 - 2hz + z^2\}$$

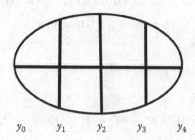

$y_0 \qquad y_1 \qquad y_2 \qquad y_3 \qquad y_4$

FIGURE 6.7   An ellipse cut into four strips.

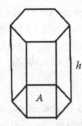 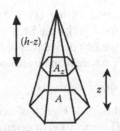

FIGURE 6.8   Volumes of a cylinder and cone.

If the cross-section is an amount δz thick, then its volume will be

$$\delta V = (A/h^2)(h^2 - 2hz + z^2)\delta z$$

If we sum all of this from the base to the apex

$$V = (A/h^2) = (A/h^2) \; [h^2z - hz^2 + z^3/3]_{z=0}^{z=h}$$

$$= (1/3) \; Ah^3/h^2 = (1/3)Ah$$

$$= \text{one-third of base area * height}$$

SUMMARY

Constant of integration: An unknown quantity that must be added after *integration*. Thus, $\int x \, dx = \frac{1}{2}x^2 + c$ where c is the constant of integration.

Definite integral: The value obtained when integrating a function between defined limits. In the case of a polynomial $y = f(x)$, it represents the area between a curve and the x-axis and the x-limits between which it is evaluated. Thus, from $x = a$ to $x = b$, it is $\int f(x)$ evaluated at b minus $\int f(x)$ evaluated at a. This eliminates the need for a *constant of integration*. If $g(x) = \int f(x)$, then the value would be $g(b) - g(a)$. The limits are written at the top and bottom of the integral sign in the form $\int_a^b$ or $[g(x)]_{x=a}^{x=b}$.

Delta (δ): A small amount, for example, δx means a very small increase in x.

Derivative: The limit of the function $\{f(x + \delta) - f(x)\}/\delta$ as δ tends to zero. Where $y = f(x)$, this is written either as dy/dx or as $f'(x)$. The derivative of a derivative (called the *second derivative*) is written as d^2y/dx^2 or as $f''(x)$. In the case of $y = x^n$, the derivative $dy/dx = nx^{(n-1)}$ then,

$d^2y/dx^2 = n(n - 1)x^{(n-2)}$
$d\{\sin(x)\}/dx = \cos(x)$
$d\{\cos(x)\}/dx = -\sin(x)$
$d\{\tan(x)\}/dx = \sec^2(x)$
$d\{\csc(x)\}/dx = -\csc(x)\cot(x)$
$d\{\sec(x)\}/dx = \sec(x)\tan(x)$
$d\{\cot(x)\}/dx = -\csc^2(x)$

The derivative of log $(x) = 1/x$. For compound functions of the form $y = f(x) * g(x)$, then $dy/dx = f(x) * g'(x) + g(x) * f'(x)$ where $f'(x)$ and $g'(x)$ are the first derivatives of $f(x)$ and $g(x)$.

If $y = u * v$ where u and v are both functions of x then,

$$dy/dx = u \, dv/dx + v \, du/dx$$

Differential calculus: The branch of mathematics that studies *derivatives*.

Differentiation: The operation of determining a lower *derivative* of a function.

Indefinite integral: The function that arises when a given function is integrated within unknown limits, hence the need to introduce an unknown quantity called the *constant of integration*.

Integral: Written with the symbol $\int$ and sometimes known as an *antiderivative*. It is the function, which when differentiated, gives the function on which the *integration* is carried out. Thus, $\int x^n \, dx = \{x^{(n+1)}\}/(n + 1) + c$ where c is a *constant of integration*.

Integral calculus: The branch of mathematics that studies *integration*.

Integration: The process of determining the *integral* of a function.

Linearization: The process of converting complex relationships into more simple linear combinations of variables.

Maclaurin's theorem: If the function "f" is infinitely differentiable then,

$$f(x) = f(0) + x^1 f'(0)/1! + x^2 f''(0)/2! + \ldots\ldots x^n f^n (0)/n! + \ldots..$$

where $f^n(0)$ is the value of the nth derivative when x is set to zero. From this

$$\sin(x) = x - x^3/3! + x^5/5! - x^7/7! + \ldots..$$
$$\cos(x) = 1 - x^2/2! + x^4/4! - x^6/6! + \ldots.$$
$$e^x = 1 + x^1/1! + x^2/2! + x^3/3! + x^4/x! + \ldots\ldots$$

Maxima and minima: Points on a curve where $dy/dx = 0$.

Newton's method: A method for solving complex equations by iteration using

$$x_{new} = x_{old} - \left\{ \frac{f(x_{old} - k)}{f'(x_{old})} \right\}$$

Normal: The line at a point on a curve perpendicular to the *tangent*.

Partial derivative: Normally written as a curly delta ∂. Differentiation of a function with two or more variables with respect to one of the variables, all the others being assumed constant. If

$$z = ax^2 + by^2 + cxy + dx + ey + f$$

then,

$$\partial z/\partial x = 2ax + cy + d$$

$$\partial z/\partial y = 2by + cx + e$$

Point of inflection: Point on a curve where $d^2y/dx^2 = 0$ and changes from positive to negative or vice versa. At this point the curve crosses from one side of its tangent to the other.

Points of maximum or minimum: Points on a curve where $dy/dx = 0$.

Second derivative (third and fourth): The *derivative* of dy/dx. Shown as d^2y/dx^2. It is the rate of change of dy/dx. The third derivative is the derivative of the second, and so on.

Simpson's rule: If a closed shape is divided into an even number of strips of width w by straight lines of length $y_0, y_1, \ldots y_n$ where n is an even number then the area of the irregular shape

$$\approx (w/3) * (y_0 + 4y_1 + 2y_2 + 4y_3 + 2\,y_4 + \ldots 2y_{(n-2)} + 4y_{(n-1)} + y_n).$$

Tangent: A line that touches a curve so that its slope $= dy/dx$.

Taylor's theorem: If $f(x)$ is a polynomial of the form $a_0 + a_1x + a_2x^2 + \cdots + a_nx^n$ and if δ is any number, then

$$f(x + \delta) = f(\delta) + x\,f'(\delta)/1! + x^2f''(\delta)/2! + x^3f'''(\delta)/3! + \ldots x^nf^n(\delta)/n!$$

For instance,

$$\log_e(1 + x) = \ln(1 + x) = x - x^2/2 + x^3/3 - x^4/4 + \ldots.$$

Trapezium rule: If a closed shape is divided into a number of strips of width w by straight lines of length $y_0, y_1, \ldots y_n$, then the area of the irregular shape is approximately

$$(w/2) * (y_0 + 2y_1 + 2y_2 + 2y_3 + 2\,y_4 + \ldots 2y_{(n-2)} + 2y_{(n-1)} + y_n).$$

Volume of pyramid or cone: Volume = one-third of base area * height.

7 Matrices and Determinants

7.1 BASIC MATRIX OPERATIONS

Matrices are a mathematical form of shorthand. At its simplest level, a *matrix* is a set of numbers aligned in rows and columns in the form of a rectangle and then enclosed in brackets. Each number within the matrix is an *element* and the whole set of numbers may be referred to as an *array*. Rather than talk about each of the elements within the array, we can simply refer to the matrix as a whole and call it **M**.

For example,

$$\mathbf{M} = \begin{pmatrix} 1 & 2 & 3 \\ 4 & 5 & 6 \end{pmatrix}$$

is a matrix with the first six integers arranged with two rows and three columns and is called a *2 * 3 matrix*. Think of **M** as six boxes or cells, each containing a number, rather like a spreadsheet:

1	2	3
4	5	6

Likewise, we may have a single column of cells with boxes stacked vertically, as in

$$\begin{pmatrix} 7 \\ 8 \\ 9 \end{pmatrix}$$

This is a column matrix with three rows and one column or 3 * 1. If the number of rows equals the number of columns then the matrix is said to be *square*. Thus,

$$\begin{pmatrix} -1 & 0 & 0 \\ 0 & 2 & 0 \\ 0 & 0 & -3 \end{pmatrix}$$

is a 3 * 3 square matrix. It has nine cells and since in this example the only numbers other than zero lie along the diagonal, it is called a *diagonal matrix*. The diagonal from top left to bottom right is called the *leading diagonal*. If all the numbers in the leading diagonal are 1 and all the other elements are zero, then the matrix is called an *identity matrix*. Thus,

$$\begin{pmatrix} 1 & 0 \\ 0 & 1 \end{pmatrix} \text{ and } \begin{pmatrix} 1 & 0 & 0 \\ 0 & 1 & 0 \\ 0 & 0 & 1 \end{pmatrix} \text{ and } \begin{pmatrix} 1 & 0 & 0 & 0 \\ 0 & 1 & 0 & 0 \\ 0 & 0 & 1 & 0 \\ 0 & 0 & 0 & 1 \end{pmatrix}$$

are all identity matrices and are often written as **I**.

The way that matrices are manipulated follows certain rules that are ideally suited to handling in a computer as the operations are in general repetitive. Two matrices can be added or subtracted if they are the same size. This is done by adding or subtracting the corresponding elements (Box 7.1). Thus,

$$\begin{pmatrix} 1 & 2 & 3 & 4 \\ 5 & 6 & 7 & 8 \end{pmatrix}$$

is a 2 * 4 matrix. If we add another 2 * 4 matrix

$$\begin{pmatrix} 6 & -5 & 4 & -3 \\ 9 & -8 & 7 & 6 \end{pmatrix}$$

we obtain

$$\begin{pmatrix} 1+6 & 2-5 & 3+4 & 4-3 \\ 5+9 & 6-8 & 7+7 & 8+6 \end{pmatrix} = \begin{pmatrix} 7 & -3 & 7 & 1 \\ 14 & -2 & 14 & 14 \end{pmatrix}$$

The matrix $\begin{pmatrix} 1 & 2 & 3 \\ 4 & 5 & 6 \end{pmatrix}$ cannot, however, be added to or subtracted from $\begin{pmatrix} 7 \\ 8 \\ 9 \end{pmatrix}$ because the two matrices are not of the same size.

BOX 7.1   MATRIX ADDITION AND SUBTRACTION

If **A** is a matrix with the number in the ith row and jth column $= a_{ij}$ and **B** is the same size with the ith row and jth column $= b_{ij}$ then

the sum of $\mathbf{A} + \mathbf{B} = \mathbf{C}$ where $c_{ij} = a_{ij} + b_{ij}$

Likewise, if $\mathbf{A} - \mathbf{B} = \mathbf{D}$ then $d_{ij} = a_{ij} - b_{ij}$

If we multiply the whole of a matrix by a number then each element must be multiplied by that number. Thus, if

$$\mathbf{M} = \begin{pmatrix} 1 & 2 & 3 & 4 \\ 5 & 6 & 7 & 8 \end{pmatrix} \text{ then } 4\mathbf{M} = \begin{pmatrix} 4 & 8 & 12 & 16 \\ 20 & 24 & 28 & 32 \end{pmatrix}$$

Note that to refer to a matrix we write it in **bold** lettering (e.g., **M** above).

To multiply one matrix by another matrix then the number of columns in the first matrix must equal the number of rows in the second. Under such an arrangement the matrices are said to be *conformable*. This is necessary because the multiplication is carried out in a special way, which creates a new matrix with the same number of rows as the first and the same number of columns as the second. Put another way, if the first matrix has p rows and q columns and the second has q rows and r columns then the product of the two will have p rows and r columns.

Consider a 2 * 2 matrix

$$\mathbf{A} = \begin{pmatrix} a & b \\ d & e \end{pmatrix} \text{ with numbers } a, b, d, \text{ and } e$$

Let **B** be a 2 * 1 matrix with numbers u and w such that

$$\mathbf{B} = \begin{pmatrix} u \\ w \end{pmatrix}$$

then

$$\mathbf{A} * \mathbf{B} = \begin{pmatrix} a & b \\ d & e \end{pmatrix} * \begin{pmatrix} u \\ w \end{pmatrix} = \begin{pmatrix} au + bw \\ du + ew \end{pmatrix}$$

For the first row, multiply each element across the row by the element that is in the same position down the column of the second matrix. Repeat for the second row of the first matrix and first column of the second. The result of multiplying a 2 * 2 matrix times a 2 * 1 matrix is a matrix with 2 rows and 1 column. Conversely, we cannot multiply a 2 * 1 matrix times a 2 * 2 since the columns in the first matrix are not as many as the rows in the second. In this case **B** * **A** does not exist.

The same procedure is followed with a 3 * 3 multiplying a 3 * 2

$$\begin{pmatrix} a & b & c \\ d & e & f \\ g & h & i \end{pmatrix} * \begin{pmatrix} u & x \\ v & y \\ w & z \end{pmatrix} = \begin{pmatrix} au+bv+cw & ax+by+cz \\ du+ev+fw & dx+ey+fz \\ gu+hv+iw & gx+hy+iz \end{pmatrix} = C$$

C is a 3 * 2 matrix.

In the above example, if in A the numbers $a = e = i = 1$ (i.e., the elements in the leading diagonal are all unity) and all the other elements of A are zero, which means that A is the identity matrix I, then

$$A = I = \begin{pmatrix} 1 & 0 & 0 \\ 0 & 1 & 0 \\ 0 & 0 & 1 \end{pmatrix} \quad \text{and} \quad \text{if } B = \begin{pmatrix} u & x \\ v & y \\ w & z \end{pmatrix}$$

then when we multiply $I * B$ to obtain C, all the elements of C will be the same as those in B, hence,

$$I * B = B.$$

Once again, if A is the 3 * 3 matrix and B is the 3 * 2, we can multiply A times B to obtain a new matrix C. The order of the matrices is important for we cannot multiply B times A since this time the number of columns in the first (B) would not equal the number of rows in the second (now A). If, however, the number of rows in A equals the number of columns in B, then we can obtain both $A * B$ and $B * A$. Thus, if A is 4 * 2 and B is 2 * 4, then $A * B$ is 4 * 4, but $B * A$ is 2 * 2 and $A * B \neq B * A$. Thus, unlike ordinary arithmetic, the order of multiplication is of critical importance (Box 7.2 and Example 7.1).

BOX 7.2   MATRIX MULTIPLICATION (1)

If A is a matrix with the element in the ith row and jth column $= a_{ij}$ and B has the same number of rows as A has columns and has its ith row and jth column $= b_{ij}$, then

$$A * B = C$$

where $c_{ij} =$ the sum of the elements of row i in A multiplied by the corresponding elements of column j in B

$$c_{ij} = \{a_{i1}b_{1j} + a_{i2}b_{2j} + a_{i3}b_{3j} + \cdots + a_{in}b_{nj}\}$$

EXAMPLE 7.1: MATRIX MULTIPLICATION

If **A** is a 4 * 2 matrix and **B** is a 2 * 4 matrix and

$$\mathbf{A} = \begin{pmatrix} 2 & 5 \\ 5 & 7 \\ 7 & 3 \\ 8 & 6 \end{pmatrix}, \mathbf{B} = \begin{pmatrix} 7 & 9 & 3 & 1 \\ 5 & 8 & 4 & 3 \end{pmatrix}; \text{ then } \mathbf{A} * \mathbf{B} = \begin{pmatrix} 39 & 58 & 26 & 17 \\ 70 & 101 & 43 & 26 \\ 64 & 87 & 33 & 16 \\ 86 & 120 & 48 & 26 \end{pmatrix}$$

$$\mathbf{B} = \begin{pmatrix} 7 & 9 & 3 & 1 \\ 5 & 8 & 4 & 3 \end{pmatrix}, \mathbf{A} = \begin{pmatrix} 2 & 5 \\ 5 & 7 \\ 7 & 3 \\ 8 & 6 \end{pmatrix}; \text{ then } \mathbf{B} * \mathbf{A} = \begin{pmatrix} 88 & 106 \\ 102 & 111 \end{pmatrix}$$

There are, however, exceptions to the rule that **A** * **B** ≠ **B** * **A**, for example, when **B** is such that **A** * **B** = **I** (the identity matrix). Consider the special case where

$$\mathbf{A} = \begin{pmatrix} a & b \\ d & e \end{pmatrix}$$

and

$$\mathbf{B} = \frac{1}{ae - bd} * \begin{pmatrix} e & -b \\ -d & a \end{pmatrix}$$

If we multiply **A** * **B**, we obtain

$$\begin{pmatrix} 1 & 0 \\ 0 & 1 \end{pmatrix}.$$

which is the identity matrix. We obtain the same result if we multiply **B** times **A** (= **B** * **A**). The matrix **B** that has this special property is said to be the *inverse* of **A** and is written as $\mathbf{A}^{-1}$. It has the property that $\mathbf{AA}^{-1} = \mathbf{A}^{-1}\mathbf{A} = \mathbf{I}$. For an illustration, see Example 7.2.

EXAMPLE 7.2: MATRIX INVERSE

If **A** is a 3 * 3 matrix $\begin{pmatrix} 1 & 4 & 3 \\ 2 & 5 & 4 \\ 7 & 6 & 7 \end{pmatrix}$ and $\mathbf{B} = \begin{pmatrix} -5.5 & 5 & -0.5 \\ -7 & 7 & -1 \\ 11.5 & -11 & 1.5 \end{pmatrix}$ then

$$\mathbf{A} * \mathbf{B} = \begin{pmatrix} 1 & 0 & 0 \\ 0 & 1 & 0 \\ 0 & 0 & 1 \end{pmatrix} = \mathbf{B} * \mathbf{A}$$

A * **B** = **I** the identity matrix and hence **B** = **A**⁻¹, the inverse of **A**.

7.2 DETERMINANTS

The number ($ae - bd$) that was used above to scale the whole multiplication to make the diagonal numbers all equal to 1 is a special number for the 2 * 2 matrix **A** that is called its *determinant*. This is usually written with straight-line brackets so that:

$$\text{The determinant of } \mathbf{A} = |\mathbf{A}| = \begin{vmatrix} a & b \\ d & e \end{vmatrix} = ae - bd.$$

The notation should not be confused with the idea of a *modulus*, which also uses straight-line brackets. The modulus (or "mod") of a value is a positive real number that is the absolute value of a quantity, that is, regardless of whether mathematically it is positive or negative. Thus, mod $(-3) = |-3| = +3$.

A determinant is always a square matrix with n rows and n columns, n being an integer such as 1, 2, 3, 4, and so on. If it happens in a 2 * 2 matrix that $ae = bd$, then $1/|\mathbf{A}| = 1/0$ which is infinite, hence the matrix **A** has no inverse. The matrix is then said to be *singular*.

For a 3 * 3 matrix the process is slightly more complicated. The calculation is broken down into three stages where

$$\begin{vmatrix} a & b & c \\ d & e & f \\ g & h & i \end{vmatrix} = \text{three parts} \begin{vmatrix} a & . & . \\ . & e & f \\ . & h & i \end{vmatrix} \text{ and } \begin{vmatrix} . & b & . \\ d & . & f \\ g & . & i \end{vmatrix} \text{ and } \begin{vmatrix} . & . & c \\ d & e & . \\ g & h & . \end{vmatrix}$$

Each of these subcomponents contains a 2 * 2 determinant that is known as the *minor* of the element in the first row. The value of the 3 * 3 determinant (Box 7.3 and Example 7.3) is then defined as:

$$|\mathbf{A}| = a(ei - fh) - b(di - fg) + c(dh - eg)$$

Note the alternating signs, plus a, minus b, plus c, and so forth.

BOX 7.3   THE DETERMINANT OF A MATRIX

If $\mathbf{A}$ is a 3 * 3 square matrix such that

$$\mathbf{A} = \begin{vmatrix} a & b & c \\ d & e & f \\ g & h & i \end{vmatrix} \text{ then } |\mathbf{A}| = a*\begin{vmatrix} e & f \\ h & i \end{vmatrix} - b*\begin{vmatrix} d & f \\ g & i \end{vmatrix} + c*\begin{vmatrix} d & e \\ g & h \end{vmatrix}$$

Thus, the determinant of $\mathbf{A} = |\mathbf{A}| = a(ei - fh) - b(di - fg) + c(dh - eg)$.

If $|\mathbf{A}| = 0$ the determinant is singular, for a 4 * 4 square matrix the determinant becomes

$$\begin{vmatrix} a & b & c & p \\ d & e & f & q \\ g & h & i & r \\ j & k & l & s \end{vmatrix} = a*\begin{vmatrix} e & f & q \\ h & i & r \\ k & l & s \end{vmatrix} - b*\begin{vmatrix} d & f & q \\ g & i & r \\ j & l & s \end{vmatrix}$$

$$+ c*\begin{vmatrix} d & e & q \\ g & h & r \\ j & k & s \end{vmatrix} - p*\begin{vmatrix} d & e & f \\ g & h & i \\ j & k & l \end{vmatrix}$$

where each of the 3 * 3 minor determinants has to be evaluated as above.

Note the overall symmetry of the operation and the alternating signs $+ - +$ $- + -$, and so forth.

If for instance two (or more) rows (or columns) differ by a common factor, then $|\mathbf{A}|$ will be zero. For instance, if in the example in Box 7.3, row 3 is a multiple of row 2 so that instead of $[g\ h\ i]$ we have $[k*d\ k*e\ k*f]$ where k is a constant, then we have

$$|\mathbf{A}| = a(e*k*f - f*k*e) - b(d*k*i - f*k*d) + c(d*k*e - e*k*d) = 0$$

The related matrix will therefore be singular.

If we change the rows in the matrix $\mathbf{A}$ into columns to form a new matrix, we call this the *transpose* of $\mathbf{A}$ and label it $\mathbf{A}^T$. In the case of a square matrix, since the calculations of the value of a determinant are symmetrical between rows and columns, the determinant of the transpose has the same numerical value as that of the original so $|\mathbf{A}| = |\mathbf{A}^T|$.

Also, if $\mathbf{A}$ has n rows and n columns, then if we multiply all the terms of $\mathbf{A}$ by a scalar quantity k, then $|k\ \mathbf{A}| = k^n |\mathbf{A}|$. Thus, if the elements of $\mathbf{A}$ are doubled in a 3 * 3 square matrix to give a new matrix $\mathbf{B} = 2\mathbf{A}$, then the value of $|\mathbf{B}| = 2^3 |\mathbf{A}| = 8|\mathbf{A}|$—see Example 7.4.

Note that if $k = -1$ then $|-\mathbf{A}| = (-1)^n |\mathbf{A}|$ and not as one might expect $-|\mathbf{A}|$ which will only occur if n is an odd number.

EXAMPLE 7.3: A 3 * 3 DETERMINANT

If the matrix $\mathbf{A}$ is $\begin{pmatrix} 1 & 4 & 3 \\ 2 & 5 & 4 \\ 7 & 6 & 7 \end{pmatrix}$ as in Example 7.2 then

$$|\mathbf{A}| = 1 * (5 * 7 - 4 * 6) - 4 * (2 * 7 - 4 * 7) + 3 * (2 * 6 - 5 * 7) = 11 + 56 - 69 = -2$$

If $\mathbf{B} = \begin{pmatrix} -5.5 & 5 & -0.5 \\ -7 & 7 & -1 \\ 11.5 & -11 & 1.5 \end{pmatrix}$ again as in Example 7.2 then

$$|\mathbf{B}| = -5.5 * \{7 * 1.5 - 1 * 11\} - 5 * \{-7 * 1.5 - (-1) * 11.5\} + (-0.5) *$$
$$\{(-7) * (-11) - (7 * 11.5)\}$$

$$= 2.75 - 5 + 1.75 = -0.5$$

Note: If the determinant of $\mathbf{A} * \mathbf{B} = |\mathbf{I}| = 1$ then $|\mathbf{A}| * |\mathbf{B}| = |\mathbf{B}| * |\mathbf{A}| = |\mathbf{A} * \mathbf{B}| = |\mathbf{B} * \mathbf{A}|$. This applies generally for determinants so long as $\mathbf{A}$ and $\mathbf{B}$ are both conformable, even though in general $\mathbf{A} * \mathbf{B} \neq \mathbf{B} * \mathbf{A}$.

If instead of multiplying all of the elements of $\mathbf{A}$ by k, we multiply only one row or one column by k' then the determinant of this matrix will only be k' times as big and hence have a value $= k' |\mathbf{A}|$. If we interchange two rows (or two columns) then the determinant will be the same size but the opposite sign. For instance, if in $|\mathbf{B}|$ in Example 7.4, we interchange columns 2 and 3 to give a new matrix $\mathbf{C}$, then

$$|\mathbf{C}| = \begin{vmatrix} 2 & 6 & 4 \\ 8 & 10 & 12 \\ 18 & 16 & 14 \end{vmatrix} = 2 * \begin{vmatrix} 10 & 12 \\ 16 & 14 \end{vmatrix} - 6 * \begin{vmatrix} 8 & 12 \\ 18 & 14 \end{vmatrix} + 4 * \begin{vmatrix} 8 & 10 \\ 18 & 16 \end{vmatrix}$$

$$= 2 * (140 - 192) - 6 * (112 - 216) + 4 * (128 - 180)$$

$$= 104 + 624 - 208 = +312 = 8 * 39$$

This is the same magnitude as $|\mathbf{B}|$ in Example 7.4.

Furthermore, consider two square matrices

$$\mathbf{A} = \begin{pmatrix} a_{11} & a_{12} & a_{13} \\ a_{21} & a_{22} & a_{23} \\ a_{31} & a_{32} & a_{33} \end{pmatrix} \text{ and } \mathbf{B} = \begin{pmatrix} b_{11} & b_{12} & b_{13} \\ b_{21} & b_{22} & b_{23} \\ b_{31} & b_{32} & b_{33} \end{pmatrix}$$

EXAMPLE 7.4: INCREASING THE ELEMENTS
IN A DETERMINANT BY A FACTOR

Let

$$|A| = \begin{vmatrix} 1 & 2 & 3 \\ 4 & 6 & 5 \\ 9 & 7 & 8 \end{vmatrix} = 1 * \begin{vmatrix} 6 & 5 \\ 7 & 8 \end{vmatrix} - 2 * \begin{vmatrix} 4 & 5 \\ 9 & 8 \end{vmatrix} + 3 * \begin{vmatrix} 4 & 6 \\ 9 & 7 \end{vmatrix}$$

$$|A| = 1 * (48 - 35) - 2 * (32 - 45) + 3 * (28 - 54) = 13 + 26 - 78 = -39 = -1 * 39$$

Let the elements in **A** be doubled to give

$$|B| = \begin{vmatrix} 2 & 4 & 6 \\ 8 & 12 & 10 \\ 18 & 14 & 16 \end{vmatrix} = 2 * \begin{vmatrix} 12 & 10 \\ 14 & 16 \end{vmatrix} - 4 * \begin{vmatrix} 8 & 10 \\ 18 & 16 \end{vmatrix} + 6 * \begin{vmatrix} 8 & 12 \\ 18 & 14 \end{vmatrix}$$

$$|B| = 2 * (192 - 140) - 4 * (128 - 180) + 6 * (112 - 160) = 104 + 208 - 624 = -312$$

$$= -8 * 39$$

A and **B** are of order $3 * 3$, hence, by doubling each of the elements in **A** to create **B**, the determinant of $B = 2^3 = 8$ times the determinant of **A**.

The product **A** * **B** =

$$\begin{pmatrix} a_{11}b_{11} + a_{12}b_{21} + a_{13}b_{31} & a_{11}b_{12} + a_{12}b_{22} + a_{13}b_{32} & a_{11}b_{13} + a_{12}b_{23} + a_{13}b_{33} \\ a_{21}b_{11} + a_{22}b_{21} + a_{23}b_{31} & a_{21}b_{12} + a_{22}b_{22} + a_{23}b_{32} & a_{21}b_{13} + a_{22}b_{23} + a_{23}b_{33} \\ a_{31}b_{11} + a_{32}b_{21} + a_{33}b_{31} & a_{31}b_{12} + a_{32}b_{22} + a_{33}b_{32} & a_{31}b_{13} + a_{32}b_{23} + a_{33}b_{33} \end{pmatrix}$$

The product **B** * **A** =

$$\begin{pmatrix} b_{11}a_{11} + b_{12}a_{21} + b_{13}a_{31} & b_{11}a_{12} + b_{12}a_{22} + b_{13}a_{32} & b_{11}a_{13} + b_{12}a_{23} + b_{13}a_{33} \\ b_{21}a_{11} + b_{22}a_{21} + b_{23}a_{31} & b_{21}a_{12} + b_{22}a_{22} + b_{23}a_{32} & b_{21}a_{13} + b_{22}a_{23} + b_{23}a_{33} \\ b_{31}a_{11} + b_{32}a_{21} + b_{33}a_{31} & b_{31}a_{12} + b_{32}a_{22} + b_{33}a_{32} & b_{31}a_{13} + b_{32}a_{23} + b_{33}a_{33} \end{pmatrix}$$

Or, on rearranging terms,

$$= \begin{pmatrix} a_{11}b_{11} + a_{21}b_{12} + a_{31}b_{13} & a_{12}b_{11} + a_{22}b_{12} + a_{32}b_{13} & a_{13}b_{11} + a_{23}b_{12} + a_{33}b_{13} \\ a_{11}b_{21} + a_{21}b_{22} + a_{31}b_{23} & a_{12}b_{21} + a_{22}b_{22} + a_{32}b_{23} & a_{13}b_{21} + a_{23}b_{22} + a_{33}b_{23} \\ a_{11}b_{31} + a_{21}b_{32} + a_{31}b_{33} & a_{12}b_{31} + a_{22}b_{32} + a_{32}b_{33} & a_{13}b_{31} + a_{23}b_{32} + a_{33}b_{33} \end{pmatrix}$$

It can be seen that the matrix product of **A** * **B** is not necessarily the same as **B** * **A**. However, since the determinant

$$|\mathbf{A}| = a_{11}(a_{22}a_{33} - a_{23}a_{32}) - a_{12}(a_{21}a_{33} - a_{23}a_{31}) + a_{13}(a_{21}a_{32} - a_{22}a_{31})$$

and

$$|\mathbf{B}| = b_{11}(b_{22}b_{33} - b_{23}b_{32}) - b_{12}(b_{21}b_{33} - b_{23}b_{31}) + b_{13}(b_{21}b_{32} - b_{22}b_{31})$$

then with a little manipulation, it can be shown that

$$|\mathbf{A} * \mathbf{B}| = |\mathbf{B} * \mathbf{A}| = |\mathbf{A}| * |\mathbf{B}| = |\mathbf{B}| * |\mathbf{A}|$$

7.3   RELATED MATRICES

The idea of changing rows into columns in a matrix **A** to form its transpose $\mathbf{A}^T$ is not restricted to square matrices. Thus, if

$$\mathbf{A} = \begin{pmatrix} a & b & c & d \\ e & f & g & h \end{pmatrix}, \text{ then } \mathbf{A}^T = \begin{pmatrix} a & e \\ b & f \\ c & g \\ d & h \end{pmatrix}$$

If **A** has m rows and n columns, then $\mathbf{A}^T$ will have n rows and m columns and we can always multiply $\mathbf{A} * \mathbf{A}^T$ and also $\mathbf{A}^T * \mathbf{A}$ although the results may differ (Box 7.4).

In the particular case where $\mathbf{A} = (x\ y\ z)$, then $\mathbf{A}^T = \begin{pmatrix} x \\ y \\ z \end{pmatrix}$ and

$$\mathbf{A} * \mathbf{A}^T = (x^2 + y^2 + z^2) \text{ which is a } 1 * 1 \text{ matrix.}$$

On the other hand, if we multiply $\mathbf{A}^T * \mathbf{A} = \begin{pmatrix} x \\ y \\ z \end{pmatrix} (x\ y\ z)$, we obtain

$$\mathbf{A}^T * \mathbf{A} = \begin{pmatrix} xx & xy & xz \\ yx & yy & yz \\ zx & zy & zz \end{pmatrix} \text{ which is a } 3 * 3 \text{ matrix.}$$

BOX 7.4   MATRIX MULTIPLICATION (2)

If $\mathbf{A} = \begin{pmatrix} a & b \\ c & d \end{pmatrix}$ and $\mathbf{B} = \begin{pmatrix} e & f \\ g & h \end{pmatrix}$ then

$$\mathbf{A}*\mathbf{B} = \begin{pmatrix} ae+bg & af+bh \\ ce+dg & cf+dh \end{pmatrix} \text{ and } \mathbf{B}*\mathbf{A} = \begin{pmatrix} ae+cf & be+df \\ ag+ch & bg+dh \end{pmatrix}$$

$$\mathbf{A}^{\mathrm{T}} \begin{pmatrix} a & c \\ b & d \end{pmatrix} \text{ hence } \mathbf{A}*\mathbf{A}^{\mathrm{T}} = \begin{pmatrix} a^2+b^2 & ac+bd \\ ac+bd & c^2+d^2 \end{pmatrix} \text{ while } \mathbf{A}^{\mathrm{T}}*\mathbf{A} = \begin{pmatrix} a^2+c^2 & ab+cd \\ ab+cd & b^2+d^2 \end{pmatrix}$$

In general, (though not in all cases)

$$\mathbf{A}*\mathbf{B} \neq \mathbf{B}*\mathbf{A} \text{ and } \mathbf{A}*\mathbf{A}^{\mathrm{T}} \neq \mathbf{A}^{\mathrm{T}}*\mathbf{A}$$

The order in which matrices are multiplied together is of crucial importance.

Earlier, we introduced the idea of labeling every element according to its row and column. Rather than saying $\mathbf{A} = (a\ b\ c\ d)$, we can say that $\mathbf{A} = (a_{11}\ a_{12}\ a_{13}\ a_{14})$.

For a 4 * 4 matrix,

$$\mathbf{A} = \begin{pmatrix} a_{11} & a_{12} & a_{13} & a_{14} \\ a_{21} & a_{22} & a_{23} & a_{24} \\ a_{31} & a_{32} & a_{33} & a_{34} \\ a_{41} & a_{42} & a_{43} & a_{44} \end{pmatrix}$$

In general, a_{ij} is the element in the ith row and jth column. In the transpose $\mathbf{A}^{\mathrm{T}}$ it becomes the jth row of the ith column or a_{ji}.

We can think of a square matrix $\mathbf{A}$ as having elements occupying the squares as in a chessboard (shown in Figure 7.1 as 8 * 8 but we will assume that it is $n * n$ where n is a positive integer). The board would contain an element such as a_{ij} for the number in the ith row and jth column. If we strip out the ith row and jth column as in Figure 7.1b, then we are left with $(n-1) * (n-1)$ matrix that is called the *sub-matrix* of a_{ij}. The determinant of this submatrix of a_{ij} is known as the *minor* of a_{ij}.

If we give the minor of a_{ij} the name $\mathbf{M}_{ij}$ then, just as we saw in Box 7.3, when we consider the determinants of $\mathbf{A}$, we need to alternate the signs (plus then minus then plus then minus then plus, etc.). We can think of black squares in the original as plus and white squares as minus or we can say that the sign is $(-1)^{(i-j)}$. We now define the *cofactor* of the element a_{ij} in $\mathbf{A}$ as the determinant of the matrix $(-1)^{(i-j)}\mathbf{M}_{ij}$ and write it as $\mathbf{A}_{ij}$. A cofactor is a determinant and reduces to a single number; it is the signed (i.e., plus or minus) minor of a_{ij}.

(a)   (b)

FIGURE 7.1   Submatrices and minors.

The overall value of the determinant of the matrix **A** is obtained from any row and its related cofactors in the form

$$|\mathbf{A}| = a_{11}\mathbf{A}_{11} + a_{12}\mathbf{A}_{12} + a_{13}\mathbf{A}_{13} + \cdots + a_{1n}\mathbf{A}_{1n}$$

Alternatively, we can use row i so that

$$|\mathbf{A}| = a_{i1}\mathbf{A}_{i1} + a_{i2}\mathbf{A}_{i2} + a_{i3}\mathbf{A}_{i3} + \cdots + a_{in}\mathbf{A}_{in}$$

This is based on the calculation using the elements in row i.

Thus, an $n * n$ matrix has a determinant that is the sum of n products between the row elements and their respective cofactors. Each cofactor can be evaluated as the sum of its minors, each minor being a $(n - 1) * (n - 1)$ matrix, and so on down to the elements of a determinant that is $2 * 2$. This is an extension of what we showed in Box 7.3.

We now define the *adjugate* or *adjoint* matrix $\mathbf{A}^*$ of a square matrix **A** as a new matrix formed by cofactors of the elements of **A** in the transposed position. Thus, for a $4 * 4$ matrix the terms are

$$\mathbf{A}^* = \begin{pmatrix} \mathbf{A}_{11} & \mathbf{A}_{21} & \mathbf{A}_{31} & \mathbf{A}_{41} \\ \mathbf{A}_{12} & \mathbf{A}_{22} & \mathbf{A}_{32} & \mathbf{A}_{42} \\ \mathbf{A}_{13} & \mathbf{A}_{23} & \mathbf{A}_{33} & \mathbf{A}_{43} \\ \mathbf{A}_{14} & \mathbf{A}_{24} & \mathbf{A}_{34} & \mathbf{A}_{44} \end{pmatrix}$$

If

$$\mathbf{A} = \begin{pmatrix} a_{11} & a_{12} & a_{13} & a_{14} \\ a_{21} & a_{22} & a_{23} & a_{24} \\ a_{31} & a_{32} & a_{33} & a_{34} \\ a_{41} & a_{42} & a_{43} & a_{44} \end{pmatrix}$$

then we can show that

$$
\mathbf{AA}^* = \begin{pmatrix} |A| & 0 & 0 & 0 \\ 0 & |A| & 0 & 0 \\ 0 & 0 & |A| & 0 \\ 0 & 0 & 0 & |A| \end{pmatrix} = |A|\,\mathbf{I}
$$

We can also show that $\mathbf{A}^* * \mathbf{A} = |A|\mathbf{I}$. We showed earlier that $\mathbf{AA}^{-1} = \mathbf{A}^{-1}\mathbf{A} = \mathbf{I}$. From this it follows that $\mathbf{A}^*/|A| = \mathbf{A}^{-1}$ which is the inverse of $\mathbf{A}$. Thus, we have a way to form the inverse of any matrix by computing its adjoint.

With a 2 * 2 matrix,

$$
\mathbf{M} = \begin{bmatrix} a & b \\ c & d \end{bmatrix}
$$

then

$$
\mathbf{M}^{-1} = \frac{1}{(ad - bc)} \begin{bmatrix} d & -b \\ -c & a \end{bmatrix}
$$

where the determinant $|\mathbf{M}| = (ad - bc)$.

If we have a 3 * 3 matrix

$$
\mathbf{M} = \begin{pmatrix} a & b & c \\ d & e & f \\ g & h & k \end{pmatrix}
$$

then

$$
|\mathbf{M}| = a(ek - fh) - b(dk - fg) + c(dh - eg)
$$

$$
\mathbf{M}^{-1} = \begin{pmatrix} a & b & c \\ d & e & f \\ g & h & k \end{pmatrix}^{-1} = \frac{1}{|\mathbf{M}|}\begin{pmatrix} A & B & C \\ D & E & F \\ G & H & K \end{pmatrix}^{\mathrm{T}} = \frac{1}{|\mathbf{M}|}\begin{pmatrix} A & D & G \\ B & E & H \\ C & F & K \end{pmatrix}
$$

where A is the cofactor of a in $\mathbf{M}$. This means that:

Determinant $|\mathbf{M}| = a(ek - fh) - b(dk - fg) + c(dh - eg)$

$A = ek - fh;$ $B = -(dk - fg);$ $C = (dh - eg);$

$D = -(bk - ch);$ $E = (ak - cg);$ $F = -(ah - bg);$

$G = (bf - ec);$ $H = -(af - cd);$ $K = (ae - bd)$

Thus,

$$\mathbf{M}^{-1} = \frac{1}{a(ek-fh)-b(dk-fg)+c(dh-eg)} \begin{bmatrix} (ek-fh)\,(ch-bk)\,(bf-ec) \\ (fg-dk)\,(ak-cg)\,(cd-af) \\ (dh-eg)\,(bg-ah)\,(ae-bd) \end{bmatrix}$$

If we have a 4 * 4 matrix $\mathbf{M} = \begin{bmatrix} a\ b\ c\ d \\ e\ f\ g\ h \\ j\ k\ l\ m \\ n\ p\ q\ r \end{bmatrix}$, then its determinant $|\mathbf{M}| =$

$$= a\begin{vmatrix} f\ g\ h \\ k\ l\ m \\ p\ q\ r \end{vmatrix} - b\begin{vmatrix} e\ g\ h \\ j\ l\ m \\ n\ q\ r \end{vmatrix} + c\begin{vmatrix} e\ f\ h \\ j\ k\ m \\ n\ p\ r \end{vmatrix} - d\begin{vmatrix} e\ f\ g \\ j\ k\ l \\ n\ p\ q \end{vmatrix}$$

or

$$= -e\begin{vmatrix} b\ c\ d \\ k\ l\ m \\ p\ q\ r \end{vmatrix} + f\begin{vmatrix} a\ c\ d \\ j\ l\ m \\ n\ q\ r \end{vmatrix} - g\begin{vmatrix} a\ b\ d \\ j\ k\ m \\ n\ p\ r \end{vmatrix} + h\begin{vmatrix} a\ b\ c \\ j\ k\ l \\ n\ p\ q \end{vmatrix}$$

or

$$= j\begin{vmatrix} b\ c\ d \\ f\ g\ h \\ p\ q\ r \end{vmatrix} - k\begin{vmatrix} a\ c\ d \\ e\ g\ h \\ n\ q\ r \end{vmatrix} + l\begin{vmatrix} a\ b\ d \\ e\ f\ h \\ n\ p\ r \end{vmatrix} - m\begin{vmatrix} a\ b\ c \\ e\ f\ g \\ n\ p\ q \end{vmatrix}$$

or

$$= -n\begin{vmatrix} b\ c\ d \\ f\ g\ h \\ k\ l\ m \end{vmatrix} + p\begin{vmatrix} a\ c\ d \\ e\ g\ h \\ j\ l\ m \end{vmatrix} - q\begin{vmatrix} a\ b\ d \\ e\ f\ h \\ j\ k\ m \end{vmatrix} + r\begin{vmatrix} a\ b\ c \\ e\ f\ g \\ j\ k\ l \end{vmatrix}$$

On expanding, this gives the determinant

$$|\mathbf{M}| = a\{f(lr-mq)-g(kr-mp)+h(kq-lp)\}-b\{e(lr-mq)-g(jr-mn)+h(jp-kn)\}$$
$$+ c\{e(kr-pm)-f(jr-mn)+h(jp-kn)\}-d\{e(kq-lp)-f(jq-ln)+g(jp-kn)\}$$

or

$$= -e\{b(lr-mq)-c(kr-mp)+d(kq-lp)\}+f\{a(lr-mq)-c(jr-mn)+d(jq-ln)\}$$
$$- g\{a(kr-mp)-b(jr-mn)+d(jp-kn)\}+h\{a(kq-lp)-b(jq-ln)+c(jp-kn)\}$$

or

$$= j\{b(gr-hq)-c(fr-hp)+d(fq-gp)\}-k\{a(gr-hq)-c(er-hn)+d(eq-gn)\}$$
$$+ l\{a(fr-hp)-b(er-hn)+d(ep-fn)\}-m\{a(fq-gp)-b(eq-gn)+c(ep-fn)\}$$

or

$$= -n\{b(gm-hl)-c(fm-hk)+d(fl-gk)\}+p\{a(gm-hl)-c(em-hj)+d(el-gj)\}$$
$$-q\{a(fm-hk)-b(em-hj)+d(ek-fj)\}+r\{a(fl-gk)-b(el-gk)+c(ek-fj)\}$$

All of these have the same value.

If the cofactors of $\mathbf{M}$ are $\begin{bmatrix} A\,B\,C\,D \\ E\,F\,G\,H \\ J\,K\,L\,M \\ N\,P\,Q\,R \end{bmatrix}$, then the adjugate of $\mathbf{M} = \mathbf{M}^* = \begin{bmatrix} A\,E\,J\,N \\ B\,F\,K\,P \\ C\,G\,L\,Q \\ D\,H\,M\,R \end{bmatrix}$

or

$$\begin{bmatrix} \{f(lr\text{-}mq)\text{-}g(kr\text{-}mp)\text{+}h(kq\text{-}lp)\} & \{\text{-}b(lr\text{-}mq)\text{+}c(kr\text{-}mp)\text{-}d(kq\text{-}lp)\} & \{b(gr\text{-}hq)\text{-}c(fr\text{-}hp)\text{+}d(fq\text{-}gp)\} & \{\text{-}b(gm\text{-}hl)\text{+}c(fm\text{-}hk)\text{-}d(fl\text{-}gk)\} \\ \{\text{-}e(lr\text{-}mq)\text{+}g(jr\text{-}mn)\text{-}h(jp\text{-}kn)\} & \{a(lr\text{-}mq)\text{-}c(jr\text{-}mn)\text{+}d(jq\text{-}ln)\} & \{\text{-}a(gr\text{-}hq)\text{+}c(er\text{-}hn)\text{-}d(eq\text{-}gn)\} & \{a(gm\text{-}hl)\text{-}c(em\text{-}hj)\text{+}d(el\text{-}gj)\} \\ \{e(kr\text{-}pm)\text{-}f(jr\text{-}mn)\text{+}h(jp\text{-}kn)\} & \{\text{-}a(kr\text{-}mp)\text{+}b(jr\text{-}mn)\text{-}d(jp\text{-}kn)\} & \{a(fr\text{-}hp)\text{-}b(er\text{-}hn)\text{+}d(ep\text{-}fn)\} & \{\text{-}a(fm\text{-}hk)\text{+}b(em\text{-}hj)\text{-}d(ek\text{-}fj)\} \\ \{\text{-}e(kq\text{-}lp)\text{+}f(jq\text{-}ln)\text{-}g(jp\text{-}kn)\} & \{a(kq\text{-}lp)\text{-}b(jq\text{-}ln)\text{+}c(jp\text{-}kn)\} & \{\text{-}a(fq\text{-}gp)\text{+}b(eq\text{-}gn)\text{-}c(ep\text{-}fn)\} & \{a(fl\text{-}gk)\text{-}b(el\text{-}gk)\text{+}c(ek\text{-}fj)\} \end{bmatrix}$$

The inverse of $\mathbf{M} = \mathbf{M}^{-1} = (1/|\mathbf{M}|) * \mathbf{M}^*$. An illustration is given in Example 7.5 and the results are summarized in Box 7.5.

With a matrix larger than 4 * 4 it becomes extremely messy when we express this in separate computational terms. The process can, however, be treated in blocks, for instance, $\mathbf{A}$, $\mathbf{B}$, $\mathbf{C}$, and $\mathbf{D}$ within a given matrix. Provided that the matrices $\mathbf{A}$

EXAMPLE 7.5: INVERTING A 4 * 4 MATRIX

$$\text{If } \mathbf{A} = \begin{bmatrix} 3 & 1 & 4 & 2 \\ 6 & 2 & 1 & 3 \\ -1 & 2 & 4 & 2 \\ 4 & 1 & 3 & 7 \end{bmatrix} \text{ then its adjoint } \mathbf{A}^* = \begin{bmatrix} 27 & 27 & -36 & -9 \\ -60 & 75 & 98 & -43 \\ 75 & -33 & -1 & -7 \\ -39 & -12 & 7 & 49 \end{bmatrix}$$

$$\mathbf{AA}^* = \begin{bmatrix} 243 & 0 & 0 & 0 \\ 0 & 243 & 0 & 0 \\ 0 & 0 & 243 & 0 \\ 0 & 0 & 0 & 243 \end{bmatrix} \text{ showing the determinant of } \mathbf{A} = |\mathbf{A}| = 243.$$

and $\mathbf{D}$ are square and that both $\mathbf{A}$ and $(\mathbf{B} - \mathbf{CA}^{-1}\mathbf{B})$ are not singular, it can then be shown that

$$\begin{bmatrix} \mathbf{A} & \mathbf{B} \\ \mathbf{C} & \mathbf{D} \end{bmatrix}^{-1} = \begin{bmatrix} \mathbf{A}^{-1} + \mathbf{A}^{-1}\mathbf{B}(\mathbf{D} - \mathbf{CA}^{-1}\mathbf{B})^{-1}\mathbf{CA}^{-1} & -\mathbf{A}^{-1}\mathbf{B}(\mathbf{D} - \mathbf{CA}^{-1}\mathbf{B})^{-1} \\ -(\mathbf{D} - \mathbf{CA}^{-1}\mathbf{B})^{-1}\mathbf{CA}^{-1} & (\mathbf{D} - \mathbf{CA}^{-1}\mathbf{B})^{-1} \end{bmatrix}$$

This reduces the operation to a more manageable form. We will, however, not prove this here, although we will discuss partitioning matrices in Section 7.6.

7.4 APPLYING MATRICES

Let us now consider some simple applications of matrices, recalling that matrices are a mathematical form of shorthand and only really come into their own when dealing with large sets of numbers. Matrices are made up of numbers such as a_{ij} and a_{kl} that will normally be subject to the four simple operations of addition, subtraction,

BOX 7.5   INVERSES AND TRANSPOSES OF SQUARE MATRICES

If $\mathbf{C} = \mathbf{AB}$, then $\mathbf{C}^T = \mathbf{B}^T\mathbf{A}^T$ and $\mathbf{C}^{-1} = \mathbf{B}^{-1}\mathbf{A}^{-1}$

Similarly, if $\mathbf{D} = \mathbf{ABC}$, then $\mathbf{D}^T = \mathbf{C}^T\mathbf{B}^T\mathbf{A}^T$ and $\mathbf{D}^{-1} = \mathbf{C}^{-1}\mathbf{B}^{-1}\mathbf{A}^{-1}$

where $\mathbf{A}^{-1}$, and so on, is the inverse of the square matrix $\mathbf{A}$, having the property that $\mathbf{AA}^{-1} = \mathbf{A}^{-1}\mathbf{A} = \mathbf{I}$.

More generally $(\mathbf{A}^{-1})^{-1} = \mathbf{A}$; $(\mathbf{A}^T)^{-1} = (\mathbf{A}^{-1})^T$; $(k\mathbf{A})^{-1} = k^{-1}\mathbf{A}^{-1}$ where k is a scalar, $\mathbf{A}^*$ is the adjugate or adjoint of the square matrix $\mathbf{A}$, made up of the cofactors of $\mathbf{A}$ in the transposed positions. $\mathbf{A}^*\mathbf{A} = |\mathbf{A}|\mathbf{I}$ or $\mathbf{A}^*/|\mathbf{A}| = \mathbf{A}^{-1}$

multiplication, and division. The answers to their manipulation are then placed in the appropriate location in the new resulting matrix.

The processes of manipulating matrices are ideally suited to handling by computer as the operations are sequential and routine. Consider a very simple example that involves the intersection of two straight lines illustrated in Box 7.6. The calculation involves the conversion of a square matrix into its inverse (A into A^{-1}) and the multiplication of two matrices ($A^{-1} * B$).

BOX 7.6   THE INTERSECTION OF TWO LINES

In Chapter 3, we calculated the intersection of two straight lines.

Let us express these in the form

$$a_{11}x + a_{12}y = b_{11}$$
$$a_{21}x + a_{22}y = b_{21}$$

or

$$\begin{pmatrix} a_{11} & a_{12} \\ a_{21} & a_{22} \end{pmatrix} \begin{pmatrix} x \\ y \end{pmatrix} = \begin{pmatrix} b_{11} \\ b_{21} \end{pmatrix}$$

or

$$AX = B$$

Multiply both sides of the expression by A^{-1} to give $A^{-1}AX = A^{-1}B$.

Since $A^{-1}A = I$, we have

$$X = A^{-1}B$$

From previously at the end of Section 7.1,

$$A^{-1} = \{1/a_{11}a_{22} - a_{12}a_{21}\} * \begin{pmatrix} a_{22} & -a_{12} \\ -a_{21} & a_{11} \end{pmatrix}$$

Hence,

$$X = \begin{pmatrix} x \\ y \end{pmatrix} = \{1/(a_{11}a_{22} - a_{12}a_{21})\} * \begin{pmatrix} a_{22} & -a_{12} \\ -a_{21} & a_{11} \end{pmatrix} \begin{pmatrix} b_{11} \\ b_{21} \end{pmatrix}$$

or

$$\begin{pmatrix} x \\ y \end{pmatrix} = \{1/(a_{11}a_{22} - a_{12}a_{21})\} * \begin{pmatrix} a_{22}b_{11} - a_{12}b_{21} \\ -a_{21}b_{11} + a_{11}b_{21} \end{pmatrix}$$

or

$$x = (a_{22}b_{11} - a_{12}b_{21})/(a_{11}a_{22} - a_{12}a_{21})$$

$$y = (a_{11}b_{21} - a_{21}b_{11})/(a_{11}a_{22} - a_{12}a_{21})$$

This is the point where the two lines intersect. For a worked example see Example 7.6.

EXAMPLE 7.6: INTERSECTING LINES

Consider the lines $3x + 4y = 10$ and $5x - 7y = 3$. We can express these as

$$\mathbf{AX = B}$$

where

$$\mathbf{A} = \begin{pmatrix} 3 & 4 \\ 5 & -7 \end{pmatrix}, \ \mathbf{X} = \begin{pmatrix} x \\ y \end{pmatrix}, \text{ and } \mathbf{B} = \begin{pmatrix} 10 \\ 3 \end{pmatrix}$$

Using the equations in Box 7.6,

$$\mathbf{A}^{-1} = \{1/(-21-20)\} \begin{pmatrix} -7 & -4 \\ -5 & 3 \end{pmatrix} \text{ hence,}$$

$$\mathbf{A}^{-1}\mathbf{B} = \{-1/41\} \begin{pmatrix} -7 & -4 \\ -5 & 3 \end{pmatrix} \begin{pmatrix} 10 \\ 3 \end{pmatrix} = \{-1/41\} \begin{pmatrix} -82 \\ -41 \end{pmatrix} = \begin{pmatrix} 2 \\ 1 \end{pmatrix} = \begin{pmatrix} x \\ y \end{pmatrix}$$

Thus, $\begin{pmatrix} x \\ y \end{pmatrix} = \begin{pmatrix} 2 \\ 1 \end{pmatrix}$. The lines intersect where $x = 2$ and $y = 1$.

In the calculations in Box 7.6 and Example 7.6, if the determinant $|\mathbf{A}| = $ zero (which happens when $a_{11}a_{22} - a_{12}a_{21} = 0$), there is no solution. This will happen when the two lines are parallel and therefore do not intersect in finite space.

7.5 ROTATIONS AND TRANSLATIONS

Matrix algebra is very useful when dealing with coordinate systems and changes in the position and orientation of the basic axes of reference. In two dimensions, let the origin be changed by an amount (a, b) from origin 1 to origin 2 (see Figure 7.2).

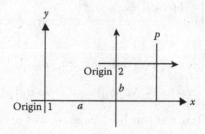

FIGURE 7.2 Shift or translation of origin.

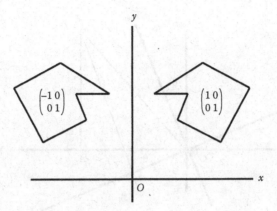

FIGURE 7.3 Reflection in the *y*-axis.

If the coordinates of any point P are $(x_1\ y_1)$ when referred to the old origin 1, then relative to the new origin 2, they are $(x_1 - a, y_1 - b)$. For a series of points:

$$\mathbf{X}_n = \mathbf{X}_o - \mathbf{A}$$

where n means new, o means old, and the matrix

$$\mathbf{X} = \begin{pmatrix} x_i \\ y_i \end{pmatrix} \quad \text{and} \quad \mathbf{A} = \begin{pmatrix} a \\ b \end{pmatrix}.$$

This relationship $\mathbf{X}_n = \mathbf{X}_o - \mathbf{A}$ will apply in any number of dimensions, especially three. It is often referred to as *translation*, meaning moving the origin from one point to another.

If we wish to obtain the reflection of an object in the *y*-axis (Figure 7.3), that is obtain the mirror image, then by applying the matrix

$$\begin{pmatrix} -1 & 0 \\ 0 & 1 \end{pmatrix} \text{ to } \begin{pmatrix} x \\ y \end{pmatrix}$$

we obtain a new x value $x_n = -x$ while y_n remains $= y$.

Similarly, by applying $\begin{pmatrix} 1 & 0 \\ 0 & -1 \end{pmatrix}$ we obtain a reflection in the *x*-axis.

If we want to change the scale of our projection by a factor s then $\mathbf{X}_n = s\,\mathbf{X}$, that is, each of the values of x and y will be multiplied by s. If we want to change the scale in the *x*-direction by a different amount than in the *y*-direction, then we need to premultiply by $\mathbf{S} = \begin{pmatrix} s_1 & 0 \\ 0 & s_2 \end{pmatrix}$ so that $\mathbf{X}_n = \mathbf{SX}$.

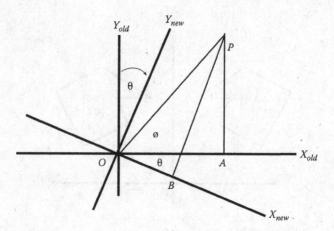

FIGURE 7.4   Rotation of axes.

Thus, if we wish to stretch the distances by a factor 2 in the x-direction and 3 in the y-direction we would apply the matrix $\begin{pmatrix} 2 & 0 \\ 0 & 3 \end{pmatrix}$. This will result in $x_n = 2x_o$ and $y_n = 3y_o$.

If we then change the origin, we obtain $\mathbf{X}_n = \mathbf{SX} - \mathbf{A}$. If we want to change the origin before we change the scale then $\mathbf{X}_n = \mathbf{S}(\mathbf{X} - \mathbf{A})$, which is not the same. Once again, the order in which we carry out operations makes a critical difference.

If we want to rotate the axes about their origin through an angle θ, then matrix algebra provides a convenient way to describe this (Figure 7.4).

Consider a point P and a set of axes OX_{old}, OY_{old}. The coordinates of P are given by $x_o = OA$ and $y_o = AP$. If the axes are rotated clockwise through an angle θ to OX_{new}, OY_{new}, then the new coordinates will be $x_n = OB$ and $y_n = BP$. Let angle $POA = \emptyset$ so that $PA = y_o = OP \sin \emptyset$ and $OA = x_o = OP \cos \emptyset$.

$$BP = y_n = OP \sin(\theta + \emptyset) = OP (\sin \theta \cos \emptyset + \cos \theta \sin \emptyset)$$

$$= OP \sin \theta \cos \emptyset + OP \cos \theta \sin \emptyset$$

Hence,

$$y_n = y_o \cos \theta + x_o \sin \theta$$

Also,

$$OB = x_n = OP \cos(\theta + \emptyset) = OP (\cos \theta \cos \emptyset - \sin \theta \sin \emptyset)$$

Hence,

$$x_n = x_o \cos \theta - y_o \sin \theta$$

Put in matrix form

$$\begin{pmatrix} x_n \\ y_n \end{pmatrix} = \begin{pmatrix} \cos\theta & -\sin\theta \\ \sin\theta & \cos\theta \end{pmatrix} \begin{pmatrix} x_o \\ y_o \end{pmatrix} \text{ or } \mathbf{X}_n = \mathbf{R}\mathbf{X}_o$$

where

$$\mathbf{R} = \begin{pmatrix} \cos\theta & -\sin\theta \\ \sin\theta & \cos\theta \end{pmatrix}$$

Note that the determinant of $\mathbf{R} = |\mathbf{R}| = \cos^2\theta + \sin^2\theta = 1$. The inverse of $\mathbf{R}$ is

$$\mathbf{R}^{-1} = \begin{pmatrix} \cos\theta & \sin\theta \\ -\sin\theta & \cos\theta \end{pmatrix}$$

Hence,

$$\mathbf{R}\mathbf{R}^{-1} = \begin{pmatrix} \cos^2\theta + \sin^2\theta & \sin\theta\cos\theta - \sin\theta\cos\theta \\ \sin\theta\cos\theta - \sin\theta\cos\theta & \sin^2\theta + \cos^2\theta \end{pmatrix} = \begin{pmatrix} 1 & 0 \\ 0 & 1 \end{pmatrix}$$

This is the identity matrix $\mathbf{I}$. Any matrix that when multiplied by its inverse gives the identity matrix is called an *orthogonal matrix*.

The processes of rotating axes but retaining their rectangularity, moving the origin, and perhaps applying a uniform scale change is called a *similarity transformation*. It is often used in photogrammetry and in computer-generated models. We will consider an example in Chapter 10.

The above derivation assumes that a positive rotation is clockwise as we look down on the plan. Since $\cos(-\theta) = \cos\theta$ and $\sin(-\theta) = -\sin\theta$, then if we were to regard an anticlockwise rotation as positive, we would have (Box 7.7 and Example 7.7)

$$\begin{pmatrix} x_n \\ y_n \end{pmatrix} = \begin{pmatrix} \cos\theta & \sin\theta \\ -\sin\theta & \cos\theta \end{pmatrix} \begin{pmatrix} x_o \\ y_o \end{pmatrix}$$

In general, when changing the direction of the axes, $\mathbf{X}_n = \mathbf{R}\mathbf{X}_o$ where $\mathbf{R}$ is called the *rotation matrix*. Because in geomatics and GIS we are often talking about bearings that are measured in a clockwise direction, we will use the convention that rotations are positive if clockwise. In photogrammetry, however, the standard convention uses the opposite, which inevitably leads to confusion. The crucial point is "the sign

---

BOX 7.7   ROTATION OF AXES

To transform old coordinates to new coordinates when rotating the x and y orthogonal axes through an angle θ measured as positive in a clockwise direction:

$$\begin{pmatrix} x_n \\ y_n \end{pmatrix} = \begin{pmatrix} \cos\theta & -\sin\theta \\ \sin\theta & \cos\theta \end{pmatrix} \begin{pmatrix} x_o \\ y_o \end{pmatrix}$$

(The same result will occur if we keep the axes fixed but rotate the object about the axes in an anticlockwise direction.)

---

of the sine." The problem arises because what is clockwise when looking upward is anticlockwise when looking downward. An example of a photogrammetric calculation is given in Chapter 10.

R as defined above is an orthogonal matrix; but there are occasions when a rectangular system needs to be skewed. For example, in Figure 7.5, the old grid that is rectangular has been skewed to form a new grid for ease of representation on a flat piece of paper. The axes are no longer at right angles although the distances parallel to the corresponding x-axis and y-axis remain the same ($AP = AP'$).

Thus, in the old grid, the coordinates of P are (x_o, y_o) or (OA, AP) while the distances for point P' on the new grid are (OA, AP') unless there is any change in scale. However, relative to the old grid the coordinates of P' have become (x', y') or (OB, BP').

$$x' = OB = OA + AP' \sin\phi = x + y \sin\phi$$

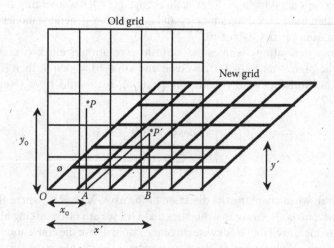

FIGURE 7.5   A skewed grid.

Matrices and Determinants

EXAMPLE 7.7: ROTATING AN OBJECT IN 2D

Consider the rectangle $ABCD$ in Figure 7.6 with coordinates A (4, 6), B (8, 6), C (8, 3), and D (4, 3). Its center is (6, 4.5).

If the coordinates (x_o, y_o) based on the origin are (x_i, y_i) with $i = 1$ to 4, and its center is $\mathbf{O} = \begin{pmatrix} 6 \\ 4.5 \end{pmatrix}$ then, based on the center of the rectangle, the new values (x_n, y_n) are

$$\mathbf{X}_n = \mathbf{X}_o - \mathbf{O} \quad \text{or} \quad \begin{pmatrix} x_n \\ y_n \end{pmatrix} = \begin{pmatrix} x_o - 6 \\ y_o - 4.5 \end{pmatrix}$$

Let us rotate the rectangle by 40° clockwise about its center, where sin 40 = 0.6428 and cos 40 = 0.7660. Applying the rotation $\begin{pmatrix} 0.7660 & 0.6428 \\ -0.6428 & 0.7660 \end{pmatrix}$ we obtain:

$$\begin{pmatrix} 0.7660 & 0.6428 \\ -0.6428 & 0.7660 \end{pmatrix}\begin{pmatrix} x_n \\ y_n \end{pmatrix} = \begin{pmatrix} 0.7660 & 0.6428 \\ -0.6428 & 0.7660 \end{pmatrix}\begin{pmatrix} x_o - 6.0 \\ y_o - 4.5 \end{pmatrix}$$

$$= \begin{pmatrix} 0.7660x_o - 4.5960 + 0.6428y_o - 2.8926 \\ -0.6428x_o + 3.8568 + 0.7660y_o - 3.4470 \end{pmatrix}$$

$$= \begin{pmatrix} 0.7660x_o + 0.6428y_o - 7.4886 \\ -0.6428x_o + 0.7660y_o + 0.4098 \end{pmatrix}$$

Note that we are rotating the object, not the axes.

Next we must return to the original origin by adding 6 to the x value and 4.5 to the y. This gives

$$\begin{pmatrix} 0.7660x_o + 0.6428y_o - 1.4886 \\ -0.6428x_o + 0.766y_o + 4.9098 \end{pmatrix}$$

for the new coordinates after rotation. A (4, 6) becomes (5.4322, 6.9346); B (8, 6) becomes (8.4962, 4.3634); C (8, 3) becomes (6.5678, 2.0654); and D (4, 3) becomes (3.5038, 4.6366).

where ø is the clockwise angle by which the y-axis has been rotated.

$$y' = BP' = AP' \cos \varnothing = y \cos \varnothing$$

Thus,

$$\begin{pmatrix} x' \\ y' \end{pmatrix} = \begin{pmatrix} 1 & \sin\phi \\ 0 & \cos\phi \end{pmatrix} \begin{pmatrix} x \\ y \end{pmatrix} \quad \text{or} \quad \mathbf{X'} = \mathbf{MX}$$

where

$$\mathbf{M} = \begin{pmatrix} 1 & \sin\phi \\ 0 & \cos\phi \end{pmatrix}$$

Here **M** is *not* an orthogonal transformation.

If we want to reduce the scale in the y-direction by a factor s_y so that y would become $s_y y$ while in the x-direction we applied a scale factor of s_x so that x becomes $s_x x$, then

$$\begin{pmatrix} x' \\ y' \end{pmatrix} = \begin{pmatrix} 1 & \sin\phi \\ 0 & \cos\phi \end{pmatrix} \begin{pmatrix} s_x & 0 \\ 0 & s_y \end{pmatrix} \begin{pmatrix} x \\ y \end{pmatrix} = \begin{pmatrix} s_x & s_y \sin\phi \\ 0 & s_y \cos\phi \end{pmatrix} \begin{pmatrix} x \\ y \end{pmatrix}$$

Note that

$$\begin{pmatrix} 1 & \sin\phi \\ 0 & \cos\phi \end{pmatrix} \begin{pmatrix} s_x & 0 \\ 0 & s_y \end{pmatrix} = \begin{pmatrix} s_x & s_y \sin\phi \\ 0 & s_y \cos\phi \end{pmatrix}$$

while

$$\begin{pmatrix} s_x & 0 \\ 0 & s_y \end{pmatrix} \begin{pmatrix} 1 & \sin\phi \\ 0 & \cos\phi \end{pmatrix} = \begin{pmatrix} s_x & s_x \sin\phi \\ 0 & s_y \cos\phi \end{pmatrix}$$

This is different. In the first case, we scale before we transform the coordinates while in the second, we apply the scale factor after the transformation.

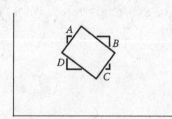

FIGURE 7.6   Rotating a rectangle.

This sequencing is especially important when considering the rotation of axes in three dimensions. The measurement of angles in a clockwise direction is sometimes referred to as the *left-hand rule* because, if your left thumb is pointing upward (in the Z-direction) and the fingers move from straight to bent, their pointing moves clockwise as you look down on your thumb. If the index finger remains straight it represents the Y-direction. See Figure 7.7. With the right-hand rule, X and Y are reversed. In each case, the fingers or thumb point in the positive direction.

In Figure 7.8a, we have the axes X (east), Y (north) in the plane of the paper and Z perpendicular to the surface. Figure 7.8b,c,d shows the effect of a clockwise turn about the X-axis, then the Y-axis then the Z-axis, as if looking along each axis in a positive direction.

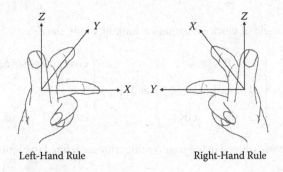

Left-Hand Rule   Right-Hand Rule

FIGURE 7.7   Left- and right-hand rules.

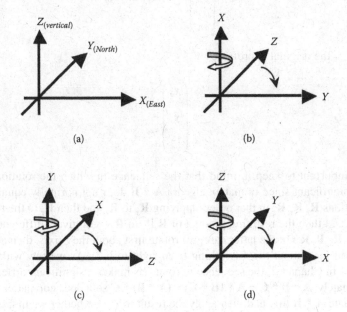

FIGURE 7.8   Positive rotations around each axis—left-hand rule.

It should however be stressed that, especially in photogrammetry, the opposite (right-hand) rule is used and positive rotations have the opposite sense. The confusion is compounded by whether an object is rotated about the axes or the axes are rotated about the object. For now, we will consider the rotation of axes; the key point is, as we have already observed, the sign of the sine.

If we are dealing in three dimensions (x, y, z), then the rotation matrix $\mathbf{R}$ about the Z-axis as shown in the two-dimensional case is

$$\mathbf{R}_z = \begin{pmatrix} \cos\phi_z & -\sin\phi_z & 0 \\ \sin\phi_z & \cos\phi_z & 0 \\ 0 & 0 & 1 \end{pmatrix}$$

where ϕ_z is the angle of clockwise rotation looking along the Z-axis. Similarly,

$$\mathbf{R}_x = \begin{pmatrix} 1 & 0 & 0 \\ 0 & \cos\phi_x & -\sin\phi_x \\ 0 & \sin\phi_x & \cos\phi_x \end{pmatrix} \text{ and } \mathbf{R}_y = \begin{pmatrix} \cos\phi_y & 0 & -\sin\phi_y \\ 0 & 1 & 0 \\ \sin\phi_y & 0 & \cos\phi_y \end{pmatrix}$$

Thus, if the axes are rotated in the positive direction as defined in Figure 7.8b,c,d, then

$$\mathbf{X}_n = \begin{pmatrix} x_n \\ y_n \\ z_n \end{pmatrix} = \mathbf{R}_x \, \mathbf{R}_y \, \mathbf{R}_z \, \mathbf{X}_o$$

where $\mathbf{X}_o$ = the original coordinates

$$\begin{pmatrix} x_o \\ y_o \\ z_o \end{pmatrix}$$

It is important to keep in mind that the sequence in which the rotations takes place is significant since in matrix algebra $\mathbf{A} * \mathbf{B}$ does not normally equal $\mathbf{B} * \mathbf{A}$. The rotations $\mathbf{R}_y \, \mathbf{R}_x \, \mathbf{R}_z$ in that order (applying $\mathbf{R}_x$ to $\mathbf{R}_z$ and then $\mathbf{R}_y$ to the result; or $\mathbf{R}_y$ to $\mathbf{R}_x$ and then the resulting product of $\mathbf{R}_y\mathbf{R}_x$ to $\mathbf{R}_z$) will give a different answer to that of $\mathbf{R}_x \, \mathbf{R}_y \, \mathbf{R}_z$. In the latter case, we rotate first about the z-axis, then about the y-axis, and then about the x (working from right to left). As we saw with dice in Figure 2.1 in Chapter 2, the sequence in rotations makes a significant difference.

Fortunately, $\mathbf{A} * \mathbf{B} * \mathbf{C} = \mathbf{A} * (\mathbf{B} * \mathbf{C}) = (\mathbf{A} * \mathbf{B}) * \mathbf{C}$ so it does not matter whether we calculate $\mathbf{A} * \mathbf{B}$ first and then apply the result to $\mathbf{C}$ or whether we first calculate $\mathbf{B} * \mathbf{C}$ and then premultiply by $\mathbf{A}$.

7.6   SIMPLIFYING MATRICES

Matrices can be *partitioned*, that is, treated in parts. Consider the two matrices **A** with 4 rows and 5 columns and **B** with 5 rows and 2 columns.

$$\mathbf{A} * \mathbf{B} = \begin{pmatrix} a_{11} & a_{12} & a_{13} & a_{14} & a_{15} \\ a_{21} & a_{22} & a_{23} & a_{24} & a_{25} \\ a_{31} & a_{32} & a_{33} & a_{34} & a_{35} \\ a_{41} & a_{42} & a_{43} & a_{44} & a_{45} \end{pmatrix} * \begin{pmatrix} b_{11} & b_{12} \\ b_{21} & b_{22} \\ b_{31} & b_{32} \\ b_{41} & b_{42} \\ b_{51} & b_{52} \end{pmatrix}$$

We can write these in the form

$$\mathbf{A} * \mathbf{B} = \left(\begin{array}{ccc|cc} a_{11} & a_{12} & a_{13} & a_{14} & a_{15} \\ a_{21} & a_{22} & a_{23} & a_{24} & a_{25} \\ a_{31} & a_{32} & a_{33} & a_{34} & a_{35} \\ a_{31} & a_{32} & a_{43} & a_{44} & a_{45} \end{array} \right) * \begin{pmatrix} b_{11} & b_{12} \\ b_{21} & b_{22} \\ \underline{b_{31}} & \underline{b_{32}} \\ b_{41} & b_{42} \\ b_{51} & b_{52} \end{pmatrix}$$

or

$$\mathbf{A} * \mathbf{B} = (\mathbf{A_1} \quad \mathbf{A_2}) * \begin{pmatrix} \mathbf{B_1} \\ \mathbf{B_2} \end{pmatrix}$$

where

$$\mathbf{A_1} = \begin{pmatrix} a_{11} & a_{12} & a_{13} \\ a_{21} & a_{22} & a_{23} \\ a_{31} & a_{32} & a_{33} \\ a_{41} & a_{42} & a_{43} \end{pmatrix}, \quad \mathbf{A_2} = \begin{pmatrix} a_{14} & a_{15} \\ a_{24} & a_{25} \\ a_{34} & a_{35} \\ a_{44} & a_{45} \end{pmatrix}$$

and

$$\mathbf{B_1} = \begin{pmatrix} b_{11} & b_{12} \\ b_{21} & b_{22} \\ b_{31} & b_{32} \end{pmatrix}, \quad \mathbf{B_1} = \begin{pmatrix} b_{41} & b_{42} \\ b_{51} & b_{52} \end{pmatrix}$$

$\mathbf{A_1}$, $\mathbf{A_2}$, $\mathbf{B_1}$, and $\mathbf{B_2}$ are partitions of the matrix. The original multiplication $\mathbf{A * B}$ is the same and can be expressed as

$$\mathbf{A * B} = \begin{pmatrix} \mathbf{A_1 B_1} & \mathbf{A_1 B_2} \\ \mathbf{A_2 B_1} & \mathbf{A_2 B_2} \end{pmatrix}$$

We can also simplify matrices by making them triangular, where $\mathbf{A_U}$ is called an *upper triangular matrix* while $\mathbf{A_L}$ is a *lower triangular matrix*:

$$\mathbf{A_U} = \begin{pmatrix} a_{11} & a_{12} & a_{13} & a_{14} & a_{15} \\ 0 & a_{22} & a_{23} & a_{24} & a_{25} \\ 0 & 0 & a_{33} & a_{34} & a_{35} \\ 0 & 0 & 0 & a_{44} & a_{45} \\ 0 & 0 & 0 & 0 & a_{55} \end{pmatrix}, \mathbf{A_L} = \begin{pmatrix} a_{11} & 0 & 0 & 0 & 0 \\ a_{21} & a_{22} & 0 & 0 & 0 \\ a_{31} & a_{32} & a_{33} & 0 & 0 \\ a_{41} & a_{42} & a_{43} & a_{44} & 0 \\ a_{51} & a_{52} & a_{53} & a_{54} & a_{55} \end{pmatrix}$$

To illustrate how this works, consider five equations for five unknowns (x_1, x_2, x_3, x_4, x_5)

$$a_{11}x_1 + a_{12}x_2 + a_{13}x_3 + a_{14}x_4 + a_{15}x_5 = b_1$$
$$a_{21}x_2 + a_{22}x_2 + a_{23}x_3 + a_{24}x_4 + a_{25}x_5 = b_2$$
$$a_{31}x_3 + a_{32}x_2 + a_{33}x_3 + a_{34}x_4 + a_{35}x_5 = b_3$$
$$a_{41}x_4 + a_{42}x_2 + a_{43}x_3 + a_{44}x_4 + a_{45}x_5 = b_4$$
$$a_{51}x_5 + a_{52}x_2 + a_{53}x_3 + a_{54}x_4 + a_{55}x_5 = b_5$$

where a_{11}, a_{12}, and so on, and b_1, b_2, and so on, are known quantities. We can express this in the form

$$\begin{pmatrix} a_{11} & a_{12} & a_{13} & a_{14} & a_{15} \\ a_{21} & a_{22} & a_{23} & a_{24} & a_{25} \\ a_{31} & a_{32} & a_{33} & a_{34} & a_{35} \\ a_{41} & a_{42} & a_{43} & a_{44} & a_{45} \\ a_{51} & a_{52} & a_{53} & a_{54} & a_{55} \end{pmatrix} * \begin{pmatrix} x_1 \\ x_2 \\ x_3 \\ x_4 \\ x_5 \end{pmatrix} = \begin{pmatrix} b_1 \\ b_2 \\ b_3 \\ b_4 \\ b_5 \end{pmatrix}$$

or as

$$\mathbf{AX = B}$$

Now premultiply both sides of $\mathbf{AX} = \mathbf{B}$ by the diagonal matrix $\mathbf{C_1}$ whose leading diagonal is made from the reciprocals of the numbers in column 1. Thus, the diagonal elements of $\mathbf{C_1}$ are $\{1/a_{11}, 1/a_{21}, 1/a_{31}, 1/a_{41}, 1/a_{51}\}$.

This gives us $\mathbf{C_1AX} = \mathbf{C_1B}$, or

$$
\begin{pmatrix}
a_{11}/a_{11} & a_{12}/a_{11} & a_{13}/a_{11} & a_{14}/a_{11} & a_{15}/a_{11} \\
a_{21}/a_{21} & a_{22}/a_{21} & a_{23}/a_{21} & a_{24}/a_{21} & a_{25}/a_{21} \\
a_{31}/a_{31} & a_{32}/a_{31} & a_{33}/a_{31} & a_{34}/a_{31} & a_{35}/a_{31} \\
a_{41}/a_{41} & a_{42}/a_{41} & a_{43}/a_{41} & a_{44}/a_{41} & a_{45}/a_{41} \\
a_{51}/a_{51} & a_{52}/a_{51} & a_{53}/a_{51} & a_{54}/a_{51} & a_{55}/a_{51}
\end{pmatrix}
*
\begin{pmatrix} x_1 \\ x_2 \\ x_3 \\ x_4 \\ x_5 \end{pmatrix}
=
\begin{pmatrix} b_1/a_{11} \\ b_2/a_{21} \\ b_3/a_{31} \\ b_4/a_{41} \\ b_5/a_{51} \end{pmatrix}.
$$

or

$$
\begin{pmatrix}
1 & a_{12}/a_{11} & a_{13}/a_{11} & a_{14}/a_{11} & a_{15}/a_{11} \\
1 & a_{22}/a_{21} & a_{23}/a_{21} & a_{24}/a_{21} & a_{25}/a_{21} \\
1 & a_{32}/a_{31} & a_{33}/a_{31} & a_{34}/a_{31} & a_{35}/a_{31} \\
1 & a_{42}/a_{41} & a_{43}/a_{41} & a_{44}/a_{41} & a_{45}/a_{41} \\
1 & a_{52}/a_{51} & a_{53}/a_{51} & a_{54}/a_{51} & a_{55}/a_{51}
\end{pmatrix}
*
\begin{pmatrix} x_1 \\ x_2 \\ x_3 \\ x_4 \\ x_5 \end{pmatrix}
=
\begin{pmatrix} b_1/a_{11} \\ b_2/a_{21} \\ b_3/a_{31} \\ b_4/a_{41} \\ b_5/a_{51} \end{pmatrix}
$$

Now subtract row 1 from the remaining rows to give

$$
\begin{pmatrix}
1 & a_{12}/a_{11} & a_{13}/a_{11} & a_{14}/a_{11} & a_{15}/a_{11} \\
0 & a_{22}/a_{21}-a_{12}/a_{11} & a_{23}/a_{21}-a_{13}/a_{11} & a_{24}/a_{21}-a_{14}/a_{11} & a_{25}/a_{21}-a_{15}/a_{11} \\
0 & a_{32}/a_{31}-a_{12}/a_{11} & a_{33}/a_{31}-a_{13}/a_{11} & a_{34}/a_{31}-a_{14}/a_{11} & a_{35}/a_{31}-a_{15}/a_{11} \\
0 & a_{42}/a_{41}-a_{12}/a_{11} & a_{43}/a_{41}-a_{13}/a_{11} & a_{44}/a_{41}-a_{14}/a_{11} & a_{45}/a_{41}-a_{15}/a_{11} \\
0 & a_{52}/a_{51}-a_{12}/a_{11} & a_{53}/a_{51}-a_{13}/a_{11} & a_{54}/a_{51}-a_{14}/a_{11} & a_{55}/a_{51}-a_{15}/a_{11}
\end{pmatrix}
*
\begin{pmatrix} x_1 \\ x_2 \\ x_3 \\ x_4 \\ x_5 \end{pmatrix}
$$

$$
=
\begin{pmatrix}
b_1/a_{11} \\
b_2/a_{21}-b_1/a_{11} \\
b_3/a_{31}-b_1/a_{11} \\
b_4/a_{41}-b_1/a_{11} \\
b_5/a_{51}-b_1/a_{11}
\end{pmatrix}
$$

We can then repeat this process ignoring the first row and first column and considering only the (4 * 4) matrix of rows 2 to 5 and columns 2 to 5 (that is, we consider the new submatrix of the original row one column one). We then repeat this again on the (3 * 3) and again on the (2 * 2) finishing up with an upper triangular matrix

$$\mathbf{B'} = \begin{pmatrix} 1 & a_{12}/a_{11} & a_{13}/a_{11} & a_{14}/a_{11} & a_{15}/a_{11} \\ 0 & 1 & a'_{23} & a'_{24} & a'_{25} \\ 0 & 0 & 1 & a'_{34} & a'_{35} \\ 0 & 0 & 0 & 1 & a'_{45} \\ 0 & 0 & 0 & 0 & 1 \end{pmatrix} * \begin{pmatrix} x_1 \\ x_2 \\ x_3 \\ x_4 \\ x_5 \end{pmatrix} = \begin{pmatrix} b_1/a_{11} \\ b'_2 \\ b'_3 \\ b'_4 \\ b'_5 \end{pmatrix}$$

The process is boring and repetitive, ideally suited to automate in a computer program that can routinely evaluate all the a' and b'. A worked example is given in Example 7.8.

The numbers $\mathbf{X}$ are found by back substitution. Thus, $x_5 = b'_5$, the number that one finishes up with in the last row of the modified matrix $\mathbf{B'}$. Moving up a row in matrix $\mathbf{B'}$, $x_4 + a'_{45} x_5 = b'_4$, or $x_4 = b'_4 - a'_{45} x_5$, and so on.

If we return for a moment to the original expression of $\mathbf{AX} = \mathbf{B}$, namely,

$$\begin{pmatrix} a_{11} & a_{12} & a_{13} & a_{14} & a_{15} \\ a_{21} & a_{22} & a_{23} & a_{24} & a_{25} \\ a_{31} & a_{32} & a_{33} & a_{34} & a_{35} \\ a_{41} & a_{42} & a_{43} & a_{44} & a_{45} \\ a_{51} & a_{52} & a_{53} & a_{54} & a_{55} \end{pmatrix} * \begin{pmatrix} x_1 \\ x_2 \\ x_3 \\ x_4 \\ x_5 \end{pmatrix} = \begin{pmatrix} b_1 \\ b_2 \\ b_3 \\ b_4 \\ b_5 \end{pmatrix}$$

The manipulation was all in terms of a and b. We could have written the equation in one solid block of numbers as

$$\begin{pmatrix} a_{11} & a_{12} & a_{13} & a_{14} & a_{15} & b_1 \\ a_{21} & a_{22} & a_{23} & a_{24} & a_{25} & b_2 \\ a_{31} & a_{32} & a_{33} & a_{34} & a_{35} & b_3 \\ a_{41} & a_{42} & a_{43} & a_{44} & a_{45} & b_4 \\ a_{51} & a_{52} & a_{53} & a_{54} & a_{55} & b_5 \end{pmatrix}$$

where the column $\mathbf{B}$ is a column *vector of constants* while the a terms form the *matrix of coefficients*. The combined expression is known as an *augmented matrix*.

EXAMPLE 7.8: THE SOLUTION OF SIMULTANEOUS EQUATIONS

Consider the four equations

$x + y + z + w = 11$; $2x - 6y + 3z + 7w = 24$; $7x + 2y + 5z - 3w = 10$; $9x - 3y + 2z + 2w = 21$

In matrix form

$$\begin{pmatrix} 1 & 1 & 1 & 1 \\ 2 & -6 & 3 & 7 \\ 7 & 2 & 5 & -3 \\ 9 & -3 & 2 & 2 \end{pmatrix} \begin{pmatrix} x \\ y \\ z \\ w \end{pmatrix} = \begin{pmatrix} 11 \\ 24 \\ 10 \\ 21 \end{pmatrix}$$

On reducing column 1 to unity, (e.g., dividing all of row 2 by 2), we obtain

$$\begin{pmatrix} 1 & 1 & 1 & 1 \\ 1 & -3 & 1.5 & 3.5 \\ 1 & 0.286 & 0.714 & -0.429 \\ 1 & -0.333 & 0.222 & 0.222 \end{pmatrix} * \begin{pmatrix} x \\ y \\ z \\ w \end{pmatrix} = \begin{pmatrix} 11 \\ 12 \\ 1.429 \\ 2.333 \end{pmatrix}$$

Then subtracting row 1 from each of the remaining rows

$$\begin{pmatrix} 1 & 1 & 1 & 1 \\ 0 & -4 & 0.5 & 2.5 \\ 0 & -0.714 & -0.286 & -1.429 \\ 0 & -1.333 & -0.778 & -0.778 \end{pmatrix} \begin{pmatrix} x \\ y \\ z \\ w \end{pmatrix} = \begin{pmatrix} 11 \\ 1 \\ -9.571 \\ -8.667 \end{pmatrix}$$

And repeating

$$\begin{pmatrix} 1 & 1 & 1 & 1 \\ 0 & 1 & -0.125 & -0.625 \\ 0 & 1 & 0.4 & 2 \\ 0 & 1 & 0.584 & 0.584 \end{pmatrix} \begin{pmatrix} x \\ y \\ z \\ w \end{pmatrix} = \begin{pmatrix} 11 \\ -0.25 \\ 13.4 \\ 6.5 \end{pmatrix}$$

Then by subtracting row 2 from rows 3 and 4

$$\begin{pmatrix} 1 & 1 & 1 & 1 \\ 0 & 1 & -0.125 & -0.625 \\ 0 & 0 & 0.525 & 2.625 \\ 0 & 0 & 0.71 & 1.21 \end{pmatrix} \begin{pmatrix} x \\ y \\ z \\ w \end{pmatrix} = \begin{pmatrix} 11 \\ -0.25 \\ 13.65 \\ 6.75 \end{pmatrix}$$

$$\begin{pmatrix} 1 & 1 & 1 & 1 \\ 0 & 1 & -0.125 & -0.625 \\ 0 & 0 & 1 & 5 \\ 0 & 0 & 1 & 1.7 \end{pmatrix} * \begin{pmatrix} x \\ y \\ z \\ w \end{pmatrix} = \begin{pmatrix} 11 \\ -0.25 \\ 26 \\ 9.5 \end{pmatrix}$$

And repeating

$$\begin{pmatrix} 1 & 1 & 1 & 1 \\ 0 & 1 & -0.125 & -0.625 \\ 0 & 0 & 1 & 5 \\ 0 & 0 & 0 & -3.3 \end{pmatrix} * \begin{pmatrix} x \\ y \\ z \\ w \end{pmatrix} = \begin{pmatrix} 11 \\ -0.25 \\ 26 \\ -16.5 \end{pmatrix}$$

And again,

$$\begin{pmatrix} 1 & 1 & 1 & 1 \\ 0 & 1 & -0.125 & -0.625 \\ 0 & 0 & 1 & 5 \\ 0 & 0 & 0 & 1 \end{pmatrix} \begin{pmatrix} x \\ y \\ z \\ w \end{pmatrix} = \begin{pmatrix} 11 \\ -0.25 \\ 26 \\ 5 \end{pmatrix}$$

And finally,

$$\begin{pmatrix} 1 & 1 & 1 & 1 \\ 0 & 1 & -0.125 & -0.625 \\ 0 & 0 & 1 & 5 \\ 0 & 0 & 0 & 1 \end{pmatrix} \begin{pmatrix} x \\ y \\ z \\ w \end{pmatrix} = \begin{pmatrix} 11 \\ -0.25 \\ 26 \\ 5 \end{pmatrix}$$

Thus, from the fourth row of this matrix,

$$w = 5$$

From the third row, $z + 5w = 26$, so $z = 1$

From the second row, $y - 0.125\,z - 0.625\,w = -0.25$, so $y = 3$

From the first row, $x + y + z + w = 11$, so $x = 2$

When reducing a matrix to diagonal form by the method outlined above, a problem can arise where one of the coefficients used for division is zero. This will happen when the five equations are not independent in which case there is no solution—there must at least be the same number of independent equations as there are unknowns in order to find a unique solution. If there are extra equations, as can happen with land survey measurements, where there are redundant measurements in order to improve the accuracy of the result, then a most probable solution can be found using statistical techniques such as the least square solution that is discussed in Chapter 14.

SUMMARY

Adjoint matrix: Also known as an *adjugate matrix*, it is the matrix $\mathbf{A}^*$ formed from a *square matrix* $\mathbf{A}$ with the *cofactor*s of $\mathbf{A}$ placed in their *transposed* positions. $\mathbf{A}^* * \mathbf{A} = |\mathbf{A}|\mathbf{I}$ where $|\mathbf{A}|$ is the *determinant* of $\mathbf{A}$. From this it follows that $\mathbf{A}^*/|\mathbf{A}| = \mathbf{A}^{-1}$ which is the *inverse* of $\mathbf{A}$.

Array: A two-dimensional rectangular arrangement of numbers or symbols in rows and columns.

Augmented matrix: A matrix that combines a set of coefficients with a column of constant values, as, for example, when solving a set of simultaneous equations.

Cofactor: The signed value of the *minor*. If the minor is $\mathbf{M}_{ij}$, then the cofactor = $(-1)^{(i-j)} * \mathbf{M}_{ij}$. Being based on a determinant, a cofactor is a single number.

Conformable: The number of elements in each row of a matrix equals the number of elements in each column of the second.

Determinant: A *scalar quantity* that is calculated from the alternating sum ($\pm$) of elements of a *square matrix*. Normally written with straight-line brackets, for example $|\mathbf{A}|$.

Diagonal matrix: A *square matrix* that has zero for every *element* except where the row number equals the column number.

Element: A number or symbol at a particular row and column within an array. The element in the ith row and jth column of the matrix $\mathbf{A}$ is written as a_{ij}.

Identity matrix: A *diagonal matrix* in which the nonzero elements are all equal to 1. Often written as $\mathbf{I}$.

$\mathbf{I} * \mathbf{A} = \mathbf{A} * \mathbf{I} = \mathbf{A}$

Inverse matrix: If a square matrix $\mathbf{A}$ and a square matrix $\mathbf{B}$ are such that

$\mathbf{A} * \mathbf{B} = \mathbf{I}$ (the *identity matrix*) then $\mathbf{B}$ is said to be the inverse of $\mathbf{A}$ and is written as $\mathbf{A}^{-1}$.

$\mathbf{A} * \mathbf{A}^{-1} = \mathbf{A}^{-1} * \mathbf{A} = \mathbf{I}$

Leading diagonal: The line of elements in a square matrix or determinant that runs from the top left corner to the bottom right.

Matrix: A set of numbers aligned in rows and columns forming a rectangular *array*. Often written in bold as $\mathbf{M}$. The number of rows is not necessarily the same as the number of columns but each row must be the same length, as must each column.

Matrix addition or subtraction: Two matrices can be added or subtracted only if they have the same number of rows and the same number of columns. If matrix $\mathbf{A}$ has elements a_{ij} and matrix $\mathbf{B}$ has elements b_{ij} then $\mathbf{A} \pm \mathbf{B}$ has elements $a_{ij} \pm b_{ij}$.

Matrix multiplication: Can only occur when the number of columns in the first matrix $\mathbf{A}$ is the same as the number of rows in the second matrix $\mathbf{B}$. The ith element of a row in $\mathbf{A}$ is used to multiply the ith element in the appropriate column in matrix $\mathbf{B}$. If A has i rows and j columns and $\mathbf{B}$ has j rows and k

columns then **A** * **B** has i rows and k columns, each element being made up from the sum of i multiplications.

In general **A** * **B** ≠ **B** * **A** unless **A** * **B** = **I**.

Minor: The *determinant* of a square *submatrix* of a given matrix.

Modulus: The absolute value of a quantity regardless of its sign.

Orthogonal matrix: A *square matrix* that when multiplied by its *inverse* gives the *identity matrix*. Thus, if $AA^{-1} = I$ then **A** is orthogonal.

Partitioning: The division of a matrix into *conformable* submatrices.

Rotation matrix: A *matrix* used to calculate the result of rotating the axes of a reference coordinate system.

Scalar quantity: A quantity that has magnitude but no direction, such as distance.

Similarity transformation: A transformation that involves translation, rotation, and uniform scale change.

Singular matrix: A square matrix that has no *inverse* since its *determinant* is zero.

Square matrix: A *matrix* that has the same number of rows as columns.

Submatrix: A matrix derived from a given matrix by removing some of its rows and columns. In particular, if a square matrix has the element a_{ij} for its ith row and jth column, then the submatrix for element a_{ij} is the matrix remaining after excluding all elements in row i and column j.

Translation: The movement of the origin of a set of coordinates.

Transpose matrix: The matrix formed by changing rows into columns. The transpose of **A** is A^T.

Triangular matrix: A matrix having zeros for all elements of the array below the *leading diagonal* in the case of an upper triangular matrix and above the leading diagonal for a lower triangular matrix.

8 Vectors

8.1 THE NATURE OF VECTORS

A matrix with one row or one column is sometimes referred to as a *vector*. Though closely related, matrix algebra should not, however, be confused with the handling of vectors. In the context of this chapter, a *vector* is a quantity such as velocity that has both magnitude and direction, while a *scalar* is a quantity that has magnitude but no direction.

The magnitude may represent a true distance such as a length or a quantity such as force or speed. In the study of geographic information systems there is a tendency to think of vectors as the lengths and directions of lines and compare them with raster representation. In practice they may represent the strength and direction of a variety of phenomena such as gravity. A bearing combined with a distance is only one example of a vector.

OP in Figure 8.1 represents a vector based on orthogonal axes that could be called x, y, and z but which, here, we will call the directions **i**, **j**, **k**. They are written in bold letters (rather like writing the symbol for the matrix **M** in bold script) to identify them as *unit vectors*, that is, vectors of unit length. If the actual length of the sides of a rectangular block of which OP is a diagonal are a_p, b_p, and c_p, then we can express the vector **OP** as:

$$\mathbf{OP} = a_p\mathbf{i} + b_p\mathbf{j} + c_p\mathbf{k}$$

What this means is that P is a distance from O of a_p in the x-direction, b_p in the y-direction, and c_p in the z-direction. If we have another vector **PQ** starting at P as in Figure 8.2 such that

$$\mathbf{PQ} = a_q\mathbf{i} + b_q\mathbf{j} + c_q\mathbf{k}$$

where a_q is the distance from P to Q in the x-direction, and so forth, then

$$\mathbf{OQ} = \mathbf{OP} + \mathbf{PQ} = (a_p + a_q)\,\mathbf{i} + (b_p + b_q)\,\mathbf{j} + (c_p + c_q)\,\mathbf{k}$$

From Q to O one needs to travel negatively in each of the directions **i**, **j**, **k** so that

$$\mathbf{OP} + \mathbf{PQ} + \mathbf{QO} = 0$$

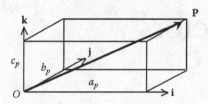

FIGURE 8.1 The axes **i**, **j**, **k** for vector **P**.

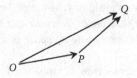

FIGURE 8.2 Vector addition.

Vectors may also be written with an arrow on top so that $\overrightarrow{OP} + \overrightarrow{PQ} = \overrightarrow{OQ}$. Thus, $\overrightarrow{OP} + \overrightarrow{PQ} + \overrightarrow{QO} = 0$.

We can also describe a vector in two parts—its length and its direction in the form

$$\mathbf{a} = |\mathbf{a}|\,\hat{\mathbf{a}}$$

This means that the vector **a** has length determined by its *modulus* $|\mathbf{a}|$ and the direction expressed as $\hat{\mathbf{a}}$, where $\hat{\mathbf{a}}$ is known as the *unit vector*. When

$$\mathbf{OP} = a_p\mathbf{i} + b_p\mathbf{j} + c_p\mathbf{k}$$

then the length of OP can be calculated by Pythagoras to be

$$|\mathbf{OP}| = \sqrt{(a_p^2 + b_p^2 + c_p^2)}$$

Note that $|\mathbf{OP}|$ (the modulus of vector **OP**) is a scalar quantity and has no direction (see Example 8.1) while $\hat{\mathbf{a}}$ has direction but is of unit length.

The cosines of the angles that **OP** makes with the reference vectors **i**, **j**, and **k** are called the *direction cosines* and are marked as α (angle **POi**), β (angle **POj**), and γ (angle **POk**) in Figure 8.3 and Example 8.2.

$$a_p = |\mathbf{OP}|\cos\alpha \qquad b_p = |\mathbf{OP}|\cos\beta \qquad c_p = |\mathbf{OP}|\cos\gamma$$

A three-dimensional vector can be defined by its length and its three direction cosines. The length of a vector can be changed simply by applying a scale factor so that, for example,

$$3\mathbf{a} = 3\,|\mathbf{a}|\,\hat{\mathbf{a}} \text{ so if } \mathbf{a} = 2\mathbf{i} + 3\mathbf{j} + 4\mathbf{k} \text{ then } 3\mathbf{a} = 6\mathbf{i} + 9\mathbf{j} + 12\mathbf{k}$$

In two dimensions we only need **i** and **j**. In Example 8.3 we calculate the intersection of two lines on a plane surface, showing the relationship between Cartesian coordinates and vectors. The procedure explained in the first part of the example could, however, be extended to three (or more dimensions).

EXAMPLE 8.1: A TRIANGLE IN SPACE

Let P (5, 9, 2), Q (3, 7, 4), and R (9, 6, 8) be the coordinates of three points in space. If $O(0, 0, 0)$ is the origin then

$$\overrightarrow{OP} = 5\mathbf{i} + 9\mathbf{j} + 2\mathbf{k}; \quad \overrightarrow{OQ} = 3\mathbf{i} + 7\mathbf{j} + 4\mathbf{k}; \quad \overrightarrow{OR} = 9\mathbf{i} + 6\mathbf{j} + 8\mathbf{k}$$

Also,

$$\overrightarrow{PQ} = -2\mathbf{i} - 2\mathbf{j} + 2\mathbf{k}: \quad \overrightarrow{QR} = 6\mathbf{i} - \mathbf{j} + 4\mathbf{k}; \quad \overrightarrow{RP} = -4\mathbf{i} + 3\mathbf{j} - 6\mathbf{k}$$

Note that

$$\overrightarrow{PQ} + \overrightarrow{QR} + \overrightarrow{RP} = 0$$

The length $PQ = \sqrt{\{(3-5)^2 + (7-9)^2 + (4-2)^2\}} = \sqrt{(2^2 + 2^2 + 2^2)}; \quad QR = \sqrt{(6^2 + 1^2 + 4^2)}; \quad RP = \sqrt{(4^2 + 3^2 + 6^2)}.$

Thus, the length $PQ = |\mathbf{PQ}| = 2\sqrt{3}; \quad |\mathbf{QR}| = \sqrt{(53)}; \quad |\mathbf{RP}| = \sqrt{(61)}.$

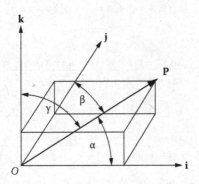

FIGURE 8.3   Direction cosines.

EXAMPLE 8.2: DIRECTION COSINES

Let P (5, 9, 2) be the coordinates of a point in space (Figure 8.3).
  If O is the origin then

$$\mathbf{OP} = 5\mathbf{i} + 9\mathbf{j} + 2\mathbf{k}$$

$$|\mathbf{OP}| = \text{the length } OP = \sqrt{(5^2 + 9^2 + 2^2)} = \sqrt{(110)} = 10.488$$

The direction cosines for **OP** are

$$\cos \alpha = 5/10.488 = 0.4767; \quad \cos \beta = 9/10.488 = 0.8581; \quad \cos \gamma = 2/10.488 = 0.1907$$

$$(\alpha \approx 61.5°; \beta \approx 30.9°; \gamma \approx 79.0°)$$

EXAMPLE 8.3: INTERSECTION OF TWO LINES IN 2D

In two dimensions, consider an origin O and four points P_1, P_2, P_3, P_4 with coordinates (x_1, y_1), and so forth. Any point P_n on the line P_1 to P_2 can be represented by the vector $\mathbf{p}_n = \mathbf{p}_1 + s(\mathbf{p}_2 - \mathbf{p}_1)$ where s is a variable quantity that equals 0 at P_1 and 1 at P_2. If P_n is also on the line P_3 to P_4 then the vector $\mathbf{p}_n = \mathbf{p}_3 + t(\mathbf{p}_4 - \mathbf{p}_3)$ where t is a variable quantity. Thus, at the point of intersection of these two lines,

$$\mathbf{p}_1 + s(\mathbf{p}_2 - \mathbf{p}_1) = \mathbf{p}_3 + t(\mathbf{p}_4 - \mathbf{p}_3)$$

Hence,

$$s = \{(\mathbf{p}_3 - \mathbf{p}_1) + t(\mathbf{p}_4 - \mathbf{p}_3)\}/(\mathbf{p}_2 - \mathbf{p}_1) \text{ or } t = \{(\mathbf{p}_1 - \mathbf{p}_3) + s(\mathbf{p}_2 - \mathbf{p}_1)\}/(\mathbf{p}_4 - \mathbf{p}_3)$$

This is true for both the set of x values and the set of y values (and z values if we were considering three dimensions). Hence,

$$t = \{(x_1 - x_3) + s(x_2 - x_1)\}/(x_4 - x_3)$$
$$= \{(y_1 - y_3) + s(y_2 - y_1)\}/(y_4 - y_3)$$

Hence,

$$s = \frac{x_1(y_4 - y_3) + x_3(y_1 - y_4) + x_4(y_3 - y_1)}{(x_4 - x_3)(y_2 - y_1) - (x_2 - x_1)(y_4 - y_3)}$$

and

$$t = \frac{x_1(y_3 - y_2) + x_2(y_3 - y_1) + x_3(y_2 - y_1)}{(x_2 - x_1)(y_4 - y_3) - (x_4 - x_3)(y_2 - y_1)}$$

From $\mathbf{p}_n = \mathbf{p}_1 + s(\mathbf{p}_2 - \mathbf{p}_1)$

$$x_n = x_1 + s(x_2 - x_1) \qquad y_n = y_1 + s(y_2 - y_1)$$

For instance, using the data in Examples 3.1 and 3.2 in Chapter 3,

$$A = P_1 = (1234.56, 2345.67) \text{ and } B = P_2 = (1296.32, 2417.38)$$
$$C = P_3 = (1300.24, 2351.77) \text{ and } D = P_4 = (1212.45, 2431.78)$$
$$s = (-5790.5758)/(-11236.8385) = 0.51532073$$
$$x_n = 1266.3 \qquad y_n = 2382.62$$

These are the same values as derived in Chapter 3.

8.2   DOT AND CROSS PRODUCTS

Vectors can be added and subtracted. They can also be multiplied or divided by a scalar quantity. Two vectors can also be multiplied together but the concept has a different meaning from ordinary multiplication. Consider two vectors **a** and **b** such that

$$\mathbf{a} = a_x\mathbf{i} + a_y\mathbf{j} + a_z\mathbf{k} \text{ and } \mathbf{b} = b_x\mathbf{i} + b_y\mathbf{j} + b_z\mathbf{k}$$

There are two forms of product, known as the *dot product* and the *cross product*. The dot product results in a scalar quantity defined as "a dot b"

$$\mathbf{a.b} = |a||b|\cos \emptyset$$

where $\emptyset$ is the difference in direction or angle between the two vectors. In particular, since $\cos(0) = 1$ and $\cos (90°) = 0$ and by definition $|i| = |j| = |k| = 1$ since they are unit vectors

$$\mathbf{i.i} = \mathbf{j.j} = \mathbf{k.k} = 1 \quad \text{while} \quad \mathbf{i.j} = \mathbf{j.k} = \mathbf{k.i} = 0$$

Thus, $\mathbf{a.b} = (a_x\mathbf{i} + a_y\mathbf{j} + a_z\mathbf{k}).(b_x\mathbf{i} + b_y\mathbf{j} + b_z\mathbf{k})$ giving $3 * 3 = 9$ relationships that contain terms such as **i.i** (which = 1) or **i.j** (which = 0) or **j.i** (which also = 0). Working this through,

$$\mathbf{a.b} = a_xb_x + a_yb_y + a_zb_z$$

which is simply a number or scalar quantity (see Example 8.4).

The cross product ("a cross b") creates a new vector perpendicular to the plane containing **a** and **b** (shown by the vector **c** in Figure 8.4). Its origins lie in the study of electromagnetism although it has relevance to a number of problems in geomatics and GIS, for instance, when determining whether any surface element of a feature, such as a building facade, is facing toward or away from the observer. It also allows levels of shading to be calculated or hidden surfaces removed. The cross product

$$\mathbf{a} \times \mathbf{b} = |a||b|\sin \emptyset \ \mathbf{c}$$

The magnitude $= |a||b| \sin \emptyset$ is a scalar quantity, while the direction is determined by convention to be a right-handed system. It follows that if we reverse the order of multiplication, the vector **c** will be pointing in the opposite direction, that is,

$$\mathbf{a} \times \mathbf{b} = -\mathbf{b} \times \mathbf{a}$$

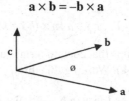

FIGURE 8.4   Dot and cross products.

EXAMPLE 8.4: THE ANGLE BETWEEN TWO VECTORS

Using the data in Example 8.1 where P (5, 9, 2) and Q (3, 7, 4) were the coordinates of two points in space and $\overrightarrow{OP} = 5\mathbf{i} + 9\mathbf{j} + 2\mathbf{k}$; $\overrightarrow{OQ} = 3\mathbf{i} + 7\mathbf{j} + 4\mathbf{k}$.

$\mathbf{OP.OQ} = 5 * 3 + 9 * 7 + 2 * 4 = 15 + 63 + 8 = 86$

$= |\mathbf{OP}|.|\mathbf{OQ}| \cos \emptyset$ where $\emptyset =$ the angle between OP and OQ

$= \sqrt{(110)} * \sqrt{(74)} \cos \emptyset$

$= 90.22 \cos \emptyset$

Hence,

$$\cos \emptyset = 86/90.22 = 0.953 \text{ or } \emptyset = 17.6°$$

Note: Using the formula $\cos \emptyset = (a^2 + b^2 - c^2)/2ab$ from Chapter 5, Box 5.2

$= (OP^2 + OQ^2 - PQ^2)/(2 * OP * OQ)$

Then

$$PQ^2 = 12, OP^2 = 110, OQ^2 = 74$$

$$\cos \emptyset = (110 + 74 - 12)/2\sqrt{(110 * 74)} = 172/180.44 = 0.953$$

which confirms the calculation.

For unit vectors, the angles between them are 90° and sin (90°) = 1 and cos (90°) = sin (0°) = 0. Hence,

$$\mathbf{i} \times \mathbf{i} = \mathbf{j} \times \mathbf{j} = \mathbf{k} \times \mathbf{k} = 0$$

Also,

$$\mathbf{i} \times \mathbf{j} = \mathbf{k}; \mathbf{j} \times \mathbf{k} = \mathbf{i}; \mathbf{k} \times \mathbf{i} = \mathbf{j} \quad \text{and} \quad \mathbf{j} \times \mathbf{i} = -\mathbf{k}; \mathbf{k} \times \mathbf{j} = -\mathbf{i} \text{ and } \mathbf{i} \times \mathbf{k} = -\mathbf{j}$$

If, as before, we have

$$\mathbf{a} = a_x\mathbf{i} + a_y\mathbf{j} + a_z\mathbf{k} \quad \text{and} \quad \mathbf{b} = b_x\mathbf{i} + b_y\mathbf{j} + b_z\mathbf{k}$$

then

$$\mathbf{a} \times \mathbf{b} = (a_x\mathbf{i} + a_y\mathbf{j} + a_z\mathbf{k}) \times (b_x\mathbf{i} + b_y\mathbf{j} + b_z\mathbf{k})$$

This gives 3 * 3 = 9 relationships that contain terms such as $\mathbf{i} \times \mathbf{i}$ (which = 0) or $\mathbf{i} \times \mathbf{j}$ (which = $\mathbf{k}$) or $\mathbf{j} \times \mathbf{i}$ (which = $-\mathbf{k}$). Working this through,

$$\mathbf{a} \times \mathbf{b} = (a_yb_z - a_zb_y)\mathbf{i} + (a_zb_x - a_xb_z)\mathbf{j} + (a_xb_y - a_yb_x)\mathbf{k}$$

EXAMPLE 8.5: THE AREA OF A TRIANGLE

Again using the data from Example 8.1, where P (5, 9, 2), Q (3, 7, 4), and R (9, 6, 8) are the coordinates of three points in space.

$$|PQ| = \sqrt{(4 + 4 + 4)} = 2\sqrt{3} \approx 3.46; \ |PR| = \sqrt{(16 + 9 + 36)} = \sqrt{(61)} \approx 7.81$$

$$PQ = -2i - 2j + 2k: \ PR = 4i - 3j + 6k; \ QR = 6i - j + 4k$$

The cross product

$$PQ * PR = \begin{vmatrix} i & j & k \\ -2 & -2 & 2 \\ 4 & -3 & 6 \end{vmatrix} = -6i + 20j + 14k \text{ the modulus of which} \approx 25.14$$

Since the area of a triangle $= \frac{1}{2}\, ab \sin C$ and the modulus of the cross product $= |PQ| * |PR| * \sin$ (angle QPR) then the modulus of $PQ * PR =$ twice the area of the triangle. The area of triangle PQR is $1/2 * 25.14 \approx 12.57$.

Since $QR = \sqrt{(53)} \approx 7.28$, this can be checked using the semiperimeter formula (see Box 4.2) where $s \approx 9.275$ giving the area as

$$\sqrt{(9.275 * 5.815 * 1.465 * 1.995)} \approx 12.56$$

The difference is due to rounding errors. If more decimal places were used, the two values would be identical.

Or, to express this in determinant form (see Example 8.5)

$$\mathbf{a \ x \ b} = \begin{vmatrix} i & j & k \\ a_x & a_y & a_z \\ b_x & b_y & b_z \end{vmatrix} = \begin{vmatrix} a_y & a_z \\ b_y & b_z \end{vmatrix} i - \begin{vmatrix} a_x & a_z \\ b_x & b_z \end{vmatrix} j + \begin{vmatrix} a_x & a_y \\ b_x & b_y \end{vmatrix} k$$

The relation $\mathbf{a}.(\mathbf{b} \times \mathbf{c})$ represents the scalar product between two vectors, the vector $\mathbf{a}$ and the vector $\mathbf{b} \times \mathbf{c}$. It is called the *scalar triple product* and gives the volume of the solid whose edges are represented by the vectors $\mathbf{a}$, $\mathbf{b}$, and $\mathbf{c}$. This solid has parallel sides and is in effect a squashed brick and is known as a *parallelepiped* (Figure 8.5).

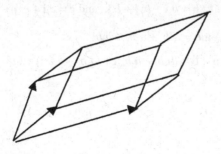

FIGURE 8.5   A parallelepiped.

If

$$\mathbf{a} = a_x\mathbf{i} + a_y\mathbf{j} + a_z\mathbf{k};\ \mathbf{b} = b_x\mathbf{i} + b_y\mathbf{j} + b_z\mathbf{k};\ \text{and}\ \mathbf{c} = c_x\mathbf{i} + c_y\mathbf{j} + c_z\mathbf{k}$$

then

$$\mathbf{b} \times \mathbf{c} = (b_yc_z - b_zc_y)\mathbf{i} + (b_zc_x - b_xc_z)\mathbf{j} + (b_xc_y - b_yc_x)\mathbf{k}$$

Hence,

$$\mathbf{a}.(\mathbf{b} \times \mathbf{c}) = \{a_x\mathbf{i} + a_y\mathbf{j} + a_z\mathbf{k}\}.\ \{(b_yc_z - b_zc_y)\mathbf{i} + (b_zc_x - b_xc_z)\mathbf{j} + (b_xc_y - b_yc_x)\mathbf{k}\}$$

$$= a_x(b_yc_z - b_zc_y) + a_y(b_zc_x - b_xc_z) + a_z(b_xc_y - b_yc_x)$$

$$= a_x(b_yc_z - b_zc_y) - a_y(b_xc_z - b_zc_x) + a_z(b_xc_y - b_yc_x)$$

Thus,

$$\mathbf{a}.(\mathbf{b} \times \mathbf{c}) = \begin{vmatrix} a_x & a_y & a_z \\ b_x & b_y & b_z \\ c_x & c_y & c_z \end{vmatrix}$$

In Box 8.1 we summarize vector multiplication.

8.3   VECTORS AND PLANES

If we have three points $A\ (x_A, y_A, z_A)$, $B\ (x_B, y_B, z_B)$, and $C\ (x_C, y_C, z_C)$, then we can represent these with three vectors (Figure 8.6) from the origin P $(0, 0, 0)$.

$$\mathbf{P}_A = (x_A\mathbf{i}, y_A\mathbf{j}, z_A\mathbf{k}),\ \mathbf{P}_B = (x_B\mathbf{i}, y_B\mathbf{j}, z_B\mathbf{k}),\ \text{and}\ \mathbf{P}_C = (x_C\mathbf{i}, y_C\mathbf{j}, z_C\mathbf{k})$$

The lines CA and CB are represented by $(\mathbf{P}_A - \mathbf{P}_C)$ and $(\mathbf{P}_B - \mathbf{P}_C)$. The cross product between these two lines is given by

$$(\mathbf{P}_A - \mathbf{P}_C) \times (\mathbf{P}_B - \mathbf{P}_C)$$

BOX 8.1   VECTOR MULTIPLICATION

If $\mathbf{a} = a_x\mathbf{i} + a_y\mathbf{j} + a_z\mathbf{k}$; $\mathbf{b} = b_x\mathbf{i} + b_y\mathbf{j} + b_z\mathbf{k}$; and $\mathbf{c} = c_x\mathbf{i} + c_y\mathbf{j} + c_z\mathbf{k}$, then

$$\mathbf{a}.\mathbf{b} = a_xb_x + a_yb_y + a_zb_z$$

$$\mathbf{a} \times \mathbf{b} = (a_yb_z - a_zb_y)\mathbf{i} + (a_zb_x - a_xb_z)\mathbf{j} + (a_xb_y - a_yb_x)\mathbf{k}$$

$$\mathbf{a}.(\mathbf{b} \times \mathbf{c}) = a_x(b_yc_z - b_zc_y) - a_y(b_xc_z - b_zc_x) + a_z(b_xc_y - b_yc_x) = \begin{vmatrix} a_x & a_y & a_z \\ b_x & b_y & b_z \\ c_x & c_y & c_z \end{vmatrix}$$

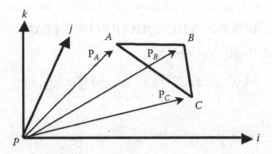

FIGURE 8.6 Vectors and a plane.

This $= \{(x_A - x_C)\mathbf{i} + (y_A - y_C)\mathbf{j} + (z_A - z_C)\,\mathbf{k}\} \times \{(x_B - x_C)\mathbf{i} + (y_B - y_C)\mathbf{j} + (z_B - z_C)\,\mathbf{k}\}$

$= \{(x_A - x_C)(y_B - y_C)\mathbf{k} - (x_A - x_C)(z_B - z_C)\mathbf{j} - (y_A - y_C)(x_B - x_C)\mathbf{k}$

$\quad + (y_A - y_C)(z_B - z_C)\mathbf{i} + (z_A - z_C)(x_B - x_C)\mathbf{j} - (z_A - z_C)(y_B - y_C)\mathbf{i}\}$

$= \{(y_A(z_B - z_C) + y_B(z_C - z_A) + y_C(z_A - z_B)\}\mathbf{i}$

$\quad + \{z_A(x_B - x_C) + z_B(x_C - x_A) + z_C(x_A - x_B)\}\mathbf{j}$

$\quad + \{x_A(y_B - y_C) + x_B(y_C - y_A) + x_C(y_A - y_B)\}\mathbf{k}$

This cross product represents a vector that is perpendicular to both the lines CA and CB in Figure 8.6, which means that it is normal to the plane of CAB. This direction can be used, for instance, to calculate the level of illumination of a surface. Note that by reversing the sequence so that we have the cross product for CB and CA (rather than CA and CB), then the direction of the normal will be reversed, that is, it will be in the opposite direction.

If we join any general point $Q\,(x, y, z)$ to C then the vector for the line CQ will be $(\mathbf{P}_Q - \mathbf{P}_C)$. If point Q lies in the plane ABC then the dot product between this vector and the normal to the plane will be zero. That is, for Q to lie in the plane ABC

$$(\mathbf{P}_A - \mathbf{P}_C) \times (\mathbf{P}_B - \mathbf{P}_C). \,(\mathbf{P}_Q - \mathbf{P}_C) = 0$$

or

$$(\mathbf{P}_A - \mathbf{P}_C) \times (\mathbf{P}_B - \mathbf{P}_C). \,\{(x - x_C)\mathbf{i} + (y - y_C)\mathbf{j} + (z - z_C)\,\mathbf{k}\} = 0$$

This gives

$$(x - x_C)\{y_A(z_B - z_C) + y_B(z_C - z_A) + y_C(z_A - z_B)\}$$

$$+ (y - y_C)\{z_A(x_B - x_C) + z_B(x_C - x_A) + z_C(x_A - x_B)\}$$

$$+ (z - z_C)\{x_A(y_B - y_C) + x_B(y_C - y_A) + x_C(y_A - y_B)\} = 0$$

This is the equation of the plane ABC (see Box 8.2 and Example 8.6).

BOX 8.2   THE EQUATION OF A PLANE

The plane that contains three points:

$$A\,(x_A, y_A, z_A),\, B\,(x_B, y_B, z_B),\, \text{and}\, C\,(x_C, y_C, z_C)$$

is given by

$$(x - x_C)\{y_A(z_B - z_C) + y_B(z_C - z_A) + y_C(z_A - z_B)\}$$
$$+ (y - y_C)\{z_A(x_B - x_C) + z_B(x_C - x_A) + z_C(x_A - x_B)\}$$
$$+ (z - z_C)\{x_A(y_B - y_C) + x_B(y_C - y_A) + x_C(y_A - y_B)\} = 0$$

which reduces to $ax + by + cz + d = 0$ where in determinant form

$$a = \begin{vmatrix} (y_A - y_C) & (z_A - z_C) \\ (y_B - y_C) & (z_B - z_C) \end{vmatrix};\; b = \begin{vmatrix} (z_A - z_C) & (x_A - x_C) \\ (z_B - z_C) & (x_B - x_C) \end{vmatrix};\; c = \begin{vmatrix} (x_A - x_C) & (y_A - y_C) \\ (x_B - x_C) & (y_B - y_C) \end{vmatrix};\; \text{and}$$

$$d = -(ax_C + by_C + cy_C)$$

This equation reduces to the form $ax + by + cz + d = 0$ where a, b, c, d are constants that define the plane through the points ABC. The relationships given for the reduced equation of the plane in Box 8.2 (see Example 8.6) can also be expressed as:

$$a = \begin{vmatrix} 1 & y_C & z_C \\ 1 & y_A & z_A \\ 1 & y_B & z_B \end{vmatrix};\; b = \begin{vmatrix} x_C & 1 & z_C \\ x_A & 1 & z_A \\ x_B & 1 & z_B \end{vmatrix};\; c = \begin{vmatrix} x_C & y_C & 1 \\ x_A & y_A & 1 \\ x_B & y_B & 1 \end{vmatrix};\; d = -(ax_C + by_C + cy_C)$$

Let us designate the vector normal to the plane as **n**, that is, the vector derived from the cross product $(\mathbf{P}_A - \mathbf{P}_C) \times (\mathbf{P}_B - \mathbf{P}_C) = \mathbf{n}$.

If the equations of the plane is given by $ax + by + cz + d = 0$, then

$$\mathbf{n} = a\mathbf{i} + b\mathbf{j} + c\mathbf{k}$$

and

$$|\mathbf{n}| = \sqrt{a^2 + b^2 + c^2}$$

If we have two planes defined by $a_1x + b_1y + c_1z + d_1 = 0$ and $a_2x + b_2y + c_2z + d_2 = 0$, then the angle between the planes is given by

$$\mathbf{n}_1.\mathbf{n}_2 = |\mathbf{n}_1| * |\mathbf{n}_2| * \cos(\emptyset)$$

EXAMPLE 8.6: THE EQUATION OF A PLANE

Once again, using the values in Example 8.1 where $P(5, 9, 2)$, $Q(3, 7, 4)$, and $R(9, 6, 8)$ and using the formulae for a, b, c, d

$$a = \begin{vmatrix} 1 & 6 & 8 \\ 1 & 9 & 2 \\ 1 & 7 & 4 \end{vmatrix} = -6;\ b = \begin{vmatrix} 9 & 1 & 8 \\ 5 & 1 & 2 \\ 3 & 1 & 4 \end{vmatrix} = 20;\ c = \begin{vmatrix} 9 & 6 & 1 \\ 5 & 9 & 1 \\ 3 & 7 & 1 \end{vmatrix} = 14;\ d = -178$$

The equation of the plane PQR is therefore

$$-6x + 20y + 14z - 178 = 0$$

The vector for the normal to the plane is

$$-6\mathbf{i} + 20\mathbf{j} + 14\mathbf{k}$$

where $\mathbf{n}_1$ and $\mathbf{n}_2$ are the vectors of the normal to each plane, that is,

$$\mathbf{n}_1 = a_1\mathbf{i} + b_1\mathbf{j} + c_1\mathbf{k} \text{ and } \mathbf{n}_2 = a_2\mathbf{i} + b_2\mathbf{j} + c_2\mathbf{k}$$

Thus, ø, the angle between the planes is such that

$$\phi = \cos^{-1}\left(\frac{\mathbf{n}_1.\mathbf{n}_2}{|\mathbf{n}_1||\mathbf{n}_2|} \right)$$

8.4 ANGLES OF INCIDENCE

We can extend this idea further by considering a vector $\mathbf{v}$ that represents a line such as a ray of light striking the plane surface defined by PQR in Figure 8.7. The angle between $\mathbf{n}$, the normal to the surface (here the plane PQR) and the vector $\mathbf{v}$ is given by the dot product

$$\mathbf{n}.\mathbf{v} = |\mathbf{n}| * |\mathbf{v}| * \cos(ø)$$

where ø is the angle that the line makes with the vertical to the plane. Thus, the angle with the normal to the plane is given by $\phi = \cos^{-1}\left(\dfrac{\mathbf{n}.\mathbf{v}}{|\mathbf{n}||\mathbf{v}|} \right)$.

Now consider the line from A to B in Figure 8.7. A is point (x_A, y_A, z_A) and B is point (x_B, y_B, z_B), and the line AB cuts the plane containing the triangle PQR at T. Let T be (x_T, y_T, z_T) and the vector $\mathbf{OT} = \mathbf{t} = x_T\mathbf{i} + y_T\mathbf{j} + z_T\mathbf{k}$. Let the vector $\mathbf{OA}$ be $\mathbf{a} = x_A\mathbf{i} + y_A\mathbf{j} + z_A\mathbf{k}$ and the vector $\mathbf{OB}$ be $\mathbf{b} = x_B\mathbf{i} + y_B\mathbf{j} + z_B\mathbf{k}$. Finally, let the vector $\mathbf{AB}$ be $\mathbf{v} = (x_B - x_A)\mathbf{i} + (y_B - y_A)\mathbf{j} + (z_B - z_A)\mathbf{k}$.

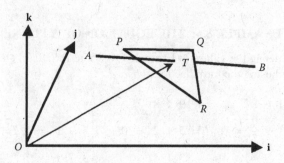

FIGURE 8.7   Angle of incidence.

T must satisfy the condition that it is both on the line AB and the plane PQR, which is defined by $ax + by + cz + d = 0$. The normal to this plane is $\mathbf{n} = a\mathbf{i} + b\mathbf{j} + c\mathbf{k}$.

$$\mathbf{n.t} = (a\mathbf{i} + b\mathbf{j} + c\mathbf{k}).(x_T\mathbf{i} + y_T\mathbf{j} + z_T\mathbf{k}) = ax_T + by_T + cz_T$$

since $\mathbf{i.i} = \mathbf{j.j} = \mathbf{k.k} = 1$ and $\mathbf{i.j} = \mathbf{j.k} = \mathbf{k.i} = 0$.

Since T lies on the plane, $ax_T + by_T + cz_T + d = 0$, then

$$\mathbf{n.t} + d = 0$$

We can also express T as the vector $\mathbf{OA}$ plus $\mathbf{AT}$ which, in turn, is a proportion of the vector $\mathbf{AB}$ in the form $\mathbf{t} = \mathbf{a} + \lambda\mathbf{v}$, where λ is some unknown scalar quantity. Thus,

$$\mathbf{n.}(\mathbf{a} + \lambda\mathbf{v}) + d = 0 \quad \text{or} \quad \mathbf{n.a} + \lambda\mathbf{n.v} + d = 0 \quad \text{or} \quad \lambda = \frac{-(\mathbf{n.a} + d)}{\mathbf{n.v}}$$

This means that we can determine the coordinates of point T based on the coordinates of A and B. We also have the value of $\mathbf{n.t}$ from which we have the cosine of the angle between the vector $\mathbf{OT}$ and the normal to the plane and hence the angle of incidence. See Example 8.7.

Note that if $\mathbf{n.v}$ is zero there is no solution as the line AB will be parallel to the plane PQR.

8.5   VECTORS AND ROTATIONS

To illustrate how vectors can be used to find the same answer as might be achieved through the use of matrices, consider an arbitrary axis OM about which we wish to rotate a point P through an angle $\emptyset$ to a new position Q (Figure 8.8).

The center for the rotation of P is M so that $MP = MQ = r =$ radius of the circle center M as P is rotated. The plane MPQ is at right angles to the axis of rotation OM.

Let the vector $\mathbf{OM} = \mathbf{m}$ and let the axis of rotation be given by the unit vector $\hat{\mathbf{m}}$ where $\hat{\mathbf{m}} = a\mathbf{i} + b\mathbf{j} + c\mathbf{k}$. The notation $\hat{\mathbf{m}}$ showing the "hat" over the $\mathbf{m}$ indicates that it is a unit vector, which means that a, b, and c have been scaled so that $\sqrt{a^2 + b^2 + c^2} = 1$.

EXAMPLE 8.7: THE INTERSECTION OF A LINE AND A PLANE

In Example 8.6 we derived the equation of the plane PQR as

$$-6x + 20y + 14z - 178 = 0$$

In Figure 8.7, let A be point $(1, 1, 2)$ and B be $(5, 7, 10)$.
The line AB is the vector

$$\mathbf{v} = 4\mathbf{i} + 6\mathbf{j} + 8\mathbf{k}$$

If $\mathbf{n}$ is the normal to the plane,

$$\mathbf{n} = -6\mathbf{i} + 20\mathbf{j} + 14\mathbf{k}$$

The dot product $\mathbf{n.v} = -24\mathbf{i}^2 + 120\mathbf{j}^2 + 112\mathbf{k}^2 = -24 + 120 + 112 = 208$;

$$|\mathbf{n}| = \sqrt{(632)} \approx 25.14; \ |\mathbf{v}| = \sqrt{(116)} \approx 10.77$$

The line AB will strike the plane at an angle ø with the normal to the plane where

$$\phi = \cos^{-1}\left(\frac{\mathbf{n.v}}{|\mathbf{n}||\mathbf{v}|}\right) \quad \text{or} \quad \phi \approx \cos^{-1}\{208/(25.14 * 10.77)\} \approx 39.8°$$

Let the line AB intersect the plane PQR at T.
Since T lies on the line AB its vector $\mathbf{OT} = \mathbf{OA} + \lambda\mathbf{AB}$ where λ is a scalar quantity. If $\mathbf{OT} = \mathbf{t}$ and $\mathbf{OA} = \mathbf{a}$ then $\mathbf{t} = \mathbf{a} + \lambda\mathbf{v}$. Also $\mathbf{n.t} + d = 0$ since T lies on the plane.
Thus,

$$\mathbf{n.a} + \lambda\,\mathbf{n.v} + d = 0 \quad \text{or} \quad \lambda = \frac{-(\mathbf{n.a} + d)}{\mathbf{n.v}}$$

Since $\mathbf{a} = 1\mathbf{i} + 1\mathbf{j} + 2\mathbf{k}$, then $\mathbf{n.a} = -6 + 20 + 28 = 42$. Hence,

$$\lambda = 136/208 = 0.653846$$

$$\mathbf{t} = (1\mathbf{i} + 1\mathbf{j} + 2\mathbf{k}) + 0.653846 * (4\mathbf{i} + 6\mathbf{j} + 8\mathbf{k})$$

$$= 3.6\mathbf{i} + 4.9\mathbf{j} + 7.2\mathbf{k} \text{ rounding to one decimal place.}$$

Thus, T has coordinates $(3.6, 4.9, 7.2)$.

Let point P be (x_P, y_P, z_P) and its new location after rotation about OM through an angle ø be (x_Q, y_Q, z_Q). Also, in the plane MPQ, let QN be such that N is the foot of the perpendicular from Q to the line PM.

By the rule of vectors, the vector $\mathbf{OQ}$ = vector $\mathbf{OM}$ + vector $\mathbf{MN}$ + vector $\mathbf{NQ}$, or

$$\overline{OQ} = \overline{OM} + \overline{MN} + \overline{NQ}$$

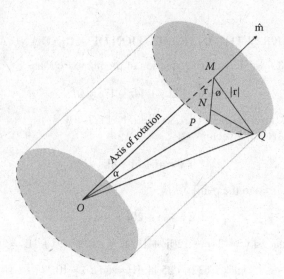

FIGURE 8.8 Rotation using vectors.

The length OM = length OP cos α where α is the angle MOP.

If **OP** is the vector **p**, then the length $OP = |\mathbf{p}|$ and the length $OM = |\mathbf{p}|$ cos α or since the angle α is built into the dot product,

$$\text{Length of } OM = \hat{\mathbf{m}} \cdot \mathbf{p}$$

The vector **OM** is therefore $\hat{\mathbf{m}}$ multiplied by its length, namely,

$$\overrightarrow{OM} = \mathbf{m} = (\hat{\mathbf{m}} \cdot \mathbf{p})\hat{\mathbf{m}}$$

The length of $MP = OP$ sin $\alpha = |\mathbf{p}|$ sin α

If we call the vector **MP** = **r**, then its length = $|\mathbf{r}| = |\mathbf{p}|$ sin α, which, of course, is also the length of MQ as P is rotated about M. Hence, the length $MN = MQ$ cos $\phi = |\mathbf{r}|$ cos ϕ and the vector for **MN** is

$$\overrightarrow{MN} = (\cos \phi)\,\mathbf{r}$$

Point P is such that the vector $\mathbf{p} = \overrightarrow{OP} = \overrightarrow{OM} + \overrightarrow{MP} = \mathbf{m} + \mathbf{r}$, or

$$\mathbf{r} = \mathbf{p} - \mathbf{m} = \mathbf{p} - (\hat{\mathbf{m}} \cdot \mathbf{p})\hat{\mathbf{m}}$$

Hence,

$$\overrightarrow{MN} = (\cos \phi)\,[\mathbf{p} - (\hat{\mathbf{m}} \cdot \mathbf{p})\hat{\mathbf{m}}]$$

To find the vector $\overrightarrow{NQ}$ we need to find a line at right angles to MN and in the plane that is at right angles to OM. This will be given by the cross product between the two vectors and $\mathbf{r}$, giving us a vector $\mathbf{s}$ for the direction $\overrightarrow{NQ}$.

$$\mathbf{s} = \hat{\mathbf{m}} \times \mathbf{r} = \hat{\mathbf{m}} \times [\mathbf{p} - (\hat{\mathbf{m}} \cdot \mathbf{p})\hat{\mathbf{m}}]$$

Or since $(\hat{\mathbf{m}} \cdot \mathbf{p})$ is a scalar and $\hat{\mathbf{m}} \times \hat{\mathbf{m}} = 0$, then

$$\mathbf{s} = \hat{\mathbf{m}} \times \mathbf{p}$$

We also know from triangle MNQ that the length $NQ = MQ \sin \phi = |\mathbf{r}| \sin \phi$.

Since $\mathbf{S} = \hat{\mathbf{m}} \times \mathbf{r}$ and $\hat{\mathbf{m}}$ is a unit vector then $|\mathbf{s}| = |\mathbf{r}|$. Hence, $\overrightarrow{NQ} = \mathbf{s} \sin \phi = (\hat{\mathbf{m}} \times \mathbf{p}) \sin \phi$

$$\overrightarrow{OQ} = \overrightarrow{OM} + \overrightarrow{MN} + \overrightarrow{NQ}$$

hence,

$$\overrightarrow{OQ} = (\hat{\mathbf{m}} \cdot \mathbf{p})\,\hat{\mathbf{m}} + [\mathbf{p} - (\hat{\mathbf{m}} \cdot \mathbf{p})\,\hat{\mathbf{m}}]\cos\phi + (\hat{\mathbf{m}} \times \mathbf{p})\sin\phi$$

$$= \mathbf{p}\cos\phi + (\hat{\mathbf{m}} \cdot \mathbf{p})\,\hat{\mathbf{m}}\,(1 - \cos\phi) + (\hat{\mathbf{m}} \times \mathbf{p})\sin\phi$$

The dot product

$$(\hat{\mathbf{m}} \cdot \mathbf{p}) = (a\mathbf{i} + b\mathbf{j} + c\mathbf{k}) \cdot (x_p\mathbf{i} + y_p\mathbf{j} + z_p\mathbf{k}) = (ax_p + by_p + cz_p)$$

Since $\mathbf{i} \times \mathbf{i} = 0$, $\mathbf{i} \times \mathbf{j} = 1$, and $\mathbf{j} \times \mathbf{i} = -1$, then the cross product

$$(\hat{\mathbf{m}} \times \mathbf{p}) = (a\mathbf{i} + b\mathbf{j} + c\mathbf{k}) \times (x_p\mathbf{i} + y_p\mathbf{j} + z_p\mathbf{k})$$

$$= [(bz_p - cy_p)\mathbf{i} + (cx_p - az_p)\mathbf{j} + (ay_p - bx_p)\mathbf{k}]$$

Hence,

$$\overrightarrow{OQ} = (x_p\mathbf{i} + y_p\mathbf{j} + z_p\mathbf{k})\cos\phi + (ax_p + by_p + cz_p)(a\mathbf{i} + b\mathbf{j} + c\mathbf{k})(1 - \cos\phi)$$

$$+ [(bz_p - cy_p)\mathbf{i} + (cx_p - az_p)\mathbf{j} + (ay_p - bx_p)\mathbf{k}]\sin\phi$$

Rearranging terms,

$$\overrightarrow{OQ} = [x_p\{a^2(1 - \cos\phi) + \cos\phi\} + y_p\{ab(1 - \cos\phi) - c\sin\phi\} + z_p\{ac(1 - \cos\phi) + b\sin\phi\}]\mathbf{i}$$

$$+ [x_p\{ab(1 - \cos\phi) + c\sin\phi\} + y_p\{b^2(1 - \cos\phi) + \cos\phi\} + z_p\{bc(1 - \cos\phi) - a\sin\phi\}]\mathbf{j}$$

$$+ [x_p\{ab(1 - \cos\phi) - b\sin\phi\} + y_p\{bc(1 - \cos\phi) + a\sin\phi\} + z_p\{c^2(1 - \cos\phi) + \cos\phi\}]\mathbf{k}$$

EXAMPLE 8.8: ROTATING A CYLINDER

Consider a cylinder centered on the z-axis (i.e., in Figure 8.8, call OM the z-axis).

Let P be a point on the cylinder with coordinates $(2, 0, z)$ so that the cylinder has a diameter of 4 units. Let us rotate the cylinder through $\phi = 45°$ so that P becomes Q.

For the z-axis, the unit vector $= \mathbf{k}$ and if $\hat{\mathbf{m}} = a\mathbf{i} + b\mathbf{j} + c\mathbf{k}$, we can write $a = b = 0$ and $c = 1$. Also, since we are considering $\phi = 45°$, then $\cos \phi = \sin \phi = 1/\sqrt{2}$.

Hence, the relationship

$$\begin{bmatrix} X_Q \\ Y_Q \\ Z_Q \end{bmatrix} = \begin{bmatrix} a^2(1-\cos\phi)+\cos\phi & ab(1-\cos\phi)-c\sin\phi & ac(1-\cos\phi)+b\sin\phi \\ ab(1-\cos\phi)+c\sin\phi & b^2(1-\cos\phi)+\cos\phi & bc(1-\cos\phi)-a\sin\phi \\ ac(1-\cos\phi)-b\sin\phi & bc(1-\cos\phi)+a\sin\phi & c^2(1-\cos\phi)+\cos\phi \end{bmatrix} * \begin{bmatrix} X_P \\ Y_P \\ Z_P \end{bmatrix}$$

reduces to

$$\begin{bmatrix} X_Q \\ Y_Q \\ Z_Q \end{bmatrix} = \begin{bmatrix} \{1/\sqrt{2}\} & \{-1/\sqrt{2}\} & \{0\} \\ \{1/\sqrt{2}\} & \{1/\sqrt{2}\} & \{0\} \\ \{0\} & \{0\} & \{1\} \end{bmatrix} \begin{bmatrix} X_P \\ Y_P \\ Z_P \end{bmatrix}$$

or

$$x_Q = x_P/\sqrt{2} - y_P/\sqrt{2} = \sqrt{2}; \ y_Q = x_P/\sqrt{2} + y_P/\sqrt{2} = \sqrt{2}; z_Q = z_P$$

Repeating through a further 45° to move P on to R,
$x_R = 0; \ y_R = 2; z_R = z_Q = z_P$, and so on.

Expressed in matrix form (See Example 8.8)

$$\begin{bmatrix} X_Q \\ Y_Q \\ Z_Q \end{bmatrix} = \begin{bmatrix} a^2(1-\cos\phi)+\cos\phi & ab(1-\cos\phi)-c\sin\phi & ac(1-\cos\phi)+b\sin\phi \\ ab(1-\cos\phi)+c\sin\phi & b^2(1-\cos\phi)+\cos\phi & bc(1-\cos\phi)-a\sin\phi \\ ac(1-\cos\phi)-b\sin\phi & bc(1-\cos\phi)+a\sin\phi & c^2(1-\cos\phi)+\cos\phi \end{bmatrix} * \begin{bmatrix} X_P \\ Y_P \\ Z_P \end{bmatrix}$$

SUMMARY

Cross product: A new vector $\mathbf{c}$ at right angles to the two generating vectors $\mathbf{a}$ and $\mathbf{b}$ that share a common start point such that "$\mathbf{a}$ cross $\mathbf{b}$" or $\mathbf{a} \times \mathbf{b} = (|\mathbf{a}| * |\mathbf{b}| * \sin \phi) \mathbf{c}$ where ϕ is the angle between the generating vectors. By convention the direction of $\mathbf{c}$ is determined by the right-hand rule (see Chapter 7). Since the two vectors **a** and **b** define a plane, **c** is normal to the plane.

Since sin (0°) = 0 and sin (90°) = 1, then for the unit vectors **i**, **j**, **k** of three orthogonal axes:

$$\mathbf{i} \times \mathbf{i} = \mathbf{j} \times \mathbf{j} = \mathbf{k} \times \mathbf{k} = 0 \quad \text{and} \quad \mathbf{i} \times \mathbf{j} = \mathbf{j} \times \mathbf{k} = \mathbf{k} \times \mathbf{i} = 1$$

Also,

$$\mathbf{j} \times \mathbf{i} = \mathbf{k} \times \mathbf{j} = \mathbf{i} \times \mathbf{k} = -1$$

Direction cosines: The cosines of angles that a vector makes with the axes used to describe the vector.

Dot product: If ø is the angle between two vectors **a** and **b**, then the dot product is a scalar quantity given by **a.b** = |**a**| * |**b**| * cos ø.

Since cos (0°) = 1 and cos (90°) = 0, then for the unit vectors **i**, **j**, **k** of three orthogonal axes

$$\mathbf{i.i} = \mathbf{j.j} = \mathbf{k.k} = 1$$

and

$$\mathbf{i.j} = \mathbf{j.k} = \mathbf{k.i} = 0$$

Modulus: The length of a *position vector*. If the vector is **v** then the modulus is written as |**v**|—not to be confused with the symbol for a determinant. If vector **OP** = $a\mathbf{i} + b\mathbf{j} + c\mathbf{k}$ where **i**, **j**, **k** are the unit vectors for the reference axes, then

$$|\mathbf{OP}| = \sqrt{a^2 + b^2 + c^2}$$

Position vector: A *vector* representing the location of a point in a coordinate system.

Scalar: A quantity that has magnitude but no direction, such as pressure or the number of people in the world.

Scalar triple product: A scalar quantity determined by combining a *dot product* with a *cross product* in the form **a.b** × **c**. It gives the volume of a solid whose parallel edges are formed by the vectors **a**, **b**, and **c**. If the scalar triple product is zero, the volume must be zero and hence the three points represented by **a**, **b**, and **c** must all lie in a plane.

Unit vector: A vector of unit length.

Vector: A quantity that has both magnitude and direction, such as velocity or the light from the sun. The term is also applied to a straight line that has length and direction.

9 Curves and Surfaces

9.1 PARAMETRIC FORMS

So far, we have seen that for points on a plane we can express a straight line as a series of values (x, y) such that

$$ax + by + c = 0$$

We can also express it in the form that

$$y = mx + n$$

where m is the slope or gradient of the line and n is a constant. These two expressions represent the same line if

$$m = -a/b \text{ and } n = -c/b$$

We also saw in Chapter 6 and can see from Figure 9.1 that if the slope of the line AB is θ (measured from the horizontal axis anticlockwise), then

$$y = x \tan \theta + c$$

Here, c is the value of y when x is zero. We can also express this as $dy/dx = \tan \theta = m$, which is constant for a straight line since

$$\int \tan \theta \, dx = x \tan \theta + c$$

The differential of a constant is zero; hence, for a line, $d^2y/dx^2 = 0$.

Also note that if we use clockwise bearings from the vertical rather than anticlockwise angles from the horizontal, the bearing of line $AB = (90 - \theta)$ and

$$y = x \cot (\text{Bearing } AB) + c$$

If the line CD is perpendicular to AB, its slope will be $(90 + \theta)$. Hence, for the line CD

$$y = m' x + c'$$

where

$$m' = -\cot (\theta) = -\tan (\text{Bearing } AB)$$

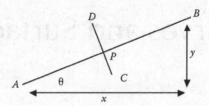

FIGURE 9.1 Orthogonal lines.

We have also seen that when two lines *AB* and *CD* are at right angles, their slopes are such that

$$m\,m' = \tan\theta * (-\cot\theta) = -1$$

If the product of the slopes of two lines $= -1$ they are said to be *orthogonal*. Thus, in Figure 9.1, the lines *AB* and *CD* are orthogonal. In Chapter 7, we also used the term *orthogonal* for a matrix where $A\,A^{-1} = I$.

In the form $y = mx + n$, *y* is the dependent variable and *x* the independent. We can however express a straight line in *parametric* form as

$$x = p + t\cos\theta;\; y = q + t\sin\theta$$

where (p, q) are the coordinates of a point on the line, θ is the fixed slope angle, and "*t*" is the parameter. Here, *t* represents the distance along the line from point (p, q). There is then only one independent variable (*t*) and two dependent variables (*x* and *y*).

In the case of curves, we have seen that the slope at any point on the curve is dy/dx and the normal to it will have a slope $(-dx/dy)$. The straight line that represents the slope at any point is called the *tangent*. In the case of a point on a circle, the tangent is normal to the radius at that point. In the case of the tangent to an ellipse, the normal does not pass through the center except when the tangent is parallel to one of the axes (Figure 9.2).

As we have already seen, the equation for a circle with center at point (x_c, y_c) can be expressed in the form $(x - x_c)^2 + (y - y_c)^2 = r^2$.

An alternative, in parametric form, would be

$$x = r\cos t + x_c \qquad y = r\sin t + y_c$$

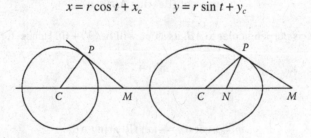

FIGURE 9.2 Tangents to a circle and an ellipse.

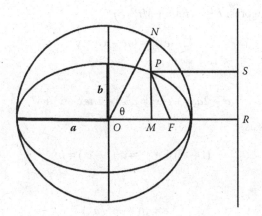

FIGURE 9.3   The ellipse and auxiliary circle.

where the variable t (the angle measured from the horizontal) is such that $0 \le t \le 2\pi$.

Consider the ellipse in Figure 9.3 with center O, semimajor axis a, and semiminor axis b. As we saw in Chapter 4, the circle of radius a centered on O is called the *auxiliary circle*. In effect the auxiliary circle is squashed or scaled down in the vertical or y-direction to give the ellipse. Thus, if N is a point on the auxiliary circle, then we can regard it as having coordinates

$$x_n = a \cos t + x_c,\ y_n = a \sin t + y_c$$

The length $MN = a \sin t$. By squashing the auxiliary circle so that OM remains the same but MN becomes scaled down to MP in the ratio $MP/MN = b/a$, then

$$MP = (b/a) * a \sin t = b \sin t$$

Hence, for point P (x_P, y_P) on the ellipse with center O at (x_c, y_c)

$$x_p = a \cos t + x_c$$

$$y_p = b \sin t + y_c$$

Then, since $\sin^2 t + \cos^2 t = 1$, we have $(x_p - x_c)^2/a^2 + (y_p - y_c)^2/b^2 = 1$.

To pursue the ellipse a little further, the distance OR to the directrix (the line RS) was stated in Chapter 4 as (a/e) with the value of e being such that $0 < e < 1$. As for any conic with center O and focus F, the distance $OF = a * e$, $OR = a/e$, and $FP = e * PS$.

For convenience, we will assume that the center of the ellipse is taken as the origin. Since in Figure 9.3, the lengths $PS = MR$ and $MR + OM = a/e$, then $PS = a/e - x$. Hence,

$$FP = e * PS = a - ex.\ FM = OF - OM = ae - x \text{ while } MP = y$$

Also, using Pythagoras, $FP^2 = FM^2 + MP^2$. So

$$(a - ex)^2 = (ae - x)^2 + y^2$$

or

$$a^2 - 2aex + e^2x^2 = a^2e^2 - 2aex + x^2 + y^2$$

We can write this as

$$(1 - e^2)x^2 + y^2 = a^2(1 - e^2) = b^2$$

where

$$e = \sqrt{(1 - b^2/a^2)}$$

Once again this confirms that for an ellipse, $x^2/a^2 + y^2/b^2 = 1$.

Expressing the equation of a circle in the parametric form

$$x = r \cos t; y = r \sin t$$

or of an ellipse as

$$x = a \cos t; y = b \sin t$$

results in our having only one variable (t). We can deduce both x and y from that variable (see Example 9.1).

**EXAMPLE 9.1: PARAMETRIC FORM
FOR A CIRCLE AND ELLIPSE**

In Figure 9.3, let the circle have radius $a = 10$ = length of semimajor axis of ellipse. Let the ellipse have semiminor axis $b = 8$ and let the center of the circle and ellipse be (0, 0).

For the first quadrant, let the coordinates for the circle be (x, y) and for the ellipse be (x, y'). Using $x = a \cos t$; $y = a \sin t$; $y' = b \sin t$. Then for:

$t =$	0°	10°	20°	30°	40°	50°	60°	70°	80°	90°
$\cos t$	1	0.98	0.94	0.87	0.77	0.64	0.5	0.34	0.17	0
$\sin t$	0	0.17	0.34	0.5	0.64	0.77	0.87	0.94	0.98	1

Given that $a = 10$ and $b = 8$, then

$x =$	10	9.8	9.4	8.7	7.7	6.4	5.0	3.4	1.7	0
$y =$	0	1.7	3.4	5.0	6.4	7.7	8.7	9.4	9.8	10
$y' =$	0	1.4	2.7	4.0	5.1	6.2	7.0	7.5	7.8	8

All intermediate values and the values for the other three quadrants can be calculated in a similar fashion.

The parabola $y^2 = 4ax$ can be expressed in parametric form as:

$$x = at^2 \quad \text{and} \quad y = 2at$$

The hyperbola can be expressed as $x^2/a^2 - y^2/b^2 = 1$. Its parametric form uses the fact that $(1 + \tan^2 \theta) = (1 + \sin^2 \theta/\cos^2 \theta) = (\cos^2 \theta + \sin^2 \theta)/\cos^2 \theta = \sec^2 \theta$. Thus,

$$\sec^2 \theta - \tan^2 \theta = 1$$

The hyperbola can therefore be expressed in the form

$$x = a \sec \theta, y = b \tan \theta$$

or using the parameter t

$$x = a \sec t; y = b \tan t$$

Parametric forms (see Box 9.1) can be used to determine the slope at any point on a curve. If, for example, we consider the circle with center (x_c, y_c) and radius r, then its parametric form is:

$$x = x_c + r \cos t \qquad y = y_c + r \sin t$$

Differentiating with respect to t

$$dx/dt = -r \sin t; \qquad dy/dt = r \cos t$$

$$(dy/dt)/(dx/dt) = -(\cos t/\sin t) = -\cot t$$

BOX 9.1 PARAMETRIC EQUATIONS FOR LINES AND CONIC SECTIONS (SECOND-DEGREE CURVES)

p, q, and θ are constant, that is, fixed quantities; t is the independent variable.
   For a line:

$$x = p + t \cos \theta ; y = q + t \sin \theta$$

For a circle:

$$x = p + r \cos t; y = q + r \sin t$$

For an ellipse:

$$x = p + a \cos t; y = q + b \sin t$$

For a parabola:

$$x = p + at^2; y = q + 2at$$

For a hyperbola:

$$x = p + a \sec t; y = q + b \tan t$$

or

$$dy/dx = -\cot t$$

This is the rate of change of the curve, or put another way, it is the measure of the slope of the curve at the point defined by t.

9.2   THE ELLIPSE

Figure 9.4 shows the auxiliary circle of radius a and the ellipse that has semiminor axis of length b. A point P on the ellipse has coordinates ($a \cos \theta$, $b \sin \theta$) referred to the center. PT is the tangent to the ellipse at point P with T being on the major axis. PQ is the normal at P, with Q being on the minor axis. The line QP makes an angle ϕ with the major axis and is known as the *geodetic* or *spheroidal latitude* of P. It is essentially the direction of the vertical at P. In practice, a plumb bob on the Earth's surface may not point exactly in this direction because of the gravitational attraction of nearby mountains, giving rise to what is known in geodesy as the *deviation of the vertical*. Given an ellipsoid of uniform density, the line PQ would be the vertical at P and, except along the equator and at the poles (D' and E'), it would not pass through the center O.

From the parametric form of a circle we can calculate the slope of the curve at any point on the circle, for example, NT in Figure 9.4. (Later, we show why the tangent at N and the tangent at P both intersect the major axis at T.) This is the slope of the tangent at that point.

Similarly for the ellipse, the tangent is PT where

$$dy/dx = (dy/d\theta)/(d\theta/dx) = (b \cos \theta)/(-a \sin \theta)$$
$$= -(b/a) \cot \theta$$

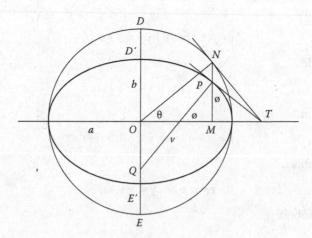

FIGURE 9.4   Normal to an ellipse.

Circles and ellipses are important figures in geomatics and GIS since, when rotated about their north/south axis, they create volumes that represent good approximations to the shape of the Earth, depending on the scale at which one is working.

The angle *POT* (not shown in the diagram) would be the *geocentric latitude* but is not normally used in geodetic calculations, although it is used in satellite position fixing (see Chapter 14) and in astronomy. The angle θ is sometimes known as the *reduced latitude*. The length *PQ* is usually referred to by the Greek letter ν (pronounced "nu" as in "new"). On an ellipsoid that is obtained by rotating the ellipse about its minor axis, the length of ν is the same for all points along a given parallel of latitude but has different values at different latitudes. The *x*-coordinate of *P* = *a* cos θ from before but also = ν cos ø from the projection of *QP* onto the *x*-axis. Hence, (see Example 9.2)

$$a \cos \theta = \nu \cos \phi$$

We mentioned earlier that in Figure 9.4, the tangents at *N* and *P* meet at *T* on the major axis. *TN* is tangential to the auxiliary circle and by squashing the auxiliary circle down to the ellipse, tangency will not be affected even though angles change. In fact, we can show this using the parametric forms of the auxiliary circle and ellipse.

For point *N* on the auxiliary circle:

$$x = a \cos t; \ y = a \sin t; \ dy/dx = (dy/dt)/(dx/dt) = -\cot t$$

If we assume that the tangent to the ellipse at *N* cuts the *x*-axis at *T*, then the equation of the line *NT* is $(y - a \sin t) = -\cot t \ (x - a \cos t)$.

This intersects the *x*-axis where $y = 0$, and $x = a \cos t + a \sin^2 t/\cos t = a \sec t$. Hence, $OT = a \sec t$ or $a \sec \theta$ in Figure 9.4.

For point *P* on the ellipse:

$$x = a \cos t; \ y = b \sin t; \ dy/dx = (dy/dt)/(dx/dt) = -(b/a) \cot t$$

EXAMPLE 9.2: THE EARTH'S RADII

The equatorial radius of the Earth = a = 6,378.1370 km. The polar radius = b = 6,356.7523 km.

For the Earth as an ellipsoid, $e^2 = 1 - (b/a)^2 = 0.00669438$ or $e = 0.081818$.

At ø = 50° north, $\tan \theta = b/a \tan \phi = 1.18775786657775$.

Hence, θ = 49.9052218256 degrees = 49° 54′ 18.8″.

Thus, there is a difference of 5 minutes 41.2 seconds of arc between the auxiliary circle latitude and the geodetic latitude. Since 1 second of arc ≈ 30 meters on the Earth's surface, this represents a difference of approximately 10.55 km on the ground.

Given that $a \cos \theta = \nu \cos \phi$, ν = 6390.7021 km at 50° north.

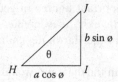

FIGURE 9.5   Theta and phi.

If we assume that the tangent to the ellipse at P cuts the x-axis at T', then the equation of the line PT' is $(y - b \sin t) = -(b/a) \cot t \, (x - a \cos t)$. This intersects the x-axis where $y = 0$, and

$$x = a \cos t + b(a/b) \sin^2 t/\cos t = a \sec t$$

Hence, $OT' = a \sec t$ demonstrating that T and T' are the same point.

Thus in Figure 9.4, if $\angle MNT = \theta$, and $\angle MPT = \phi$, then $NM = a \sin \theta$, $OM = x$ value of $P = a \cos \theta$, and $PM = y$ value of $P = b \sin \theta$.

In triangle MNT, $MT = NM \tan \theta = a \sin \theta \tan \theta$ while in triangle MPT,

$$MT = PM \tan \phi = b \sin \theta \tan \phi$$

Hence, $a \tan \theta = b \tan \phi$. We can express this as:

$$\tan \theta = (b \sin \phi)/(a \cos \phi)$$

Since angles are just ratios, we can represent this relationship by the right-angled triangle HIJ in Figure 9.5 in which $IJ = b \sin \phi$ and $HI = a \cos \phi$ and $\angle JHI = \theta$. Then by suitable manipulation as shown in the Box 9.2, we can show that:

$$x = \nu \cos \phi; \; y = \nu \, (1 - e^2) \sin \phi; \; \nu = a/\sqrt{(1 - e^2 \sin^2 \phi)}$$

BOX 9.2   THE RADIUS ν

From Figure 9.5

$$HI = a \cos \phi, \qquad IJ = b \sin \phi, \quad \text{and} \quad \angle JHI = \theta$$

Thus,

$IJ/HI = \tan \theta = (b/a) \tan \phi$. Using Pythagoras, $HJ^2 = a^2 \cos^2\phi + b^2 \sin^2\phi$

Since $\cos^2\phi + \sin^2\phi = 1$, we can write this as

$$HJ^2 = a^2 - a^2 \sin^2 \phi + b^2 \sin^2\phi = a^2 \left(1 - \frac{(a^2 - b^2)}{a^2} \sin^2 \phi \right)$$

$$= a^2(1 - e^2 \sin^2 \phi)$$

---

where

$e^2 = (1 - b^2/a^2)$ as before. Hence, $\quad HJ = a\sqrt{1 - e^2 \sin^2 \text{ø}}$

In triangle *HIJ*,

$$\cos \theta = HI/HJ = a \cos \text{ø} / \{a\sqrt{1 - e^2 \sin^2 \text{ø}}\}$$

or

$$a \cos \theta = a \cos \text{ø} / \sqrt{1 - e^2 \sin^2 \text{ø}}$$

We have previously shown that:

$$a \cos \theta = v \cos \text{ø}$$

So $a \cos \theta = v \cos \text{ø} = a \cos \text{ø} / \sqrt{(1 - e^2\sin^2\text{ø})}$ and $v = a/\sqrt{(1 - e^2\sin^2\text{ø})}$, or

$$a = v \sqrt{(1 - e^2\sin^2\text{ø})}$$

But,

$$x^2/a^2 + y^2/b^2 = 1 \quad \text{and} \quad x = v \cos \text{ø}$$

Thus,

$$y^2 = b^2 - (b^2/a^2)x^2 = a^2(1 - e^2) - (1 - e^2)x^2 = (1 - e^2)(a^2 - x^2)$$

Since $a^2 = v^2 (1 - e^2\sin^2\text{ø})$ and $x^2 = v^2 \cos^2\text{ø} = v^2 (1 - \sin^2\text{ø})$, then

$(a^2 - x^2) = v^2 (1 - e^2) \sin^2\text{ø}$. As a result $y^2 = v^2 (1 - e^2)^2 \sin^2\text{ø}$.

Thus,

$$x = v \cos \text{ø}; y = v (1 - e^2) \sin \text{ø}; v = a/\sqrt{(1 - e^2\sin^2\text{ø})}$$

These relationships are used in geodetic computations.

The quantities v and ø are of particular importance in the field of geodesy, the scientific study of the size and shape of the Earth and, for example, in calculations of position using the satellite global positioning system (GPS).

9.3 THE RADIUS OF CURVATURE

Before leaving the ellipse, there is one further quantity that is required. It is known as the *radius of curvature* at *P* along the line of the ellipse and is normally referred to as "ρ," the Greek letter "rho" (pronounced "row" as in a row of beans). *Curvature*

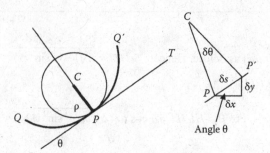

FIGURE 9.6 Radius of curvature.

is the rate of change of the direction of a tangent in relation to the length of arc; the *radius of curvature* at P is the radius of a circle with curvature equal to that of the curve at P or, put another way, the radius of the circle that just touches the curve. At the point of contact, the tangent to the circle is also the tangent to the curve. This applies to all curves, not just ellipses.

Consider the curve QPQ' in Figure 9.6 with the circle center C just touching the curve at P. At the point of contact the radius $CP (= \rho)$ will be normal to the tangent PT that has a slope value that is here called θ. Consider also the two nearby points P and P' shown on the inset. Their coordinates will differ by δx and δy while the length along the curve is denoted by δs. If the slope at P is θ and at P' is $\theta + \delta\theta$, then the angle $PCP' = \delta\theta$ (exaggerated in Figure 9.6).

By measuring the angle θ in radians in triangle PCP', then $\delta s = CP\delta\theta = \rho\,\delta\theta$ or $(\delta\theta/\delta s) = 1/\rho$. The slope of the line $PP' = \theta$, therefore $\delta y/\delta x = \tan\theta$ or

$$dy/dx = \tan\theta$$

The second derivative or rate of change of the slope is the curvature $= d^2y/dx^2 = d(dy/dx)/dx = d(\tan\theta)/dx = \{d\theta/dx\} * \{d(\tan\theta)/d\theta\} = (d\theta/dx)\sec^2\theta$ since the differential of $\tan\theta = \sec^2\theta$. Also, $\delta\theta/\delta x = (\delta\theta/\delta s) * (\delta s/\delta x)$ and $\delta s/\delta x = \sec\theta$ as can be seen from Figure 9.6. So

$$d^2y/dx^2 = (1/\rho)\sec^3\theta$$

Since $\sec^2 = 1 + \tan^2$ and $\tan\theta = dy/dx$, therefore

$$d^2y/dx^2 = (1/\rho)(1 + (dy/dx)^2)^{3/2}$$

On rearranging, the radius of curvature of any curve is given by

$$\rho = \{(1 + (dy/dx)^2)^{3/2}\}/(d^2y/dx^2)$$

If this is applied to an ellipse, then it can be shown that

$$\rho = v(1 - e^2)(1 - e^2\sin^2\phi)^{-1} = a(1 - e^2)(1 - e^2\sin^2\phi)^{-3/2}$$

BOX 9.3   RADII OF CURVATURE

For any curve the radius of curvature

$$\rho = \{(1 + (dy/dx)^2)^{3/2}\}/(d^2y/dx^2)$$

For an ellipse

$$\rho = v(1 - e^2)(1 - e^2\sin^2\phi)^{-1} = a(1 - e^2)(1 - e^2\sin^2\phi)^{-3/2}$$

$$x = v \cos \phi$$

$$y = v (1 - e^2) \sin \phi$$

$$v = a/\sqrt{(1 - e^2 \sin^2 \phi)}$$

where ϕ is the geodetic latitude.

The value "ρ" (Box 9.3) is used extensively in geodetic computations; but these are beyond the scope of the present text. For now, we will consider more about curvature. Every second-degree (quadratic) curve bends only one way. The circle and ellipse, for example, bend all the way around until they close back on themselves while the parabola and hyperbola go off to infinity before turning back.

From Figure 9.6 we can also note that $\delta s^2 = \delta x^2 + \delta y^2$ or $\delta s/\delta x = \sqrt{1 + (\delta y / \delta x)^2}$. From this, we obtain

$$ds/dx = \sqrt{1 + (dy / dx)^2}$$

or

$$s = \int \sqrt{1 + (dy / dx)^2} \; dx$$

This provides a method for calculating the length along a curve. In the particular case of the ellipse, the integration is not straightforward; but for some functions it provides an elegant way to determine such lengths.

9.4   FITTING CURVES TO POINTS

The simplest curve that bends forward and backward is the cubic. If we take the general case of the cubic

$$y = a + bx + cx^2 + dx^3$$

Then

$$dy/dx = b + 2cx + 3dx^2$$

There can be two values for which $dy/dx = 0$. These are where the maxima and minima for the curve are located. Also,

$$d^2y/dx^2 = 2c + 6dx$$

There is only one value for which $d^2y/dx^2 = 0$, namely, where $x = -c/(3d)$. This is the point of inflection. (See Figure 6.3 in Chapter 6.)

Because there are four unknown constants (a, b, c, d) in the cubic equation for y, we need four independent equations to calculate them. Put another way, a cubic can be made to pass through four points just as a quadratic curve such as a circle can be made to pass through any three points (albeit with infinite radius for three points on a straight line). Five points are needed to define a quartic (fourth degree), six for a quintic (fifth degree), and so forth.

Given a cubic in the form $y = f(x)$ with four known points (x_1, y_1), (x_2, y_2), (x_3, y_3), and (x_4, y_4), then the equation:

$$y = y_1 \frac{(x-x_2)(x-x_3)(x-x_4)}{(x_1-x_2)(x_1-x_3)(x_1-x_4)} + y_2 \frac{(x-x_1)(x-x_3)(x-x_4)}{(x_2-x_1)(x_2-x_3)(x_2-x_4)}$$

$$+ y_3 \frac{(x-x_1)(x-x_2)(x-x_4)}{(x_3-x_1)(x_3-x_2)(x_3-x_4)} + y_4 \frac{(x-x_1)(x-x_2)(x-x_3)}{(x_4-x_1)(x_4-x_2)(x_4-x_3)}$$

is a cubic in x and passes through all four points. Note that none of the elements in the denominator can equal zero because there is only one value for y for a given value of x in the equation that we are considering. Note also that the sequence can be extended to quartics, quintics, and any other order and can be expressed in mathematical shorthand as

$$y = \sum_{i=1}^{n} y_i \prod_{i \neq j} \frac{(x-x_j)}{(x_i-x_j)}$$

where $\sum_{i=1}^{n} y_i$ means the sum of all the y_i from $i = 1$ to $i = n$, and $\prod_{i \neq j}$ means the product of the following expression, subject to the constraint that i does not equal j ($i \neq j$).

But what if we want to fit a cubic to five points? One solution would be to make the best-fit approximation to all five. Another would be to ensure that the curve goes exactly through all five points but that is normally not possible with only one cubic. The solution is to fit the curve step by step in sections, or what is called *piecewise*. This means fitting several cubic curves so that if we have points A, B, C, D, E,...N then we fit the curve to AB, then BC, then CD, then DE, and so on, until the last point N is reached. The result is what is called a *spline* curve, the points A, B, C, and so forth, being known as *nodes* or *knots*.

Before considering piecewise polynomials, let us return to the way in which we express a straight line between two points in parametric form, t being the independent variable.

$$x = a + bt \qquad y = c + dt$$

For a quadratic

$$x = a + bt + ct^2 \quad y = d + et + ft^2$$

For a cubic

$$x = a + bt + ct^2 + dt^3 \quad y = e + ft + gt^2 + ht^3$$

The reason for expressing (x, y) in this form is that with a computer plotter we can increment t from 0 to 1 in, for example, 10 steps of 0.1. The computer calculates (x, y) for $t = 0$, then for $t = 0.1$, then for $t = 0.2$, and so on, until $t = 0.9$, and then $t = 1.0$. We would then have 11 values for (x, y) from the start point (x_s, y_s) where $t = 0$ to the end point (x_e, y_e) where $t = 1$. A computer-driven plotter can draw a series of straight lines joining these points consecutively so as to portray the curve from its beginning to its end. By taking finer increments of t (for example, every 0.01 or 0.05 rather than 0.1) we can create the appearance of a smoother curve but at a cost of longer processing time.

If we have two points A (x_1, y_1) and B (x_2, y_2), then any point on the line AB can be expressed in the form

$$x = x_1 + (x_2 - x_1) t \quad \text{and} \quad y = y_1 + (y_2 - y_1) t$$

(where $t = 0$ at point A and $t = 1$ at B).

Consider three points A (x_A, y_A), B (x_B, y_B), and C (x_C, y_C) in Figure 9.7. A second-degree curve can be made to pass exactly through three points but here we will divide it into two separate but continuous sections. For each section AB and BC there is a quadratic of the form:

$$x = a + bt + ct^2 \qquad dx/dt = b + 2ct \qquad d^2x/dt^2 = 2c$$

$$y = e + ft + gt^2 \qquad dy/dt = f + 2gt \qquad d^2y/dt^2 = 2g$$

Let us consider two sets of equations for the two sections of the curve from A to B and from B to C. Let any point on the first section of the curve be (x_1, y_1) and on the second section (x_2, y_2). For the first section, these coordinates must satisfy the quadratic expressions:

$$x_1 = a_1 + b_1 t_1 + c_1 t_1^2 \qquad dx/dt = b_1 + 2c_1 t_1 \qquad d^2x/dt^2 = 2c_1$$

$$y_1 = e_1 + f_1 t_1 + g_1 t_1^2 \qquad dy/dt = f_1 + 2g_1 t_1 \qquad d^2y/dt^2 = 2g_1$$

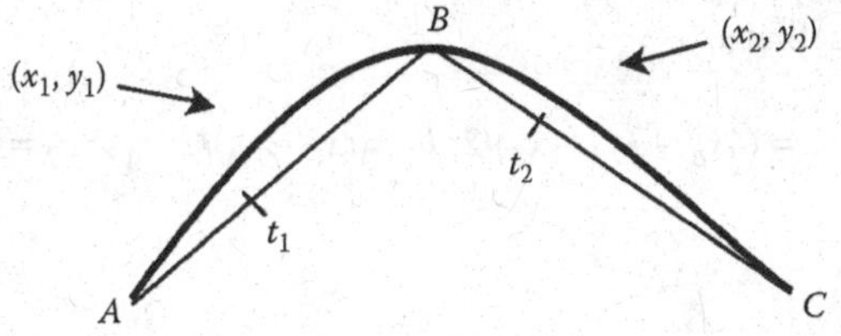

FIGURE 9.7 Fitting a second-degree (quadratic) curve.

and for the second section:

$$x_2 = a_2 + b_2 t_2 + c_2 t_2^2 \qquad dx/dt = b_2 + 2c_2 t_2 \qquad d^2x/dt^2 = 2c_2$$
$$y_2 = e_2 + f_2 t_2 + g_2 t_2^2 \qquad dy/dt = f_2 + 2g_2 t_2 \qquad d^2y/dt^2 = 2g_2$$

For section 1 from A to B:

$$\text{At } A, t_1 = 0 \text{ and } x_1 = a_1 \text{ and } y_1 = e_1$$

Thus,

$$a_1 = x_A \text{ and } e_1 = y_A$$
$$\text{At } B, t_1 = 1 \text{ and } x_1 = x_B = a_1 + b_1 + c_1 \text{ or}$$

hence,

$$b_1 + c_1 = x_B - x_A$$

Likewise,

$$f_1 + g_1 = y_B - y_A$$

For section 2 from B to C:

$$\text{At } B, t_2 = 0 \text{ and } x_2 = a_2 \text{ and } y_2 = e_2$$

Since this is point B, $a_2 = x_B$ and $e_2 = y_B$

$$\text{At } C, t_2 = 1 \text{ and } x_2 = x_C = a_2 + b_2 + c_2 \text{ or } b_2 + c_2 = x_C - x_B$$

Likewise,

$$f_2 + g_2 = y_C - y_B$$

For the curve to be continuous at point B the first and second derivatives must be the same. For the first derivative this means that $b_1 + 2c_1 = b_2$, and $f_1 + 2g_1 = f_2$. For the second derivative, $c_1 = c_2$, and $g_1 = g_2$.

Putting this together

$$a_1 = x_A; a_2 = x_B; b_1 = (4x_B - 3x_A - x_C)/2; b_2 = (x_C - x_A)/2; c_1 = c_2 = (x_A - 2x_B + x_C)/2$$

Similarly,

$$e_1 = y_A; e_2 = y_B; f_1 = (4y_B - 3y_A - y_C)/2; f_2 = (y_C - y_A)/2; g_1 = g_2 = (y_A - 2y_B + y_C)/2$$

EXAMPLE 9.3: FITTING A QUADRATIC PIECEWISE

As an example of a piecewise quadratic, consider the points A (3, 5), B (23, 25), and C (40, 15) shown in Figure 9.8. Using the equations derived in Section 9.4,

$$a_1 = x_A = 3; \ a_2 = x_B = 23; \ b_1 = (4x_B - 3x_A - x_C)/2 = 21.5; \ b_2 = (x_C - x_A)/2 = 18.5$$

$$c_1 = c_2 = (x_A - 2x_B + x_C)/2 = -1.5$$

$$e_1 = y_A = 5; \ e_2 = y_B = 25;$$

$$f_1 = (4y_B - 3y_A - y_C)/2 = 35; \ f_2 = (y_C - y_A)/2 = 5$$

$$g_1 = g_2 = (y_A - 2y_B + y_C)/2 = -15$$

We can now compile a table based on $x_1 = a_1 + b_1 t_1 + c_1 t_1^2$, and so forth, using x_1, y_1 for section AB and x_2, y_2 for section BC:

$t =$	0	0.1	0.2	0.3	0.4	0.5	0.6	0.7	0.8	0.9	1
x_1	3	5.1	7.2	9.3	11.4	13.4	15.5	17.3	19.2	21.1	23
y_1	5	8.4	11.4	14.2	16.6	18.8	20.6	22.2	23.4	24.4	25
x_2	23	24.8	26.6	28.4	30.2	31.9	33.6	35.2	36.8	38.4	40
y_2	25	25.3	25.4	25.2	24.6	23.8	22.6	21.2	19.4	17.4	15

The (x, y) values can then be plotted as shown in Figure 9.8.

With these values, we can calculate $x = a_1 + b_1 t + c_1 t_2$ and $y = e_1 + f_1 t + g_1 t_2$ for t incremented from 0 to 1. From these values we can then plot the first section of the curve; and similarly for the second section as illustrated in Example 9.3.

The trouble with a quadratic is that it bends only one way—it cannot twist left and then right, clockwise then counterclockwise, or vice versa. A cubic can do this

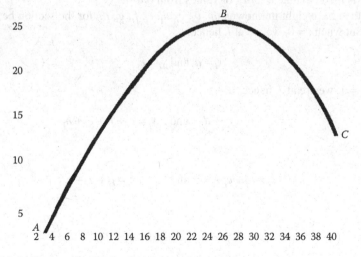

FIGURE 9.8 Two quadratics fitted to three points.

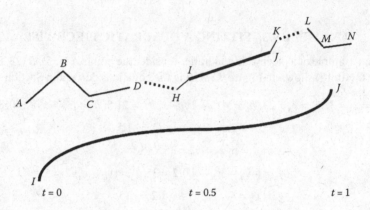

FIGURE 9.9   A piecewise cubic.

since it can have a maximum and minimum and a point of inflection in between. As a result we can fit a set of cubic curves piecewise through any number of points.

Consider the series of points $A, B, C, D, \ldots M, N$ in Figure 9.9. The upper part of Figure 9.9 shows a series of points joined by lines. Let these be:

$$A\,(x_0, y_0),\; B\,(x_1, y_1),\; C\,(x_2, y_2),\; D\,(x_3, y_3),\; \ldots\; M\,(x_{n-1}, y_{n-1}),\; N\,(x_n, y_n).$$

The series of lines $A - B$, $B - C$, and so forth, is said to form a *string*.

Now consider the two successive points along this string, $I\,(x_i, y_i)$ and $J\,(x_{i+1}, y_{i+1})$. For a cubic curve between I and J, we can let

$$x = a_i + b_i t + c_i t^2 + d_i t^3$$

$$y = e_i + f_i t + g_i t^2 + h_i t^3$$

where t is made to take a series of values from 0 to 1.

We then have eight unknowns $(a_i, b_i, c_i, d_i, e_i, f_i, g_i, h_i)$ for the section between I and J. But when $t = 0$, we are at I, hence,

$$x_i = a_i \text{ and } y_i = e_i$$

When $t = 1$, we are at J, hence,

$$x_j = a_j + b_j + c_j + d_j \quad \text{and} \quad y_j = e_j + f_j + g_j + h_j$$

or

$$x_j - x_i = b_j + c_j + d_j \quad \text{and} \quad y_j - y_i = f_j + g_j + h_j$$

Also,

$$dx/dt = b_i + 2c_i t + 3d_i t^2 \qquad dy/dt = f_i + 2g_i t + 3h_i t^2$$

$$d^2x/dt^2 = 2c_i + 6d_i t \qquad d^2y/dt^2 = 2g_i + 6h_i t$$

To obtain a smooth curve at I, there must be continuity in both the first and second differentials so that both the slope and rate of change of the slope, namely the curvature are continuous. That means that at I (where $t = 0$), $dx/dt = b_i$ and must have the same value as the previous section HI where $t = 1$. This would have been $b_h + 2c_h + 3d_h$ using the same notation. Hence,

$$b_i = b_h + 2c_h + 3d_h$$

Similarly,

$$f_i = f_h + 2g_h + 3h_h$$

At point I on the section IJ, where $t = 0$, $d^2x/dt^2 = 2c_i$ and this must equal the value from the previous section of the curve when $t = 1$, namely, $2c_h + 6d_h$. Hence, we can calculate c_i and similarly g_i and since we have

$$d_i = (x_j - x_i) - b_j - c_j \quad \text{and} \quad h_i = (y_j - y_i) - f_j - g_j$$

we have all the parameters for the section IJ provided that we have established the parameters for HI. Thus, for all the sections of the curve $BC, CD,....IJ,.....MN$ we can fit a cubic in parametric form, based on values obtained from the previous section. The trouble is how to get started, since we need to know the initial slope and curvature in order to calculate the section AB. There are an infinite number of possibilities.

Of the many practical solutions, the most common is to set the second derivative to zero at the very start and very end of the string. An alternative is to use the values from the quadratic described above as the start for the line AB and carry through the values at B as the initial values for fitting a cubic to BC, and so on; or fit a cubic exactly through the first four points but use only the first section.

There is, however, a fundamental problem with the approach described in that although it works in many circumstances, it can lead to a curve that crosses over itself as the angle of slope increases from zero to 2π—see Figure 9.10.

Such a loop is unacceptable if it is meant to represent a contour line through a set of interpolated points. It is not the intention of this text to discuss curve-fitting algorithms but rather to explain the basic mathematics behind them. Readers should consult relevant books on computer graphics for more details.

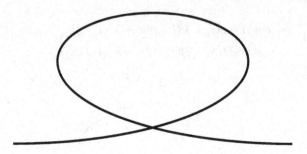

FIGURE 9.10   Looping curve.

9.5 THE BEZIER CURVE

There is, however, another approach that will be mentioned, since it is commonly referred to in computer drawing packages. Known as the *family of Bezier curves*, they use Bernstein polynomials (named after the Russian mathematician, Sergei Bernstein) and take the form:

$$x = f_x(t) = \sum_{i=0}^{i=n} {}_nC_i t^i (1-t)^{n-i} x_i$$

$$y = f_y(t) = \sum_{i=0}^{i=n} {}_nC_i t^i (1-t)^{n-i} y_i$$

where $f(t)$ means the function for which t takes a series of steps from 0 to 1 calculated separately for each increment of x and y and

$$_nC_i = n!/\{i!(n-i)!\}$$

$_nC_i$ is a symbol often used for the number of distinct *combinations* of i elements chosen from a set of n objects. It is sometimes written as $\binom{n}{i}$. Thus, for $n = 3$ then, given that $0! = 1$,

$$_3C_0 = 3!/\{0!3!\} = 1 = {}_3C_3 \quad \text{and} \quad _3C_1 = 3!/\{1!2!\} = 3 = {}_3C_2$$

Thus,

$$\sum_{i=0}^{i=3} {}_3C_i t^i (1-t)^{3-i} = (1-t)^3 + 3t(1-t)^2 + 3t^2(1-t) + t^3$$

The curve starts at A (x_o, y_o) and finishes at B (x_n, y_n) with $(n-1)$ guiding points P_1 (x_1, y_1), P_2 (x_2, y_2) through to $P_{(n-1)}$ (x_{n-1}, y_{n-1}). With a quadratic $(n = 2)$, there would be one guiding point; with a cubic $(n = 3)$ there would be two guiding points as in Figure 9.11. A and B can be regarded as P_0 and P_n.

The curve starts tangential to the line $P_0 P_1$ (AP_1 in Figure 9.11) and finishes tangential to the line $P_{n-1} P_n$ (here $P_2 B$). For any point t along a Bezier curve of the third degree, we have

$$x_t = (1-t)^3 x_0 + 3t(1-t)^2 x_1 + 3t^2(1-t)x_2 + t^3 x_3$$

$$y_t = (1-t)^3 y_0 + 3t(1-t)^2 y_1 + 3t^2(1-t)y_2 + t^3 y_3$$

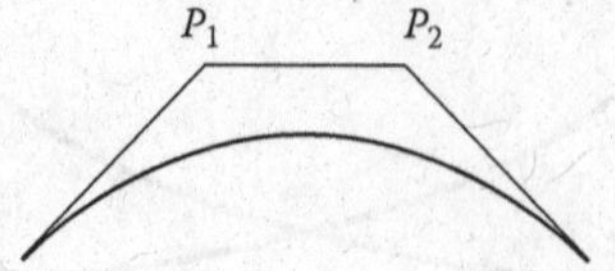

FIGURE 9.11   A Bezier curve with two control points.

EXAMPLE 9.4: BEZIER CURVES

Figure 9.11 shows two versions of a cubic curve from A to B with two intermediate points P_1 and P_2. In the upper curve we have A (20, 20), B (40, 25), P_1 (25, 25), and P_2 (35, 35). In the lower curve, the same values are used except that the y value of P_2 has been changed to P'_2 (35, 28).

$$x_t = (1 - t)^3 x_0 + 3t(1 - t)^2 x_1 + 3t^2(1 - t)x_2 + t^3 x_3$$

$$y_t = (1 - t)^3 y_0 + 3t(1 - t)^2 y_1 + 3t^2(1 - t)y_2 + t^3 y_3$$

t	$1-t$	$(1-t)^3$	$3t(1-t)^2$	$3t^2(1-t)$	t^3	x_t	y_t	y'_t
0	1	1	0	0	0	20	20	20
0.1	0.9	0.729	0.243	0.027	0.001	21.64	21.625	21.409
0.2	0.8	0.512	0.384	0.096	0.008	23.52	23.4	22.632
0.3	0.7	0.343	0.441	0.189	0.027	25.58	25.175	23.663
0.4	0.6	0.216	0.432	0.288	0.064	27.76	26.8	24.496
0.5	0.5	0.125	0.375	0.375	0.125	30	28.125	25.125
0.6	0.4	0.064	0.288	0.432	0.216	32.24	29	25.544
0.7	0.3	0.027	0.189	0.441	0.343	34.42	29.275	25.747
0.8	0.2	0.008	0.096	0.384	0.512	36.48	28.8	25.728
0.9	0.1	0.001	0.027	0.243	0.729	38.36	27.425	25.481
1	0	0	0	0	1	40	25	25

In the case of P' the curve is closer to the straight line AB. In both cases, the curve starts off tangential to AP_1 and finishes tangential to P_2B (or P'_2B) (Figure 9.12).

We then increment t in small steps from 0 to 1 and plot the points (x_t, y_t) to obtain the curve as in Example 9.4.

Thus, the selection of the control points P_1, P_2, and so forth, is critical to the shape of the curve. By choosing various controlling points, a graphic designer can adjust a curve to meet any required design. In general, in matrix notation, if

$$\mathbf{P}_i = \begin{bmatrix} x_i \\ y_i \end{bmatrix} \quad \text{and} \quad \mathbf{P}(t) = \begin{bmatrix} p_x(t) \\ p_y(t) \end{bmatrix} \quad \text{where } p_x(t) \text{ means the } x \text{ value of } \mathbf{P} \text{ for the value } t.$$

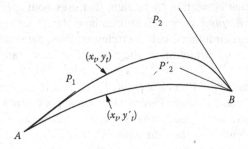

FIGURE 9.12   Two versions of a Bezier curve.

Then

$$\mathbf{P}(t) = \sum_{r=0}^{r=n} {}_nC_r t^r (1-t)^{n-r} P_r$$

The idea of curve fitting can be extended to three dimensions. Thus, if we are fitting the cubic in the form

$$x = a_i + b_i t + c_i t^2 + d_i t^3$$

and

$$y = e_i + f_i t + g_i t^2 + h_i t^3$$

then we can add

$$z = l_i + m_i t + n_i t^2 + o_i t^3$$

Similarly, the Bezier function can be extended so that

$$\mathbf{P}(t) = \begin{bmatrix} p_x(t) \\ p_y(t) \\ p_z(t) \end{bmatrix}$$

Curves and surfaces may be made to go exactly through a series of fixed points or can be constructed as the best mean fit. The latter requires a statistical approach and is discussed in Chapters 12 and 13.

9.6   B-SPLINES

In Section 9.4 we considered a set of cubic curves that passed through a series of points while in Section 9.5 we fitted a cubic to end points using points off the curve as guides to the shape. We can combine both ideas by drawing a smooth curve piecewise, that is, section by section, that uses both approaches in what is known as a set of *B-spline curves*. B-splines have the advantage that they can be used to describe regular figures such as circles, ellipses, parabolas, and hyperbolas and their shapes are invariant under the rotation of axes, translations, and scale changes.

Consider a series of "$n + 1$" control points $P_0, P_1, P_2, \dots P_i, \dots P_n$ whose coordinates are (X_0, Y_0), (X_1, Y_1), and so forth (Figure 9.13). These points are sometimes referred to as *knots*. The curve will be made up from "$n - 2$" sections $S_0, S_1, S_2, \dots, S_i, \dots S_{(n-3)}$ with S_i being influenced by the four points P_i, $P_{(i+1)}$, $P_{(i+2)}$, and $P_{(i+3)}$. Thus, when $i = n - 3$ we will have reached the final point P_n.

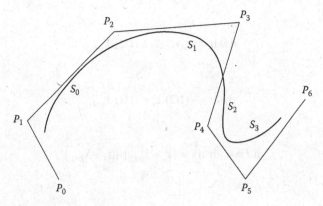

FIGURE 9.13 B-spline and knots.

Each section is then defined as

$$S_i(t) = \sum_{k=0}^{k=3} P_{i+k} Q_k(t)$$

where t is incremented from 0 to 1 in an appropriate number of steps (for example, $t = 0, 0.01, 0.02, \ldots 1$) and the function Q is given by

$$Q_0(t) = (1 - t)^3/6$$
$$Q_1(t) = (3t^3 - 6t^2 + 4)/6$$
$$Q_2(t) = (-3t^3 + 3t^2 + 3t + 1)/6$$
$$Q^3(t) = t^3/6$$

This means that for the section S_i:

$$x_i(t) = \{(1 - t)^3/6\}X_i + \{(3t^3 - 6t^2 + 4)/6\}X_{(i+1)} + \{(-3t^3 + 3t^2 + 3t + 1)/6\}X_{(i+2)} + \{t^3/6\}X_{(i+3)}$$
$$y_i(t) = \{(1 - t)^3/6\}Y_i + \{(3t^3 - 6t^2 + 4)/6\}Y_{(i+1)} + \{(-3t^3 + 3t^2 + 3t + 1)/6\}Y_{(i+2)} + \{t^3/6\}Y_{(i+3)}$$

Note that the section starts where

$$x = (1/6)X_i + (4/6)X_{(i+1)} + (1/6)X_{(i+2)}$$
$$y = (1/6)Y_i + (4/6)Y_{(i+1)} + (1/6)Y_{(i+2)}$$

It ends where

$$x = (1/6)X_{(i+1)} + (4/6)X_{(i+2)} + (1/6)X_{(i+3)}$$
$$y = (1/6)Y_{(i+1)} + (4/6)Y_{(i+2)} + (1/6)Y_{(i+3)}$$

Thus, the coordinates of the end point are the start of the next section so there is continuity. Furthermore:

$$dx/dt = \{-3(1 - t)^2/6\}X_i + \{(9t^2 - 12t)/6\}X_{(i+1)} + \{(-9t^2 + 6t + 3)/6\}X_{(i+2)} + \{3t^2/6\}X_{(i+3)}$$

When $t = 0$,

$$dx/dt = -(1/2) X_i + (1/2) X_{(i+2)}$$

Similarly,

$$dy/dt = -(1/2) Y_i + (1/2) Y_{(i+2)}$$

Hence,

$$\text{at } t = 0, dy/dx = \{Y_i - Y_{(i+2)}\}/\{X_i - X_{(i+2)}\}$$

When $t = 1$,

$$dx/dt = -(1/2) X_{(i+1)} + (1/2) X_{(i+3)}$$

$$dy/dt = -(1/2) Y_{(i+1)} + (1/2) Y_{(i+3)}$$

Hence,

$$\text{at } t = 1, dy/dx = \{Y_{(i+1)} - Y_{(i+3)}\}/\{X_{(i+1)} - X_{(i+3)}\}$$

The slope at the end of the section is the same as the slope at the start of the next section. Similarly,

$$d^2x/dt^2 = (1 - t)X_i + (3t - 2)X_{(i+1)} + (-3t + 1)X_{(i+2)} + tX_{(i+3)}$$

$$\text{at } t = 0, d^2x/dt^2 = X_i - 2X_{(i+1)} + X_{(i+2)}$$

$$\text{at } t = 1, d^2x/dt^2 = X_{(i+1)} - 2X_{(i+2)} + X_{(i+3)}$$

and similarly for y. Thus, the rate of change of the slope at the end of one section is the same as the rate of change at the start of the next section. There is therefore continuity for both the first and second derivatives and hence a smooth curve throughout.

The ideas behind the B-spline can be extended to surfaces and beyond and can also be generated by more complex curves than the cubic. Such procedures are beyond the present text. Readers who are interested should consult textbooks on computer graphics for more information.

SUMMARY

Bezier curve: A curve built from the family of polynomials of the form

$$x = \mathbf{f}_x(t) = \sum_{i=0}^{i=n} {}_nC_i t^i (1-t)^{n-i} x_i; \quad y = \mathbf{f}_y(t) = \sum_{r=0}^{r=n} {}_nC_r t^r (1-t)^{n-r} y_i$$

Combination: The number of distinct ways in which r objects can be chosen from a set of n objects. Written as ${}_nC_r$.

Curvature: The rate of change of a tangent to a curve.

Geocentric latitude: The latitude as measured from the notional center of the Earth.

Geodetic latitude: Also known as the *spheroidal latitude*. The latitude determined by the direction of the normal to an ellipsoid at any point.

Nodes: A point at which two or more branches of a graph meet.

Orthogonal lines: Lines which are perpendicular to each other.

Parametric form: A set of equations that relate the variables such as coordinates to a set of parameters, for instance,

For a line: $x = p + lt$; $y = q + mt$

For a circle: $x = p + r \cos t$; $y = q + r \sin t$

For an ellipse: $x = p + a \cos t$; $y = q + b \sin t$

For a hyperbola: $x = p + a \sec t$; $y = q + b \tan t$

Piecewise: The fitting of a curve section by section using lower-order polynomial equations.

Radius of curvature: The radius of the circle that has the same *curvature* as that of a point on a curve. Known as ρ (rho).

$$\rho = \{(1 + (dy/dx)^2)^{3/2}\}/(d^2y/dx^2)$$

Reduced latitude: The latitude determined by the normal to the auxiliary circle to an ellipse.

Spline: A simplified way of representing a complex curve using low-order polynomials.

String: A series of elements connected in a specific order, as with points along a line.

10 2D/3D Transformations

10.1 HOMOGENEOUS COORDINATES

In this chapter we will consider the two-dimensional (2D) representation of three-dimensional (3D) objects in what is sometimes called *visualization*. We often need to display the landscape in perspective on a screen or on paper. The problem is how best to transfer the 3D coordinates of a point in space onto a flat surface. In Chapter 11, we will look at the specific case of points on the surface of the Earth and how they can be displayed on a map, for instance, in an atlas. Here, we are looking at 3D Euclidean space in general and how that may be transformed into two dimensions.

First, we need to extend the idea of 3D coordinates (x, y, z) to homogeneous coordinates (x, y, z, w) where in effect the w is a scaling factor and not a fourth dimension. The actual 3D coordinates in normal Euclidean space are then $(x/w, y/w, z/w)$. In the two-dimensional case, the homogenous coordinates would be (x, y, w) with the traditional values with which we normally deal being $(x/w, y/w)$. The reason for introducing this extra complexity is to allow us to describe points that are at infinity.

Consider the line AB and a point C in Figure 10.1. Every line through C cuts the line AB somewhere (e.g., CD, CE, or CF); but when the line through C is parallel to AB, it is an infinite distance away, although in a known direction. It is not just anywhere but specifically in the direction in which AB is pointing.

If in conventional terms AB is the line $y = mx + c$, then, if we use homogeneous coordinates, we can express the point at infinity as $(x, mx, 0)$ or even as $(1, m, 0)$ and still retain the relationship that y is basically m times x as shown in Example 10.1. We can use this relationship to draw perspective, as parallel lines are made to appear to converge on the horizon as illustrated in Figure 10.2 at what are called the *vanishing points*.

In Box 9.1 in Chapter 9, we showed various equations in parametric form; for instance, the straight line was given as

$$x = p + t \cos \theta; y = q + t \sin \theta$$

where (p, q) is a point on the line of slope θ, and t is the distance along that line from (p, q).

If $l = \cos \theta$ and $n = \sin \theta$, then $x = p + lt$ and $y = q + nt$. If we let $w = 1/t$, then in homogeneous coordinates, any point on the line can be represented as

$$(pw + l, qw + n, w)$$

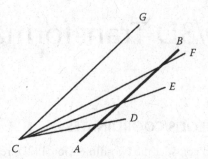

FIGURE 10.1 Points at infinity.

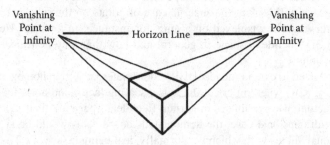

FIGURE 10.2 Vanishing points for a rectangular block.

As the number t goes off to infinity, w tends to zero and we move to point $(l, n, 0)$.

Note that if k is any constant, then the point $\{k(pw + l), k(qw + n), kw)\}$ represents the same Euclidean value as $(pw + l, qw + n, w)$ and hence $(kl, kn, 0)$ is the same point as $(l, n, 0)$. This is the point at infinity in the direction of the slope angle θ, namely, $\tan^{-1}(n/l)$.

EXAMPLE 10.1: HOMOGENEOUS COORDINATES

Consider the lines (a) $y = 2x + 3$ and (b) $y = 2x + 5$. Point $(1, 5)$ lies on line (a) since $5 = 2 * 1 + 3$ but does not satisfy line (b) since $5 \neq 2 * 1 + 5$.

The two lines are parallel, and therefore in Euclidean space in terms of ordinary coordinates, they do not meet.

If we express any point on the lines as (x, y, w) in homogeneous coordinates, then we can express the first equation as $(y/w) = 2(x/w) + 3$ or equation (a) as $y = 2x + 3w$.

Similarly, equation (b) takes the form $y = 2x + 5w$.

In homogeneous coordinates, point $(1, 2, 0)$ satisfies both equation (a) and equation (b) and hence that is a point at which the two parallel lines meet. It is known as the point at infinity on the line $y = 2x + c$.

(In the nonhomogeneous version, $(1, 2)$ represented point $(1, 2, 1)$ which does not lie on the line $y = 2x + 3$.)

10.2 ROTATING AN OBJECT

Consider a point whose 3D homogeneous coordinates are (x, y, z, w). In Chapter 7, we discussed the rotation of axes. In order to rotate an object or the axes that determines its position, we can apply a 4 * 4 matrix of the form:

$$
\begin{pmatrix}
\cos\theta & -\sin\theta & 0 & 0 \\
\sin\theta & \cos\theta & 0 & 0 \\
0 & 0 & 1 & 0 \\
0 & 0 & 0 & 1
\end{pmatrix}
*
\begin{pmatrix}
x \\ y \\ z \\ w
\end{pmatrix}
=
\begin{pmatrix}
x\cos\theta - y\sin\theta \\
x\sin\theta + y\cos\theta \\
z \\
w
\end{pmatrix}
$$

This results in new homogeneous coordinates $\{(x\cos\theta - y\sin\theta), (x\sin\theta + y\cos\theta), z, w\}$.

As we saw in Chapter 7, this represents a rotation about the z-axis and we can call the matrix $\mathbf{R}_z$. Similarly, about the y-axis, we have

$$
\mathbf{R}_y =
\begin{pmatrix}
\cos\phi & 0 & -\sin\phi & 0 \\
0 & 1 & 0 & 0 \\
\sin\phi & 0 & \cos\phi & 0 \\
0 & 0 & 0 & 1
\end{pmatrix}
$$

For a rotation about the x-axis,

$$
\mathbf{R}_x =
\begin{pmatrix}
1 & 0 & 0 & 0 \\
0 & \cos\omega & -\sin\omega & 0 \\
0 & \sin\omega & \cos\omega & 0 \\
0 & 0 & 0 & 1
\end{pmatrix}
$$

In Chapter 7, we called these two matrices $\mathbf{R}_y$ and $\mathbf{R}_x$ and discussed how we can combine them into one 3 * 3 matrix, which we will show here as a 4 * 4 matrix

$$
\mathbf{R} =
\begin{pmatrix}
a & b & c & 0 \\
d & e & f & 0 \\
g & h & i & 0 \\
0 & 0 & 0 & 1
\end{pmatrix}
$$

The values of a, b, c, and so forth, will depend on the sequence in which the rotations take place, but so long as the original rotations represent orthogonal transformations, the effect will be to keep right angles as right angles and straight lines as straight lines. If $\mathbf{R}$ is not an orthogonal matrix, the axes will become skewed.

Consider next the transformation

$$
\begin{pmatrix}
1 & 0 & 0 & t \\
0 & 1 & 0 & u \\
0 & 0 & 1 & v \\
0 & 0 & 0 & 1
\end{pmatrix}
*
\begin{pmatrix}
x \\
y \\
z \\
w
\end{pmatrix}
=
\begin{pmatrix}
x+tw \\
y+uw \\
z+vw \\
w
\end{pmatrix}
$$

This results in new coordinates $x' = x/w + t$, $y' = y/w + u$, $z' = z/w + v$. This is a simple translation of the origin from its original point to a new origin at $(-t, -u, -v)$.

Now consider

$$
\begin{pmatrix}
1 & 0 & 0 & 0 \\
0 & 1 & 0 & 0 \\
0 & 0 & 1 & 0 \\
0 & 0 & 0 & s
\end{pmatrix}
*
\begin{pmatrix}
x \\
y \\
z \\
w
\end{pmatrix}
=
\begin{pmatrix}
x \\
y \\
z \\
sw
\end{pmatrix}
$$

This results in a new set of coordinates that in more conventional form are:

$$x' = x/sw, \; y' = y/sw, \text{ and } z' = z/sw$$

This means that all dimensions have been changed in scale by an amount $= 1/s$.

Finally, consider:

$$
\begin{pmatrix}
1 & 0 & 0 & 0 \\
0 & 1 & 0 & 0 \\
0 & 0 & 1 & 0 \\
p & q & r & s
\end{pmatrix}
*
\begin{pmatrix}
x \\
y \\
z \\
w
\end{pmatrix}
$$

This results in

$$
\begin{pmatrix}
x \\
y \\
z \\
px + qy + rz + sw
\end{pmatrix}
$$

The expression $px + qy + rz + sw$ is the equation of a plane and all values will be reduced to a plane that is skew to our original axes, the amount of skew being dependent on the values p, q, and r. So taken overall, if we apply the matrix

$$
\mathbf{M} =
\begin{pmatrix}
a & b & c & t \\
d & e & f & u \\
g & h & i & v \\
p & q & r & s
\end{pmatrix}
$$

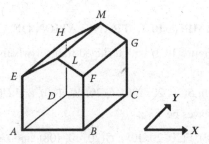

FIGURE 10.3   A barn.

to the set of coordinates $(x, y, z, w)^T$ then we will change the origin, rotate the axes, change the overall scale, and project the whole scene onto a plane surface. Thus, by suitable choice of the 16 elements, we can project any string of coordinates from three dimensions into two.

In Examples 10.2 through 10.5 we show the translation and rotation of a solid figure—in this case the barn shown in Figure 10.3. First, in Example 10.2 we move the origin to point A.

Next we will rotate the object about the z-axis and then about the x-axis, as shown in Example 10.3.

In Examples 10.2 and 10.3 we carried out a series of operations one at a time. These can be combined into a single operation as shown in Example 10.4.

It is important to note that the sequence of operations is critical and that $\mathbf{R}_x * \mathbf{R}_z *$ $\mathbf{N}'$ as shown in Example 10.3 is not the same as $\mathbf{R}_z * \mathbf{R}_x * \mathbf{N}'$. This is demonstrated in Example 10.5 where the rotation about the x-axis is applied first, and then there is a rotation about the z-axis. Likewise, if the origin is moved after, rather than before the rotations, there will be a different solution.

The matrix $\mathbf{N}'_{xz}$ in Example 10.3 represents the coordinates in space of the 10 points after the rotation of the axes. There is no change in shape and no change in scale. The lengths of the sides are the same as before and right angles remain as right angles in three-dimensional space. This is a *similarity transformation* in that it is a simple rotation and translation in which orthogonal lines remain orthogonal. A similarity transformation can also incorporate a uniform scale change. It is also *affine* in that it preserves collinearity and parallelism, straight lines remaining straight lines.

The image shown in Figure 10.4 treats the y values as zero and the image is displayed on the plane $y = 0$. From $\mathbf{N}'_{xz}$ in Examples 10.3 and 10.4, the nonhomogeneous coordinates are (x, z), namely,

A (0, 0), B (8.7, –0.9), C (–1.3, –3.9), D (–10, –3), E (0, 7.9), F (8.7, 7), G (–1.3, 4), H (–10, 4.9), L (4.3, 11.4), M (–5.7, 8.4)

EXAMPLE 10.2: TRANSLATION OF AXES

Consider a barn (Figure 10.3) with a ridged roof for which the coordinates of the floor are

$A(10, 20, 100)$, $B(20, 20, 100)$, $C(20, 40, 100)$, and $D(10, 40, 100)$.

Let the line of the eaves be

$E(10, 20, 108)$, F(20, 20, 108), $G(20, 40, 108)$, and $H(10, 40, 108)$

and the ridge be

$$L(15, 20, 112), M (15, 40, 112)$$

The AB direction is x, the AD is y, and the AE is z.

Let us express all these 10 points as homogeneous coordinates in a matrix N where the columns represent the four coordinates of the corner points.

$$
N = \begin{array}{cccccccccc}
A & B & C & D & E & F & G & H & L & M
\end{array}
$$

$$
N = \begin{pmatrix}
10 & 20 & 20 & 10 & 10 & 20 & 20 & 10 & 15 & 15 \\
20 & 20 & 40 & 40 & 20 & 20 & 40 & 40 & 20 & 40 \\
100 & 100 & 100 & 100 & 108 & 108 & 108 & 108 & 112 & 112 \\
1 & 1 & 1 & 1 & 1 & 1 & 1 & 1 & 1 & 1
\end{pmatrix}.
$$

Now let us change the origin to point $A(10, 20, 100, 1)$ by applying the matrix T where

$$
T = \begin{pmatrix}
1 & 0 & 0 & -10 \\
0 & 1 & 0 & -20 \\
0 & 0 & 1 & -100 \\
0 & 0 & 0 & 1
\end{pmatrix}
$$

$T * N =$

$$
\begin{pmatrix}
1 & 0 & 0 & -10 \\
0 & 1 & 0 & -20 \\
0 & 0 & 1 & -100 \\
0 & 0 & 0 & 1
\end{pmatrix} *
\begin{pmatrix}
10 & 20 & 20 & 10 & 10 & 20 & 20 & 10 & 15 & 15 \\
20 & 20 & 40 & 40 & 20 & 20 & 40 & 40 & 20 & 40 \\
100 & 100 & 100 & 100 & 108 & 108 & 108 & 108 & 112 & 112 \\
1 & 1 & 1 & 1 & 1 & 1 & 1 & 1 & 1 & 1
\end{pmatrix}
$$

Call this

$$
N' = \begin{pmatrix}
0 & 10 & 10 & 0 & 0 & 10 & 10 & 0 & 5 & 5 \\
0 & 0 & 20 & 20 & 0 & 0 & 20 & 20 & 0 & 20 \\
0 & 0 & 0 & 0 & 8 & 8 & 8 & 8 & 12 & 12 \\
1 & 1 & 1 & 1 & 1 & 1 & 1 & 1 & 1 & 1
\end{pmatrix}
$$

This is a simple translation of axes by moving the location of the origin.

EXAMPLE 10.3: ROTATION OF AN OBJECT

Using the data in Example 10.2, now let us rotate the building about the z-axis of the new origin, by an amount of 30° (with cos 30 ≈ 0.866 and sin 30 = 0.5).

$$\mathbf{R}^* \, \mathbf{N}' = \begin{pmatrix} 0.866 & -0.5 & 0 & 0 \\ 0.5 & 0.866 & 0 & 0 \\ 0 & 0 & 1 & 0 \\ 0 & 0 & 0 & 1 \end{pmatrix} * \mathbf{N}' = \mathbf{N}'_z$$

Our 10 points then, subject to rounding errors, become:

$$\mathbf{N}'_z = \begin{array}{ccccccccccc} A & B & C & D & E & F & G & H & L & M \\ \begin{pmatrix} 0 & 8.66 & -1.34 & -10 & 0 & 8.66 & -1.34 & -10 & 4.33 & -5.67 \\ 0 & 5 & 22.32 & 17.32 & 0 & 5 & 22.32 & 17.32 & 2.5 & 19.82 \\ 0 & 0 & 0 & 0 & 8 & 8 & 8 & 8 & 12 & 12 \\ 1 & 1 & 1 & 1 & 1 & 1 & 1 & 1 & 1 & 1 \end{pmatrix} \end{array}$$

Hence, A has become $(0, 0, 0, 1)$, B $(8.66, 5, 0, 1)$, and so forth. Now let us tip the image backward by rotating the shape by 10° about the x-axis (clockwise for right-handed systems; cos 10 = 0.985 and sin 10 = 0.174). The transformation matrix is then

$$\mathbf{R}_x * \mathbf{N}'_z = \begin{pmatrix} 1 & 0 & 0 & 0 \\ 0 & 0.985 & 0.174 & 0 \\ 0 & -0.174 & 0.985 & 0 \\ 0 & 0 & 0 & 1 \end{pmatrix} * \mathbf{N}'_z = \mathbf{N}'_{xz}$$

from where

$$\mathbf{N}'_{xz} = \begin{pmatrix} 0 & 8.66 & -1.34 & -10 & 0 & 8.66 & -1.34 & -10 & 4.33 & -5.67 \\ 0 & 4.92 & 21.99 & 17.06 & 1.39 & 6.31 & 23.38 & 18.45 & 4.55 & 21.61 \\ 0 & -0.87 & -3.88 & -3.01 & 7.88 & 7.01 & 4.00 & 4.87 & 11.38 & 8.37 \\ 1 & 1 & 1 & 1 & 1 & 1 & 1 & 1 & 1 & 1 \end{pmatrix}$$

EXAMPLE 10.4: COMBINING ROTATIONS

In Example 10.3 we could have combined all three operations by applying one matrix **R** where

$$
\mathbf{R} = \begin{pmatrix} 1 & 0 & 0 & 0 \\ 0 & 0.985 & 0.174 & 0 \\ 0 & -0.174 & 0.985 & 0 \\ 0 & 0 & 0 & 1 \end{pmatrix} * \begin{pmatrix} 0.866 & -0.5 & 0 & 0 \\ 0.5 & 0.866 & 0 & 0 \\ 0 & 0 & 1 & 0 \\ 0 & 0 & 0 & 1 \end{pmatrix} * \begin{pmatrix} 1 & 0 & 0 & -10 \\ 0 & 1 & 0 & -20 \\ 0 & 0 & 1 & -100 \\ 0 & 0 & 0 & 1 \end{pmatrix}
$$

$$
= \begin{pmatrix} 0.866 & -0.5 & 0 & 1.34 \\ 0.492 & 0.853 & 0.174 & -39.38 \\ -0.087 & -0.151 & 0.985 & -94.62 \\ 0 & 0 & 0 & 1 \end{pmatrix}
$$

What we would then have, subject to rounding errors, is: **R * N =**

$$
\begin{pmatrix} 0.866 & -0.5 & 0 & 1.34 \\ 0.492 & 0.853 & 0.174 & -39.38 \\ -0.087 & -0.151 & 0.985 & -94.62 \\ 0 & 0 & 0 & 1 \end{pmatrix} * \begin{pmatrix} 10 & 20 & 20 & 10 & 10 & 20 & 20 & 10 & 15 & 15 \\ 10 & 20 & 40 & 40 & 20 & 20 & 40 & 40 & 20 & 40 \\ 100 & 100 & 100 & 100 & 108 & 108 & 108 & 108 & 112 & 112 \\ 1 & 1 & 1 & 1 & 1 & 1 & 1 & 1 & 1 & 1 \end{pmatrix}
$$

$$
= \begin{pmatrix} 0 & 8.66 & -1.34 & -10 & 0 & 8.66 & -1.34 & -10 & 4.33 & -5.67 \\ 0 & 4.92 & 21.99 & 17.06 & 1.39 & 6.31 & 23.38 & 18.45 & 4.55 & 21.61 \\ 0 & -0.87 & -3.88 & -3.01 & 7.88 & 7.01 & 4.00 & 4.87 & 11.38 & 8.37 \\ 1 & 1 & 1 & 1 & 1 & 1 & 1 & 1 & 1 & 1 \end{pmatrix}
$$

which is the same value for $\mathbf{N'_{xz}}$ as the one derived earlier in Example 10.3.

EXAMPLE 10.5: THE SEQUENCE OF ROTATIONS

The sequence in which rotations are carried out is important. If in Example 10.3 we had carried out the rotation about the x-axis before we carried out the rotation about the z, we would obtain a different answer **R'**.

$$
\mathbf{R'} = \begin{pmatrix} 0.866 & -0.5 & 0 & 0 \\ 0.5 & 0.866 & 0 & 0 \\ 0 & 0 & 1 & 0 \\ 0 & 0 & 0 & 1 \end{pmatrix} * \begin{pmatrix} 1 & 0 & 0 & 0 \\ 0 & 0.985 & 0.174 & 0 \\ 0 & -0.174 & 0.985 & 0 \\ 0 & 0 & 0 & 1 \end{pmatrix} * \begin{pmatrix} 1 & 0 & 0 & -10 \\ 0 & 1 & 0 & -20 \\ 0 & 0 & 1 & -100 \\ 0 & 0 & 0 & 1 \end{pmatrix}
$$

$$
= \begin{pmatrix} 0.866 & -0.492 & -0.082 & 9.89 \\ 0.5 & 0.853 & 0.151 & -37.13 \\ 0 & -0.174 & 0.985 & -95.02 \\ 0 & 0 & 0 & 1 \end{pmatrix}
$$

This is not the same as the original **R** shown in Example 10.4.

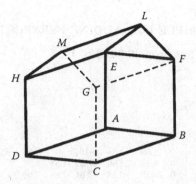

FIGURE 10.4 The barn after two rotations.

In Figure 10.4, the 3D coordinates show that the lengths DH, AE, and FB are all the same as are DA, HE, and ML, and so forth. EF and AB are parallel and meet at a point on the x-axis that is an infinite distance away. This point can be given the coordinates $(1, 0, 0, 0)$. If we multiply this by the skew matrix:

$$
\mathbf{S}_x = \begin{pmatrix} 1 & 0 & 0 & 0 \\ 0 & 1 & 0 & 0 \\ 0 & 0 & 1 & 0 \\ 0.025 & 0 & 0 & 1 \end{pmatrix} \quad \text{then} \quad \mathbf{S}_x * \begin{pmatrix} 1 \\ 0 \\ 0 \\ 0 \end{pmatrix} = \begin{pmatrix} 1 \\ 0 \\ 0 \\ 0.025 \end{pmatrix}
$$

The point at infinity along the x-axis becomes the point $(1, 0, 0, 0.025)$ or in conventional coordinates $(40, 0, 0)$ since $1/(0.025) = 40$. Similarly, lines in the y-direction can be drawn to the point at infinity in the y-direction, namely, point $(0, 1, 0, 0)$. Thus,

$$
\mathbf{S}_{xy} = \begin{pmatrix} 1 & 0 & 0 & 0 \\ 0 & 1 & 0 & 0 \\ 0 & 0 & 1 & 0 \\ 0.025 & 0.025 & 0 & 1 \end{pmatrix}
$$

will bring both the points at infinity in the x- and y-directions into a range that can be plotted on a sheet of paper. We then have a perspective view (see Figure 10.5) as shown in Example 10.6.

Figure 10.5 shows the barn after translation and rotation before and after applying perspective. The results from Example 10.6 provide the coordinates of the barn with a perspective view, again as in Figure 10.4, treating the y value as zero.

EXAMPLE 10.6: ADDING PERSPECTIVE

Using the data from Examples 10.3 and 10.4,

$$
\begin{pmatrix}
1 & 0 & 0 & 0 \\
0 & 1 & 0 & 0 \\
0 & 0 & 1 & 0 \\
0.025 & 0.025 & 0 & 1
\end{pmatrix} * \mathbf{N}'_{xz}
$$

$$
= \begin{pmatrix}
1 & 0 & 0 & 0 \\
0 & 1 & 0 & 0 \\
0 & 0 & 1 & 0 \\
0.025 & 0.025 & 0 & 1
\end{pmatrix} * \begin{bmatrix}
0 & 8.66 & -1.34 & -10 & 0 & 8.66 & -1.34 & -10 & 4.33 & -5.67 \\
0 & 4.92 & 21.99 & 17.06 & 1.39 & 6.31 & 23.38 & 18.45 & 4.55 & 21.61 \\
0 & -0.87 & -3.88 & -3.01 & 7.88 & 7.01 & 4.00 & 4.87 & 11.38 & 8.37 \\
1 & 1 & 1 & 1 & 1 & 1 & 1 & 1 & 1 & 1
\end{bmatrix}
$$

$$
= \begin{pmatrix}
0 & 8.66 & -1.34 & -10 & 0 & 8.66 & -1.34 & -10 & 4.33 & -5.67 \\
0 & 4.92 & 21.99 & 17.06 & 1.39 & 6.31 & 23.38 & 18.45 & 4.55 & 21.61 \\
0 & -0.87 & -3.88 & -3.01 & 7.88 & 7.01 & 4.00 & 4.87 & 11.38 & 8.37 \\
1 & 1.34 & 1.52 & 1.17 & 1.03 & 1.37 & 1.55 & 1.21 & 1.22 & 1.40
\end{pmatrix}
$$

$$
= \begin{pmatrix}
0 & 6.46 & -0.88 & -8.55 & 0 & 6.32 & -0.86 & -8.26 & 3.55 & -4.05 \\
0 & 3.67 & 14.47 & 14.58 & 1.35 & 4.61 & 15.08 & 15.25 & 3.73 & 15.43 \\
0 & -0.65 & -2.55 & -2.57 & 7.65 & 5.12 & 2.58 & 4.02 & 9.33 & 5.98 \\
1 & 1 & 1 & 1 & 1 & 1 & 1 & 1 & 1 & 1
\end{pmatrix}
$$

Prior to applying the perspective transformation, the nonhomogeneous coordinates were

A (0, 0, 0), B (8.7, 4.9, −0.9), C (−1.3, 22, −3.9), D (−10, 17, −3), E (0, 1.4, 7.9),

F (8.7, 6.3, 7), G (−1.3, 23.4, 4), H (−10, 18.4, 4.9), L (4.3, 4.6, 11.4), M (−5.7, 21.6, 8.4)

After applying the perspective transformation they became:

A (0, 0, 0), B (6.5, 3.7, −0.6), C (−0.9, 14.5, −2.5), D (−8.5, 14.6, −2.6), E (0, 1.3, 7.6),

F (6.3, 4.6, 5.1), G (−0.9, 15.1, 2.6), H (−8.3, 15.2, 4.0), L (3.5, 2.7, 9.3),

M (−4.0, 15.4, 6.0)

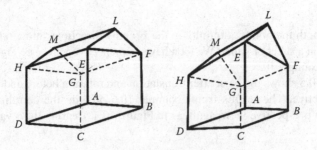

FIGURE 10.5   The affine and perspective projections of the barn.

Note that the two points at infinity $(1, 0, 0, 0)$ and $(0, 1, 0, 0)$ under the original transformation became

$$
\begin{pmatrix}
0.866 & -0.5 & 0 & 1.34 \\
0.492 & 0.853 & 0.174 & -39.38 \\
-0.087 & -0.151 & 0.985 & -94.62 \\
0 & 0 & 0 & 1
\end{pmatrix}
*
\begin{pmatrix}
1 & 0 \\
0 & 1 \\
0 & 0 \\
0 & 0
\end{pmatrix}
=
\begin{pmatrix}
0.866 & -0.5 \\
0.492 & 0.853 \\
-0.087 & -0.151 \\
0 & 0
\end{pmatrix}
$$

and were still at infinity. However, by applying the matrix $\mathbf{S}_{xy}$ we obtain

$$
\begin{pmatrix}
1 & 0 & 0 & 0 \\
0 & 1 & 0 & 0 \\
0 & 0 & 1 & 0 \\
0.025 & 0.025 & 0 & 1
\end{pmatrix}
*
\begin{pmatrix}
0.8666 & -0.5 \\
0.492 & 0.853 \\
-0.087 & -0.151 \\
0 & 0
\end{pmatrix}
=
\begin{pmatrix}
0.866 & -0.5 \\
0.492 & 0.853 \\
-0.087 & -0.151 \\
0.034 & 0.009
\end{pmatrix}
$$

$$
=
\begin{pmatrix}
25.47 & -55.6 \\
14.47 & 94.78 \\
-2.54 & -16.78 \\
1 & 1
\end{pmatrix}
$$

Thus, the lines AB, EF, HG, and DC converge at P $(25.5, 14.5, -2.6)$ as shown in Figure 10.6, while AD, EH, FG, BC, and LM converge at $(-55.6, 94.8, -16.8)$ (not shown).

Transformations of the kind illustrated above can be used, for example, to simulate a three-dimensional landscape including the effects of changing viewing positions, for instance, for "fly-by" presentations. Each point or line can be transformed and then displayed at its new coordinates. Straight lines remain straight but parallel lines can be made to converge.

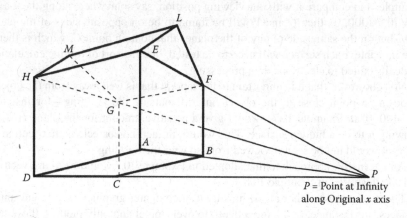

P = Point at Infinity along Original x axis

FIGURE 10.6   Transformation into a perspective view.

10.3   HIDDEN LINES AND SURFACES

Reverting to Euclidean space and nonhomogeneous coordinates, in Chapter 3 we noted that for two points on a line, A (x_A, y_A) and B (x_B, y_B), then

$$(y - y_A) = \frac{(y_B - y_A)}{(x_B - x_A)} * (x - x_A)$$

Given a point P (x_P, y_P), if we calculate

$$y = y_A + \frac{(y_B - y_A)}{(x_B - x_A)} * (x_P - x_A)$$

and if $(y_P > y)$, then P is above the line AB while if $y_P < y$, then P lies below the line. If $y_P = y$, then P is on the line AB.

We can extend this idea to any plane through points ABC, which in Chapter 8 (Section 8.3) we showed to have the form

$$(x - x_C)\{y_A(z_B - z_C) + y_B(z_C - z_A) + y_C(z_A - z_B)\}$$
$$+ (y - y_C)\{z_A(x_B - x_C) + z_B(x_C - x_A) + z_C(x_A - x_B)\}$$
$$+ (z - z_C)\{x_A(y_B - y_C) + x_B(y_C - y_A) + x_C(y_A - y_B)\} = 0$$

When we insert values for x, y, and z this quantity will be either positive or negative unless the point lies on the plane. Two points on the same side of the plane will share the same sign (i.e., both will give a positive value or both will give a negative).

We can use this fact to determine which lines are "hidden" when we view a solid object projected onto a flat surface. Every plane surface can be defined by three points so, for example in Figure 10.6, for the surface defined by the four points A, D, H, and E, we only need the coordinates of three of these (see Example 10.7). If we take a coordinate value inside the building, say a point Q at $(0, 5, 0)$ in the above example, and compare it with our viewing position, say somewhere along the y-axis at V $(0, -1000, 0)$, then Q and V will be found to be on opposite sides of the plane ADE but on the same side of any of the planes that contain point G, which is therefore invisible. Each surface will need to be tested, but this sort of routine calculation is ideally suited to electronic data processing.

Note, however, that it is important to use a point that is well away from the object. Choosing a point close to the object but still outside the building—for instance $(0, -100, 0)$ in Example 10.7—would give a negative amount for the plane $HELM$, showing it to be a hidden surface. This would, in fact, be correct, as that section of the roof would not be seen if viewed from so near the building.

An alternative approach to that adopted in Example 10.7 would be to use vectors. This is illustrated in Example 10.8.

There are various other algorithms used in computer graphics to hide unwanted surfaces, for instance, by giving a depth to every pixel and only plotting those that

EXAMPLE 10.7: HIDDEN SURFACES USING PLANES

There are seven faces in Figure 10.6. From the outside and in an anticlockwise direction, these are *ABFLE, BCGF, CDHMG, DAEH, BADC, HELM,* and *FGML.*

From Example 10.6, the perspective coordinates are: *A* (0, 0, 0), *B* (6.46, 3.67, −0.65), *C* (−0.88, 14.47, −2.55), *D* (−8.55, 14.58, −2.57), *E* (0, 1.35, 7.65), *F* (6.32, 4.61, 5.12), *G* (−0.86, 15.08, 2.58), *H* (−8.26, 15.25, 4.02), *L* (3.55, 3.73, 9.33), and *M* (−4.05, 15.43, 5.98).

The plane of *ABF* is given by $ax + by + cz + d = 0$. Using the formulae for deriving a, b, c, d, then for *ABF* we have $21.79x - 36.88y + 6.31z = 0$ or dividing through by

$$\sqrt{(21.79^2 + 36.88^2 + 6.31^2)} \text{ to reduce } \sqrt{(a^2 + b^2 + c^2)} \text{ to } 1$$

ABF	$0.5x - 0.85y + 0.15z = 0$

Similarly,

BCG	$0.830x + 0.553y - 0.069z - 7.4 = 0$
CDH	$0.01x + 0.996y - 0.09z - 14.7 = 0$
DAE	$-0.866x - 0.492y + 0.087z = 0$
BAD	$-0.174y - 0.985z = 0$
HEL	$-0.421x - 0.013y + 0.907z - 6.9 = 0$
FGM	$0.631x + 0.562y + 0.534z - 9.3 = 0$

Now consider a point within the building (for instance (0, 5, 0) and a point well outside (0, −1000, 0). Inserting these values in the equations for the planes, we obtain

	Point Inside	**Point Outside**
ABF	− 4.2	+ 850
BCG	− 4.6	− 560
CDH	− 9.7	−1010
DAE	− 0.2	+ 492
BAD	− 0.1	+ 174
HEL	− 7.0	+ 6
FGM	− 6.6	− 572

Here, those planes with opposite signs will be visible when viewed from a distance, whereas those with the same sign (that is, *BCG, CDH,* and *FGM*) will be hidden.

EXAMPLE 10.8: HIDDEN SURFACES USING VECTORS

Using the same data as in Example 10.7, the equation of the plane *ABF* is

$$0.5x - 0.85y + 0.15z = 0$$

The normal to the plane is therefore the unit vector $0.5\mathbf{i} - 0.85\mathbf{j} + 0.15\mathbf{k}$.

If our viewing point is (0, –50, 0), which has a unit vector (0, –1, 0) or $0 * \mathbf{i} - 1 * \mathbf{j} + 0 * \mathbf{k}$ then the cosine of the angle of incidence (cos ø) is given by the dot product with the plane *ABFLE*.

Thus,

$$\cos ø = 0 * 0.5 + 1 * 0.85 + 0 * 0.15 = +0.85$$

For *BCGF*, cos ø = –0.553; for *CDHMG*, cos ø = –0.996; for *DAEH*, cos ø = +0.492; for *BADC*, cos ø = +0.174; for *HELM*, cos ø = +0.013; and for *FGML*, cos ø = –0.562.

The cosine of an angle that is greater than 90° is negative thus the normals to the planes *BCGF*, *CDHMG*, and *FGML* all point away from the viewer, and thus the surfaces are hidden.

are of minimum depth. The reader should consult books on computer graphics for more information.

10.4 PHOTOGRAMMETRIC MEASUREMENTS

Most topographic maps these days are produced using vertical aerial photography and a series of techniques known as *stereophotogrammetry*. In Figure 10.7, a camera takes a photograph of the ground from a point known as a base station and then flies forward to the next base station and takes a second photograph, the ground in

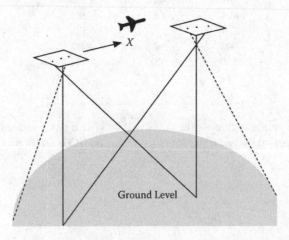

FIGURE 10.7   Stereo pair of aerial photographs.

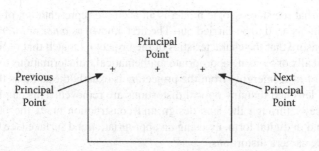

FIGURE 10.8 Principal point in the center of an aerial photograph.

between being covered by both photographs. The distance flown between the taking of each photograph is known as the *air base* and is approximately 40% of the width of each resulting aerial photograph (Figure 10.8). Rays of light from the ground pass through the camera lens and register as a negative image. If the focal length of the camera lens is f, the flying height is H, and the terrain height h (so that the height above the ground is $H - h$), then the scale of the photography will $\approx \frac{f}{H-h}$. Since h varies across the terrain, this means that the scale varies from one point on the photograph to the next.

The center of each photograph is known as the *principal point* and mapping takes place in the overlap area between successive principal points. Here a virtual 3D model of the terrain is formed and precise measurements within the model can be taken giving, with suitable photography and appropriate technology, ground value accuracies of the order of 1 centimeter or so. The coordinates of points visible on each photograph are measured and transformed, either by analogue or digital means, into points on a map.

The axis of the camera is said to be the z-axis (Figure 10.9); with the negative it points downward but with a positive print it points upward. The direction of flight for a series of aerial photographs is called the *x-axis* while the horizontal line at right angles to the direction of flight is called the *y-axis*.

The photogrammetric process analyzes the rays of light by projecting them back out so that corresponding rays intersect, in effect turning the camera in Figure 10.7 into a projector. Using the two camera stations, a three-dimensional model can be formed, the scale of which will depend on the distance between the two projectors, which in turn represents the length of the air base.

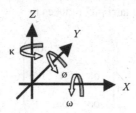

FIGURE 10.9 Photogrammetric rotations.

In order that the stereoscopic model is an accurate representation of the ground, three operations need to be carried out. The first, known as *inner* or *interior orientation*, is to ensure that the characteristics of the projectors match that of the cameras, either physically or by using appropriate mathematical transformations to ensure that the bundle of rays emerging from the projectors is exactly the same as that entering the camera lens and that any optical distortions are removed. This can be achieved either optically through the lens design and construction or, if the photographic image is held in digital form, by using an appropriate trend surface (see Chapter 13) to model the camera distortions.

The second operation is known as *relative orientation*. This means that the two projectors are adjusted so that their relative tips and tilts match those of the original camera positions.

Rotations of the camera about the x-axis or the roll of the aircraft are designated *omega*—the Greek letter "ω." A rotation about the z-axis, which for an aircraft is known as the *yaw*, is said to be *kappa*—the Greek letter "κ." Rotations about the y-axis are known as the *pitch* of the aircraft and are called *phi*—the Greek letter "ϕ." The rotations are right handed in that if your right-hand thumb points along the direction of the arrows, your fingers curl to point as shown in Figure 7.7 and Figure 10.9.

If we rotate a camera/projector first about the x-axis (known as the *primary axis*) then about the y-axis, and finally the z-axis, the transformation will be of the form:

$$\mathbf{M} = \begin{pmatrix} \cos\kappa & \sin\kappa & 0 \\ -\sin\kappa & \cos\kappa & 0 \\ 0 & 0 & 1 \end{pmatrix} * \begin{pmatrix} \cos\phi & 0 & \sin\phi \\ 0 & 1 & 0 \\ -\sin\phi & 0 & \cos\phi \end{pmatrix} * \begin{pmatrix} 1 & 0 & 0 \\ 0 & \cos\omega & \sin\omega \\ 0 & -\sin\omega & \cos\omega \end{pmatrix}$$

$$= \begin{pmatrix} \cos\kappa\cos\phi & -\cos\kappa\sin\phi\sin\omega+\sin\kappa\cos\omega & \cos\kappa\sin\phi\cos\omega+\sin\kappa\sin\omega \\ -\sin\kappa\cos\phi & \sin\kappa\sin\phi\sin\omega+\cos\kappa\cos\omega & -\sin\kappa\sin\phi\cos\omega+\cos\kappa\sin\omega \\ -\sin\phi & -\cos\phi\sin\omega & \cos\phi\cos\omega \end{pmatrix}$$

In particular, if the angles through which we make the rotations are small and we call these $\delta\omega$, $\delta\phi$, and $\delta\kappa$, then we have shown in Chapter 6 that $\cos(\delta\omega)$, $\cos(\delta\phi)$, and $\cos(\delta\kappa)$ each equal 1, while the sine functions become $\delta\omega$, $\delta\phi$, and $\delta\kappa$ in radians. Thus, for small rotations the matrix $\mathbf{M}$ reduces to

$$\begin{pmatrix} 1 & \delta\kappa & \delta\phi \\ -\delta\kappa & 1 & \delta\omega \\ -\delta\phi & -\delta\omega & 1 \end{pmatrix}$$

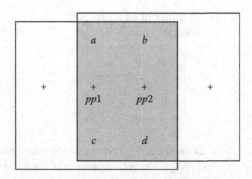

FIGURE 10.10   Model area for overlapping stereo pair of photographs.

Since there are two cameras or projectors (*L* for left and *R* for right), there are six unknowns, namely, $\delta\omega_L$, $\delta\phi_L$, $\delta\kappa_L$, and $\delta\omega_R$, $\delta\phi_R$, $\delta\kappa_R$. However, if we rotate the projectors by the same amount about the *x*-axis, the rays will still intersect and the values for $\delta\omega$ cannot at this stage be determined absolutely. Relative orientation allows us to determine the five values $\delta\phi_L$, $\delta\kappa_L$, $\delta\phi_R$, $\delta\kappa_R$, and the relative value $\delta\omega$.

This is achieved by choosing clearly identifiable points in the general area of the model (*a*, *b*, *c*, *d*, and *pp*1 and *pp*2), as shown in Figure 10.10 and measuring their locations on each flat-surface photograph and transforming their (*x*, *y*) values so that they match, using the rotation matrix **M** with the appropriate values for each projector. Finally, to bring the model into agreement with the ground coordinate values, we apply a 4 * 4 matrix transformation of the form discussed in Section 10.2.

The stereoscopic model must be leveled and adjusted for scale. This process, known as the *absolute orientation*, removes the uncertainties about tip and tilt. Three points are chosen that have known height and are suitably located to adjust the level of the model—such as *a* and *b* for roll and *b* and *c* for pitch in Figure 10.10. Likewise, the distance between two points (such as *a* to *d*) must be known in order that we can adjust for scale. These known points may have been established either by ground surveys or through what is known as *aerotriangulation*. Further details can be found in standard books on photogrammetry. Our interest here is in the mathematics behind the treatment of photogrammetric measurements.

SUMMARY

Affine: A transformation that preserves collinearity and parallelism so that points in a straight line remain in a straight line.

Hidden lines: Lines in a 3D object that cannot be seen by the viewer in 2D.

Homogeneous coordinates: The introduction of an additional element in a set of coordinates that provides scale and can be used to describe points at infinity.

Orthogonality: The preservation of right angles as right angles.

Principal point: The optical center of an aerial photograph.

Rotation matrix: In homogeneous coordinates, a 4 * 4 matrix of the form

$$\begin{pmatrix} a & b & c & t \\ d & e & f & u \\ g & h & i & v \\ p & q & r & s \end{pmatrix}$$

in which the nine elements $a, b \dots i$ represent the rotations, t, u, v the translation, and p, q, r the projection onto a plane; s is the scale change.

Similarity transformation: A transformation by translation and rotation and uniform scale change that preserves *orthogonality*.

Stereophotogrammetry: The measurement of position from overlapping pairs of photographs.

Vanishing point: The point at infinity where parallel lines meet. It can be transformed into a finite point in space.

11 Map Projections

11.1 MAP PROJECTIONS

In this chapter we will consider ways in which the location of points on the surface of the Earth can be represented in what are commonly referred to as *map projections*. The assumption behind the transformations considered in Chapter 10 has been that we have been dealing with straight lines or flat surfaces, in effect, a flat Earth. When we wish to represent the curved surface of the Earth, we need to take a different approach.

Consider a point A on the Earth's surface with latitude ϕ_A and longitude λ (Figure 11.1). The length of the arc from the equator (Q) to $A \doteq R\,\phi_A$ where R is the radius of the Earth, assumed here to be a sphere, and ϕ is measured in radians. Let C be a point with latitude ϕ_C and longitude λ_C. Let the difference in longitude between A and $C = \Delta\lambda = \lambda_C - \lambda_A$ and the difference in latitude $= \Delta\phi = \phi_C - \phi_A$. Let B be on the same parallel of latitude as C and the same meridian of longitude as A. In triangle ABC then in terms of physical length: $AB = R\,\Delta\phi$. Since the radius of the circle for the parallel of latitude $\phi = PB$ and $PB = PC = R\cos\phi$, then

$$BC = PB\,\Delta\lambda = R\cos\phi_C\,\Delta\lambda$$

These quantities form the basis for plotting on a flat surface with a rectangular grid (see Figure 11.2). AB on the sphere becomes $A'B' = \Delta N$ and BC becomes $B'C' = \Delta E$, the differences in northings and eastings between A' and C'.

In the simplest map projection, known as the *simple cylindrical* or *plate carrée* projection, the distance north is plotted as $N = R\phi$ and the eastings as $R\lambda$ so that there is a rectangular grid. This means, in effect, that the distances north–south are treated as correct but distances east–west must be stretched from ($R\cos\phi$) λ on the globe to $R\lambda$ on the grid. East–west distances on the globe must be increased by a scale factor of ($1/\cos\phi$) or $\sec\phi$ in order to plot them on the rectangular grid.

Another approach might be to plot the northings correctly as $R\phi$ and the eastings as ($R\cos\phi$) λ so that both north–south and east–west had a scale factor of 1. But consider what happens to triangle ABC. If for the sake of argument we set the latitude of $A = 30°$ and $C = 40°$ so $\Delta\phi = 10° = 0.1745$ radians and we set $\Delta\lambda = 15° = 0.2618$ radians and we have a globe of radius $R = 100$ units then $A'B'$ on the flat $= R\Delta\phi = 17.45$ and $B'C' = R\Delta\lambda * \cos 40 = 20.05$. Tan ($\angle B'A'C'$) = 20.05/17.45 or $\angle B'A'C' = 49°$ while the distance $A'C' = \sqrt{(17.45^2 + 20.05^2)} = 26.58$.

On the globe, however, we have to use the spherical triangle formulae on the triangle ANC in Figure 11.1. In this case, $NC = 50°$, $NA = 60°$, angle $ANC = 15°$, and if we call the angle $AC = b$, then by using the cosine formula for spherical triangles

$$\cos b = \cos 50 \cos 60 + \sin 50 \sin 60 \cos 15 = 0.9622, \text{ or}$$

$$b = 15.8° = 0.2758 \text{ radians}$$

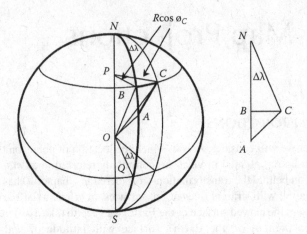

FIGURE 11.1   Points on a globe.

Thus, on the sphere, the distance is 27.58 but on the grid it is 26.58. Likewise, using the sine formula the angle *NAC* on the globe = 53.7° rather than 49° as previously derived. Thus, although we have preserved the scale in both the north–south and east–west directions we have changed the scale along the diagonal and the angle at *A*. We would also have the meridians bending toward the center and not crossing the parallels at right angles. Some distortion is inevitable because a curved surface must be stretched or shrunk in order to represent it on a plane.

Now let us consider the triangle *ABC* in Figure 11.1 as being very small, so small as to allow us to use differential calculus. We will call this small triangle the *elemental triangle* with sides *AB* and *BC* on the map (i.e., on a flat surface) being represented by δN and δE while on the sphere they are $R\delta\phi$ and $R\cos\phi\ \delta\lambda$ (Figure 11.3). The question is, "How are δN and δE related to $R\delta\phi$ and $R\cos\phi\ \delta\lambda$?" There are many possible answers, each one trying to preserve the angles, the distances, or the area,

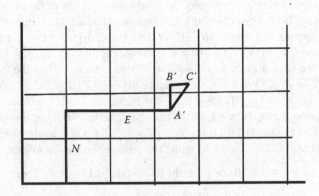

FIGURE 11.2   The simple cylindrical projection.

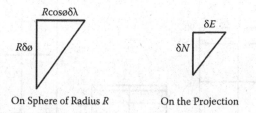

On Sphere of Radius R On the Projection

FIGURE 11.3   The elemental triangle on a sphere and on a plane surface.

or somehow compromise between these three quantities. There are in general three approaches, one based on a cylinder wrapped around the sphere, another based on a cone touching or cutting it, and a third on a plane. Both the cylinder and cone can be cut along a line and opened up into a flat surface (Figure 11.4).

Projections based around these three approaches are called *cylindrical*, *conical*, and *azimuthal* (or *zenithal*). They are normally considered to have the axis of rotation of the Earth as vertical but they can be applied obliquely. In much of what

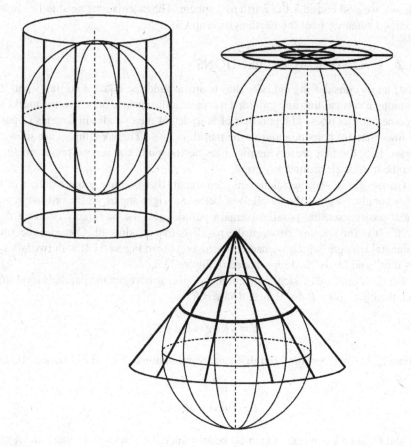

FIGURE 11.4   Cylinder, cone, and plane.

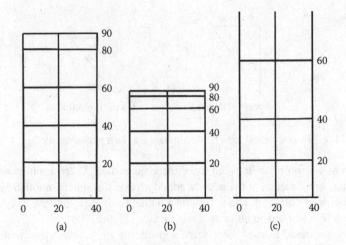

FIGURE 11.5   Cylindrical equidistant (a), equal area (b), and conformal (c).

follows, we will consider the Earth as a sphere. The calculations need to be slightly modified when we treat the Earth as an ellipsoid.

11.2   CYLINDRICAL PROJECTIONS

First, let us consider the cylinder that wraps around the equator so that when it is cut along a vertical line and unfolded it creates a flat surface on which the meridians become parallel lines. The parallels of latitude cut the cylinder in a series of parallel lines that are perpendicular to the meridian lines. Three examples are shown in Figure 11.5, the first preserving distances north–south, the second preserving area, and the third scale around any point.

On the globe, the meridians are convergent, that is, they come together at the poles but always cross the parallels of latitude at right angles. In the ordinary cylindrical projections, the parallels remain parallel and the meridians intersect them at 90°. It is the spacing between the parallels that is adjusted. Consider the small elemental triangle (which we had enlarged as ABC in Figure 11.1) with two sides "$R \cos \phi \, \delta\lambda$" and "$R \, \delta\phi$" and reproduced in Figure 11.3.

For the simple cylindrical projection, the spacing between the parallels is retained and, therefore, $\delta N = R \, \delta\phi$. Hence, $dN/d\phi = R$, or

$$N = \int R \, d\phi = R \phi + C_n$$

where C_n is some constant. For differences in longitude, $\delta E = R \, \delta\lambda$. Hence, $dE/d\lambda = R$, or

$$E = \int R d\lambda = R \lambda + C_e$$

C_n and C_e are the constants that must be introduced on integration, and ϕ and λ must be measured in radians. Since $N = 0$ when $\phi = 0$ and $E = 0$ when $\lambda = 0$, then $C_n = C_e = 0$.

Hence, for the simple cylindrical projection

$$N = R\,\phi \quad \text{and} \quad E = R\,\lambda$$

To preserve the area of the small element, if we retain $E = R\,\lambda$ when it really ought to be $(R\cos\phi)\,\lambda$, then we have, in effect, increased the scale west–east by sec ϕ. Thus, to preserve area we must reduce the south–north distances by an amount cos ϕ. Put another way, since the area of a triangle is half its base * height, the area of the elemental triangle on the sphere is

$$(1/2)R^2\cos\phi\ \delta\phi\ \delta\lambda$$

The area on the projection is $(1/2)\ \delta N\ \delta E$. But,

$$\delta E = R\ \delta\lambda$$

hence,

$$\delta N = R\cos\phi\ \delta\phi \quad \text{and} \quad N = \int R\cos\phi\ d\phi = R\sin\phi$$

Thus, for the cylindrical equal area projection,

$$E = R\,\lambda \text{ and } N = R\sin\phi$$

Projections that preserve area are called *equal-area* or *equivalent* or *authalic*.

If we want to maintain the shape of the elemental triangle, then since the west–east scale factor is "sec ϕ," then we must do the same for the south–north and thus, $\delta N = R\sec\phi\ \delta\phi$.

The integral of sec ϕ was quoted in Box 6.4 to be $\log_e$ (sec ϕ + tan ϕ). This can also be expressed as $\log_e \tan(\pi/4 + \phi/2)$. Hence, if $dN = R\sec\phi\ d\phi$, then ignoring the constant term,

$$N = \int R\sec\phi\ d\phi = R\log_e\tan(\pi/4 + \phi/2)$$

By maintaining the scale around any point, the shape of small areas is preserved. Projections that maintain this property are known as *orthomorphic* or *conformal*. They have the advantage that angles around any point are preserved so that angles measured in the field with a theodolite can be used directly in the projection. This is particularly important in topographic mapping.

One consequence is that at the North or South Pole, the value for N then becomes $\log_e \tan(90°)$ and that is infinitely large. Recalling that this is based on a cylinder wrapped around the equator, there is no way that the position of the poles can be plotted on a map when using this form of cylindrical projection.

The cylindrical orthomorphic projection is also known as *Mercator's projection*. Since angles around any point are preserved and since any straight line drawn on the map cuts all the meridians at the same angle, it means that the bearing along the line is always the same. A line of constant bearing is known as a *rhumb line* or *loxodrome*. Such lines have been particularly important to navigators who can draw a straight line on a chart using the Mercator's projection and know that by following the bearing of that line they will reach their destination. This line may not be the shortest route; but it may be the easiest to navigate. The scale at any point on the Mercator's projection is given by "sec ϕ." This becomes large at high latitudes; but around the equator it is close to 1.

By wrapping a cylinder around a meridian rather than the equator, we have a *transverse cylindrical projection*.

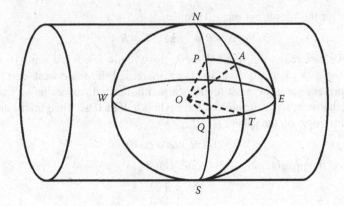

FIGURE 11.6   The Transverse Mercator.

In Figure 11.6, the cylinder is wrapped around the meridian $NPQS$, known as the *central meridian*, with points E and W being 90° east and west of the central meridian, in effect becoming the poles in the standard case. For any point A, $EAPW$ is a great circle and the coordinates of A can be expressed in terms of QP (the equivalent of λ in the normal cylindrical case but here being ϕ_P) and PA (the equivalent of the latitude in the nontransverse case). Let us assume that its value is α so angle $POA = \alpha$. $NPQS$ is the central meridian to which we will give the value $\lambda_M\ (= \lambda_Q)$.

Let NA cut the equator at T. QT is the difference in longitude between the central meridian and the meridian through A. AT = latitude of $A = \phi_A$; $NA = 90° - \phi_A$; angle PNA = difference in longitude between A and the central meridian $= \lambda_A - \lambda_M$. Let us call this λ_A so that angle $ANE = 90° - \lambda_A$.

In triangle ATE, angle $ATE = 90°$; side $AT = \phi_A$; side TE = angle $ANE = 90° - \lambda_A$; $AE = 90° - \alpha$. Using the cosine formula on triangle ATE

$$\cos (AE) = \cos (AT) \cos (TE) + \sin (AT) \sin (TE) \cos (90)$$

or, since $\cos (90) = 0$,

$$\sin \alpha = \cos \phi_A \sin \lambda_A = \cos \phi_A \sin(\lambda_A - \lambda_M) = \cos \phi_A \sin(\lambda_A - \lambda_Q)$$

Also, using the sine formula on triangle ATE where $ATE = 90°$

$$\sin (AET)/\sin (AT) = \sin (ATE)/\sin (AE)$$

or, since angle $AET = PQ = $ the latitude of $P = \phi_P$.

$$\sin \phi_P = \sin \phi_A/\cos \alpha, \text{ or}$$

$$\cos \alpha = \sin \phi_A/\sin \phi_P$$

Using the cosine formula in triangle PNA,

$$\cos PA = \cos PN \cos NA + \sin PN \sin NA \cos (\lambda_A - \lambda_Q)$$

$$\cos \alpha = \sin \phi_P \sin \phi_A + \cos \phi_P \cos \phi_A \cos (\lambda_A - \lambda_Q) \text{ and from above,}$$

$$= \sin \phi_A/\sin \phi_P$$

Thus,

$$\sin \phi_A = \sin \phi_A \sin^2 \phi_P + \sin \phi_P \cos \phi_P \cos \phi_A \cos (\lambda_A - \lambda_Q)$$

$$\sin \phi_P \cos \phi_P \cos \phi_A \cos (\lambda_A - \lambda_Q) = \sin \phi_A (1 - \sin^2 \phi_P) = \sin \phi_A (\cos^2 \phi_P)$$

Rearranging, $\tan \phi_P = \tan \phi_A \sec (\lambda_A - \lambda_Q)$, which with $\sin \alpha = \cos \phi_A \sin(\lambda_A - \lambda_Q)$ means that we can calculate the values of $QP (= \phi_P)$ and $PA (= \alpha)$.

EXAMPLE 11.1: THE MERCATOR AND TRANSVERSE MERCATOR

With the Mercator projection,

$$E = R \lambda \quad \text{and} \quad N = R \log_e \tan (\pi/4 + \phi/2)$$

using the 100° west meridian and the equator as the origin, for a point A at latitude 40° north and longitude 103° west and a globe of radius 100 units

$$\lambda = 3° = 0.05236 \text{ radians}; \qquad \tan (\pi/4 + \phi/2) = 2.1445$$

Hence, the coordinates of A would be

$$E = 5.236 \qquad N = 76.291$$

The scale at A would be $\sec (40°) = 1.3054$.

With the Transverse Mercator, in Figure 11.6, angle $AOT = \phi_A = 40°$; angle $PNA = (\lambda_A - \lambda_P) = 3°$. Let us call angle $POA = \alpha$.

We have shown that

$$\sin \alpha = \cos \phi_A \sin \lambda = 0.04009. \text{ Hence, } \alpha = 2.3°$$

Also,

$$\tan \phi_P = \tan \phi_A \sec (\lambda_A - \lambda_P) = 0.\,84025. \text{ Hence, } \phi_P \approx 40.\,04°$$

From Box 11.1,

$$E = R \log_e (\sec \alpha + \tan \alpha) \qquad N = R \phi_P$$

Hence, the coordinates of A would be

$$E = 4.011 \qquad N = 69.881$$

The scale at $A = \sec \alpha = 1.0008$.

Note: Along the central meridian the scale is $\sec (0) = 1$ exactly. Our example shows that at 3° east or west the scale is 1.0008. If we apply an overall scale factor of 0.9995 to all our measurements, this reduces to 1.0003. In the Universal Transverse Mercator (UTM) where the Earth is mapped in strips 6° wide, this results in scales always being close to unity.

For the transverse cylindrical equidistant projection (also known as the *Cassini projection*), then the northings of point A are plotted as $N = R \, \emptyset_P$ and while for the eastings, $E = R \, \alpha$. The Cassini projection was much used in early days for topographic mapping. Similarly, if we let $E = R \sin \alpha$, then we have the transverse cylindrical equal area. If, however, we make

$$E = R \log_e (\sec \alpha + \tan \alpha) = R \log_e \tan(\pi/4 + \alpha/2)$$

then we have an orthomorphic projection, known as the *Transverse Mercator* (see Example 11.1).

By extending this principle to an ellipsoid rather than to a sphere and then mapping the globe in strips 3° either side of the central meridians that are chosen every 6° of longitude, a picture of the globe can be built up in what is known as the *Universal Transverse Mercator* or UTM. This was originally developed in the 1940s by the U.S. military and has been widely adopted around the world, especially for topographic mapping.

The Transverse Mercator is the most commonly used projection in surveying and is normally related to the spheroid, rather than the sphere, as this gives the best mathematical representation of the shape of the Earth. As we saw in Chapter 9, the geometry of the ellipse and ellipsoid is more complex than that of the circle and sphere. In essence, there are two radii of curvature that must be considered namely, ρ and ν. Further discussions of this, and how they affect map transformations, is outside the scope of the present book. Likewise, the summary in Box 11.1 relates to spherical rather than ellipsoidal transformations.

BOX 11.1   CYLINDRICAL PROJECTIONS FOR A SPHERE

For the cylindrical equidistant or *plate carrée*

$$E = R \, \lambda \quad \text{and} \quad N = R \, \emptyset$$

For the cylindrical equal area

$$E = R \, \lambda \quad \text{and} \quad N = R \sin \emptyset$$

For the cylindrical orthomorphic (Mercator)

$$E = R \, \lambda \quad \text{and} \quad N = R \log_e (\sec \emptyset + \tan \emptyset)$$

For the transverse cylindrical, for point A and central meridian M

$$\tan \emptyset_P = \tan \emptyset_A \sec (\lambda_A - \lambda_M), \sin \alpha = \cos \emptyset_A \sin(\lambda_A - \lambda_M)$$

For the transverse cylindrical equidistant (the Cassini projection)

$$E = R \, \alpha \quad \text{and} \quad N = R \, \emptyset_P$$

For the transverse cylindrical orthomorphic (the Transverse Mercator)

$$E = R \log_e (\sec \alpha + \tan \alpha) \quad \text{and} \quad N = R \, \emptyset_P$$

11.3   AZIMUTHAL PROJECTIONS

An azimuthal projection is one where the surface onto which points are projected is a plane that is tangential to the sphere. One of the easiest azimuthal projections to imagine is the polar case (known as the *zenithal projection*, shown in Figure 11.7) in which either pole (here designated N) becomes the center, with the meridians radiating outward as straight lines. The parallels become circles centered around N, their radius depending on the characteristics that we wish to preserve.

On the sphere, the distance along the surface from the pole to the parallel = $R(\pi/2 - \phi) = R\chi$ where ϕ is the latitude of point P and χ = colatitude (the angle as measured from the pole, rather than from the equator). This distance plots on the flat as the length r, the value of which depends on the characteristics of the projection. The angles λ representing the meridians are plotted as their values on the Earth's surface, that is, they are true to scale.

Consider a small triangle of width $\delta\lambda$ and height $\delta\phi$ on the sphere and on the projection. For the zenithal equidistant, circles are drawn around the pole with radius $r = R(\pi/2 - \phi) = R\chi$. For the equal area projection, we have to make the two elemental triangles in Figure 11.7 the same area: hence,

$$(1/2) R^2 \sin\chi \; \delta\chi \; \delta\lambda = (1/2) \; r \; \delta r, \quad \text{or} \quad \int R^2 \sin\chi \; d\chi = \int r \; dr$$

This gives

$$C - R^2 \cos \chi = (1/2)r^2$$

where C is some constant.

Since $r = 0$ at the center where $\chi = 0$ and $\cos 0 = 1$, then $C = R^2$. Rearranging, $r^2 = 2R^2(1 - \cos \chi)$ and this equals $4R^2 \sin^2 (\chi/2)$ as shown in Chapter 5: hence,

$$r = 2R \sin (\chi/2) \text{ for the equal area projection.}$$

For the zenithal orthomorphic projection, in order to preserve shape then from Figure 11.7

$$R \sin\chi \; \delta\lambda/R\delta\chi = r \; \delta\lambda/\delta r$$

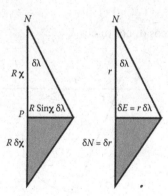

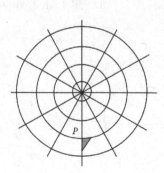

On Sphere of Radius R with
Colatitude $\chi = \pi/2 - \phi$

Projection of Elemental Triangle for Latitude
ϕ and Radius r

FIGURE 11.7   Zenithal projection.

or

$$(1/\sin \chi) \, d\chi = (1/r) \, dr$$

Integrating,

$$\log_e (\tan \chi/2) + \text{constant} = \log_e r$$

If we call the constant $\log_e C$, then

$$\log_e (C \tan \chi/2) = \log_e r$$

or

$$r = C \tan(\chi/2)$$

The value of C determines the scale overall and is commonly set so that $C = 2R$. This projection is often known as the *stereographic projection*.

The relationships established above can also be applied to the projection onto a plane that touches the sphere at any point P, not just the pole N. The formulae modified as χ and λ used above take on different meanings. The angle NPA in Figure 11.8a,b represents the difference in longitude so we must replace λ by Ω (omega) while the angle PA is now what was the colatitude, so we must replace χ by Ψ (psi). The values can be calculated from the sine and cosine formulae for spherical triangles.

In Figure 11.8b, from the sine formula:

$$\sin \Omega/\sin (\pi/2 - \phi_A) = \sin \lambda/\sin \Psi \text{ or, since } \sin (\pi/2 - \phi) = \cos \phi, \text{ then}$$

$$\sin \Omega \sin \Psi = \sin \lambda \cos \phi_A$$

From the cosine formula

$$\cos \Psi = \cos (\pi/2 - \phi_A) \cos (\pi/2 - \phi_P) + \sin (\pi/2 - \phi_A) \sin (\pi/2 - \phi_P) \cos \lambda$$

or

$$\cos \Psi = \sin \phi_A \sin \phi_P + \cos \phi_A \cos \phi_P \cos \lambda$$

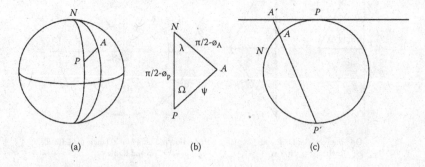

(a)   (b)   (c)

FIGURE 11.8   Oblique azimuthal.

BOX 11.2   ZENITHAL PROJECTIONS

If the colatitude is $\chi = 90 - \phi$, then

For the zenithal equidistant, $r = R \chi$

For the zenithal equal-area, $r = 2R \sin (\chi/2)$

For the zenithal orthomorphic (or stereographic), $r = 2R \tan (\chi/2)$.

Hence, we can find Ψ and from that we can find Ω. The results of all azimuthal projections for a sphere are summarized in Box 11.2.

The stereographic projection is equivalent to projecting onto a plane from the point on the sphere diametrically opposite point P shown as P' in Figure 11.8c. See Example 11.2. Although the plane onto which the projection is made is assumed in Example 11.2 to be tangential to the sphere at P, it can be any plane parallel to the tangent plane at P (for instance, through the center O) provided it is not through the point diametrically opposite to P. It has the characteristic that a circle on the surface of the Earth plots as a circle on the projection. Hence, when looking for the epicenter of an earthquake, for example, we can plot the time when shock waves were recorded and this will give an indication as to where the center of the earthquake lay.

EXAMPLE 11.2: OBLIQUE STEREOGRAPHIC PROJECTION

For the stereographic projection of a sphere based on point P at 40° North and 100° West, (see Figure 11.8a,b), then for a point A at 45° North and 103° West, using a sphere of radius $R = 100$

$$\phi_A = 45; \qquad \phi_P = 40; \qquad \lambda = 3$$

Using

$$\cos \Psi = \sin \phi_A \sin \phi_P + \cos \phi_A \cos \phi_P \cos \lambda = 0.99545, \quad \text{then} \quad \Psi \approx 5.466°$$

Using

$$\sin \Omega \sin \Psi = \sin \lambda \cos \phi_A, \text{ then } \sin \Omega = 0.38848 \text{ and } \Omega \approx 22.86°$$

Thus, the distance $PA = 2R \tan (\Psi/2) = 9.5478$.

Relative to the origin P,

$$E = PA \sin \Omega = 3.709; \qquad N = PA \cos \Omega = 8.798$$

Now consider point P that is diametrically opposite P on a sphere in Figure 11.8c. Let the vector PA meet the tangent plane at P at point A. In geocentric coordinates

$$P \text{ is } (R\cos \phi_P, 0, R\sin\phi_P) = (76.60444, 0, 64.27876)$$

$$A \text{ is } (R \cos \phi_A \cos \lambda, R \cos \phi_A \sin \lambda, R\sin \phi_A) = (70.61377, 3.70071, 70.71068).$$

Now rotate about the whole configuration about the y-axis by an amount $(90° - \phi_P)$ so that point P in effect becomes the North Pole. This means applying the matrix

$$\begin{pmatrix} \cos \phi_P & 0 & \sin \phi_P \\ 0 & 1 & 0 \\ -\sin \phi_P & 0 & \cos \phi_P \end{pmatrix}$$

The coordinates of P become $(R, 0, 0)$ while A becomes

$$x_A = R \cos \phi_A \cos \phi_P \cos \lambda + R\sin \phi_A \sin \phi_P; = 99.545$$

$$y_A = R \cos \phi_A \sin \lambda; = 3.701$$

$$z_A = -R \cos \phi_A \sin \phi_P \cos \lambda + R\sin \phi_A \cos \phi_P = 8.778$$

The center O remains $(0, 0, 0)$ while P' becomes $(-100, 0, 0)$.

We now project point A onto the tangent plane at P from P' so that the x value for A' becomes 100, as shown in Figure 11.8c. This means extending $P'A$ to $P'A'$ or from 199.545 to 200.

Thus,

$$y_A \text{ becomes } E = 3.701 * 200/199.545 = 3.709$$

$$z_A \text{ becomes } N = 8.778 * 200/199.545 = 8.798$$

These are the same values as we achieved earlier, showing that the stereographic projection is a view of the world from P', which is called the *antipode* of point P.

11.4 CONICAL PROJECTIONS

The third group of projections is known as *conical*. They are based on the idea of a cone wrapped around the midlatitudes, it being possible to cut a cone so that its surface can be uncurled onto a flat piece of paper. The cone may touch the sphere—in which case the parallel of latitude where this happens is known as the *standard parallel*—or cut the sphere in which case there will be *two standard parallels*.

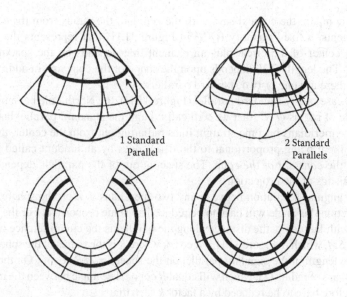

FIGURE 11.9 Conical projection with one or two standard parallels.

When the cone is unwrapped and laid flat (Figure 11.9), each parallel of latitude becomes a part of a set of concentric arcs while the meridians are lines that radiate from the center. The pole also becomes a circle, whose radius will depend upon the extent to which the cone is pointed.

In fact, the cylindrical and zenithal projections are special cases of the conical with the angle of the cone being zero for the cylindrical and 180° for the zenithal; if QN in Figure 11.10 is zero, the cone will have become a plane while if QN is infinite in length, then the cone will have become a cylinder.

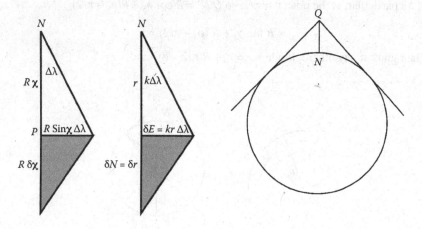

FIGURE 11.10 Elemental triangles for conical projections.

If we compare the conical case with the zenithal, the radius from the center has two elements, a fixed quantity ($Q'N'$ in Figure 11.11) that represents the distance from the center of the circle plus an element that depends on the spacing of the parallels. The length $Q'N'$ depends upon the shape of the cone and in addition upon whether there are one or two standard parallels.

In the case of any standard parallel (Figure 11.11), the North Pole becomes an arc of a circle of radius $Q'N'$ (that we will call r_0) and the standard parallel has correct scale. The meridians become straight lines radiating out from the center, the angles between them being proportional to the true values by an amount called k, which is called the *constant of the cone*. The spacing out of the parallels depends on the characteristics of the projection.

If the angular separation between any two meridians is λ on the sphere, then on the projection this angle will have to be reduced by some proportion. For the standard parallel with latitude ϕ_s the distance along the parallel is the circumference of a circle of radius SM, which in Figure 11.11 $= R \cos \phi_S$ where R is the radius of the sphere. Thus, the whole length of the standard parallel on the sphere $= 2\pi R \cos \phi_S$. On the flat this will be an arc of radius QS, which will equal $R \cot \phi_S$. The angle between the meridians will therefore have to be reduced by a factor k such that

$$K * 2\pi * R \cot \phi_S = 2\pi R \cos \phi_S = \text{the length of the standard parallel}$$

Thus, $k = \sin \phi_S = \cos \chi_S$ where χ is the colatitude or $(\pi/2 - \phi)$. A difference in longitude of λ becomes $k \lambda$ on the projection so $\theta = k \, \delta\lambda$ (Figure 11.11).

For the conical equidistant with one standard parallel, we need to keep the parallels correctly spaced. Then for any parallel of colatitude χ, the radius of the circular arc that appears on the projection will be QS + the distance along the meridian from the standard parallel $= QS + R(\phi_S - \phi)$. But, $QS = R \cot \phi_s$. Thus, if the distance from the center $Q' = r$, then

$$r = R \cot \phi_S + R(\phi_S - \phi) = R \tan \chi_S + R(\chi - \chi_S)$$

In particular, at the pole, $\phi = \pi/2$, so $Q'N' = R \cot \phi_S + R(\phi_S - \pi/2)$

$$= R \tan \chi_S + R (\phi_S - \pi/2) = r_0$$

The radius at any other latitude $= r = r_0 + R(\pi/2 - \phi)$.

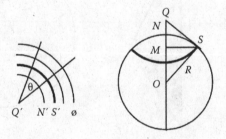

FIGURE 11.11   Conical equidistant with one standard parallel.

For the case where there are two standard parallels ϕ_1 and ϕ_2, then the length of each parallel on the sphere will be such that $k\, r = R \cos \phi = R \sin \chi$ where k is the *constant of the cone* and χ is the colatitude. For the two standard parallels

$$k\{r_0 + R(\chi_1)\} = R \sin \chi_1$$

$$k\{r_0 + R(\chi_2)\} = R \sin \chi_2$$

After suitable manipulation, we obtain

$$r_o = R\, (\chi_2 \sin \chi_1 - \chi_1 \sin \chi_2)/(\sin \chi_2 - \sin \chi_1)$$

$$r = r_0 + R\chi$$

For the conical equal area projection, the two elemental triangles must be of equal area; hence,

$$(1/2)\, R\, d\chi\, R \sin \chi\, \delta\lambda = (1/2)\, k\, r\, \delta\lambda\, \delta r \quad \text{or} \quad R^2 \sin \chi\, d\chi = k\, r\, dr$$

Integrating

$$C - 2R^2 \cos\chi = kr^2$$

where C is the constant of integration and k is not yet determined. Thus,

$$kC - 2kR^2 \cos \chi = k^2 r^2$$

Along a standard parallel, distances on the sphere are the same as on the projection. Hence, along a standard parallel

$$kr = R \sin \chi$$

If there are two standard parallels (1 and 2), then

$$kC - 2kR^2 \cos \chi_1 = R^2 \sin^2\chi_1 \quad \text{and} \quad kC - 2kR^2 \cos \chi_2 = R^2 \sin^2\chi_2$$

Subtracting the two equations, and dividing,

$$2k = (\sin^2\chi_2 - \sin^2\chi_1)/(\cos \chi_1 - \cos \chi_2)$$

But,

$$\sin^2 = 1 - \cos^2; \text{ so } (\sin^2\chi_2 - \sin^2\chi_1) = (\cos^2\chi_1 - \cos^2\chi_2)$$

$$= (\cos \chi_1 + \cos \chi_2)\, (\cos \chi_1 - \cos \chi_2)$$

Hence,

$$k = (1/2)\, (\cos \chi_1 + \cos \chi_2) = (1/2)\, (\sin \phi_1 + \sin \phi_2)$$

where ϕ_1 and ϕ_2 are the latitudes rather than colatitudes of the standard parallels. Substituting for k in the $C - 2R^2 \cos \chi = kr^2$, then

$$C = R^2(1 + \sin \phi_1 \sin \phi_2)/k = R^2(1 + \cos \chi_1 \sin \chi_2)/k$$

From this, r^2 can be calculated and hence r derived.

The derivation of the values of C and r for the conical orthomorphic follows the same principles as those for the stereographic or zenithal orthomorphic projection. From the elemental triangles in Figure 11.10 the angles must be preserved: hence,

$$R \, d\chi/(R \sin \chi \, \delta\lambda) = \delta r/(k \, r \, \delta\lambda)$$

or

$$k \, \text{cosec} \, \chi \, d\chi = (1/r) \, dr$$

On integrating $k \log_e \{\tan(\chi/2)\} + C = \log_e r$.

After suitable manipulation and the insertion of the values for the standard parallels χ_1 and χ_2, it can be shown that

$$k = (\log_e \sin \chi_1 - \log_e \sin \chi_2)/\{\log_e \tan(\chi_1/2) - \log_e \tan(\chi_2/2)\}$$

$$r = R\left(\frac{\sin \chi_1}{k}\right) * \left(\tan\left(\frac{\chi}{2}\right) \Big/ \tan\left(\frac{\chi_1}{2}\right)\right)^k$$

The results are summarized in Box 11.3.

BOX 11.3   CONICAL PROJECTIONS OF THE SPHERE

For the conical equidistant with one standard parallel ϕ_s, the radius of each arc of the projection = r where

$$r = r_0 + R(\pi/2 - \phi)$$

where

$$r_0 = R \{\cot \phi_s + (\phi_s - \pi/2)\}$$

The meridians converge at a point and if the angle of longitude between two meridians is λ on the sphere and θ on the projection:

$$\theta = (\sin \phi_s) \, \lambda$$

For the conical equidistant, with two standard parallels whose colatitudes are χ_1 and χ_2,

$$r = R(\chi_2 \sin \chi_1 - \chi_1 \sin \chi_2)/(\sin \chi_2 - \sin \chi_1) + R\chi$$

For the conical equal-area,

$$k = (1/2) (\cos \chi_1 + \cos \chi_2); \quad C = R^2(1 + \cos \chi_1 \sin \chi_2)/k; \quad r = \sqrt{\{C/k - 2R^2 \cos \chi/k\}}$$

For the conical orthomorphic,

$$k = (\log_e \sin \chi_1 - \log_e \sin \chi_2)/(\log_e \tan(\chi_1/2) - \log_e \tan(\chi_2/2))$$

$$r = R\{(\sin \chi_1)/k\}\{\tan(\chi/2)/\tan(\chi_1/2)\}^k$$

$$= R\left(\frac{\sin\chi_1}{k}\right) * \left(\tan\left(\frac{\chi}{2}\right) \Big/ \tan\left(\frac{\chi_2}{2}\right)\right)^k$$

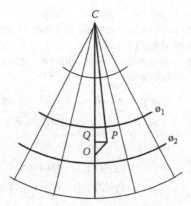

FIGURE 11.12 A conical projection with two standard parallels.

When working on the ellipsoid, the principles remain the same although the value of R, the radius of the Earth, will be replaced by v, derived from the normal to the ellipse (see Figure 9.4 in Chapter 9) and the latitudes will be geodetic latitudes rather than geocentric.

In Figure 11.12, O is the origin for the coordinates of the projection, here located between the two standard parallels $ø_1$ and $ø_2$ (or colatitude χ_1 and χ_2). On the map, let the coordinates of O be (x_O, y_O) and of point P be (x_P, y_P), where P is any point to be plotted on the map. C is the point where the meridians converge in a conical projection.

On the ellipsoid, let O have latitude $ø_O$ and longitude λ_O, and P have latitude $ø_P$ and longitude λ_P.

On the projection, let $CP = r$; $CO = r_O$; and angle $OCP = k(\lambda - \lambda_O)$ where λ is the longitude of P relative to the origin (here OC). If Q is the foot of the perpendicular from P onto the line OC in the projection, then

$$x_P = x_O + QP = x_O + r \sin \{k(\lambda - \lambda_O)\}$$

$$y_P = y_O + OQ = y_O + r_O - r \cos \{k(\lambda - \lambda_O)\}$$

To plot our point P on the conical projection we need to know the formulae for r and k. If we want our projection to be conformal (i.e., orthomorphic), then if the Earth were a sphere:

$$k = (\log_e \sin \chi_1 - \log_e \sin \chi_2)/(\log_e \tan(\chi_1/2) - \log_e \tan(\chi_2/2))$$

$$r = R\left(\frac{\sin \chi_1}{k}\right) * \left(\tan\left(\frac{\chi}{2}\right) \Big/ \tan\left(\frac{\chi_2}{2}\right)\right)^k$$

for which R is the radius of the Earth to which a suitable scale factor must be applied.

If we are working on the best-fit ellipsoid that represents the surface of the Earth, then the formulae need some modification. The actual values will depend on which ellipsoid is chosen and that tends to be a matter of national preference. The key elements that need defining are the radius at the equator R and the flattening that is either defined as the ratio (a/b) or as the value of e (see Chapter 4, Section 4.6). Here we will use e with $R = a$ on the ellipsoid.

The formulae for the conformal conical with two standard parallels, known as *Lambert's projection*, then become:

For k, instead of $\quad\quad$ $\sin \chi_1$ we need $\cos \phi_1/\sqrt{(1 - e^2\sin^2\phi_1)}$

$$\text{Call this} = n_1 = \cos \phi_1/\sqrt{(1 - e^2\sin^2\phi_1)}$$

$$\sin \chi_2 \quad\quad\quad n_2 = \cos \phi_2/\sqrt{(1 - e^2\sin^2\phi_2)}$$

$$\tan(\chi_1/2) \quad\quad n_3 = \tan(\chi_1/2)/\{(1 - e\sin \phi_1)/(1 + e\sin \phi_1)\}^{e/2}$$

$$\tan(\chi_2/2) \quad\quad n_4 = \tan(\chi_2/2)/\{(1 - e\sin \phi_2)/(1 + e\sin \phi_2)\}^{e/2}$$

$$k = (\ln n_1 - \ln n_2)/(\ln n_3 - \ln n_4)$$

(Here, we write "ln" to mean the natural logarithm, which is the same as $\log_e$ which we avoid, so as not to confuse the "e" of natural logarithms with the "e" for the ellipsoid.)

The formula for r must be modified in a similar way:

$$\sin \chi_1 \text{ is still} \quad\quad n_1 = \cos \phi_1/\sqrt{(1 - e^2\sin^2\phi_1)}$$

$$\tan(\chi_1/2) \quad\quad n_3 = \tan(\chi_1/2)/\{(1 - e\sin \phi_1)/(1 + e\sin \phi_1)\}^{e/2}$$

$$\tan(\chi/2) \quad\quad n = \tan(\chi/2)/\{(1 - e\sin \phi)/(1 + e\sin \phi)\}^{e/2}$$

$$r = R * (n_1/k) * \{n/n_3\}^k$$

The formulae are summarized in Box 11.4. For a worked example, see Example 11.3.

BOX 11.4   LAMBERT'S PROJECTION OF THE ELLIPSOID

Let the radius at the equator be R, the radius of the arcs for the conical projection of the parallels of latitude by r from the central point, and let the longitudes be modified by the factor k. Let the standard parallels be ϕ_1 and ϕ_2 and their colatitudes be χ_1 and χ_2.

Let $n_1 = \cos \phi_1/\sqrt{(1 - e^2\sin^2\phi_1)}$; $n_3 = \tan(\chi_1/2)/\{(1 - e\sin \phi_1)/(1 + e\sin \phi_1)\}^{e/2}$

$n_2 = \cos \phi_2/\sqrt{(1 - e^2\sin^2\phi_2)}$; $n_4 = \tan(\chi_2/2)/\{(1 - e\sin \phi_2)/(1 + e\sin \phi_2)\}^{e/2}$

Then

$$k = (\ln n_1 - \ln n_2)/(\ln n_3 - \ln n_4)$$

Let $n = \tan(\chi/2)/\{(1 - e\sin \phi)/(1 + e\sin \phi)\}^{e/2}$

Then

$$r = R * (n_1/k) * \{n/n_3\}^k$$

EXAMPLE 11.3: LAMBERT'S CONICAL ORTHOMORPHIC WITH TWO STANDARD PARALLELS

Consider an ellipsoid with $a = R = 6378206.400$ meters or 20925832.16 U.S. geodetic feet and $e = 0.08227185$. (This is known as the *Clarke 1866 ellipsoid*.) Let the origin O be at latitude 45° N and 100° W, and the standard parallels be 40° N and 50° N. Let point P be at 47° N and 95° W.

Then using the formulae $k = (\ln n_1 - \ln n_2)/(\ln n_3 - \ln n_4)$

$$r = R * (n_1/k) * \{n/n_3\}^k$$

$n_1 = 0.64406800504$ $\qquad$ $n_2 = 0.76711787277$

$n_3 = 0.36586487239$ $\qquad$ $n_4 = 0.46834279637$

$n_P = 0.39586764826$ $\qquad$ $n_O = 0.41620305983$

From these values

$$k = 0.708020879$$

$r_P = 20128131.243$ feet $\qquad$ $r_o = 20854828.952$ feet

$$\{k(\lambda - \lambda_o)\} = 0.06178648 \text{ radians}$$

Let the coordinates of the origin in Figure 11.12 be (2000000, 0) using U.S. geodetic feet.

Then,

$$x_P = 3242855.245 \qquad y_P = 765105.750$$

In addition to these basic projections, there are various alternatives that compromise between the preservation of various angles, small shapes, areas, and distances. The reader should consult other texts for the details (see "Further Reading").

SUMMARY

Antipode: The point diametrically opposite any given point on the surface of the Earth.

Authalic projection: An equal area projection.

Azimuthal projection: A *map projection* based on a plane tangential to the surface of the Earth.

Cassini projection: The transverse cylindrical equidistant projection.

Central meridian: The meridian around which a *transverse cylindrical projection* is wrapped.

Conformal: Another name for an orthomorphic projection.

Conical projection: A *map projection* based on a cone that touches or cuts the surface of the Earth and whose vertex is usually taken as being above the North or South Pole.

Constant of the cone: The factor used in *conical projections* to determine the radius of circles representing the parallels of latitude.

Cylindrical projection: A *map projection* based on a cylinder wrapped around the equator of the Earth.

Elemental triangle: A small triangle on the surface of the globe defined by the increments $\delta\lambda$ and $\delta\phi$, both of which tend to zero.

Equal-area projection: Also known as *equivalent* or *authalic*. A projection that preserves area by contracting distances in one direction and stretching them in the direction at right angles.

Equidistant projection: Also known as *equirectangular*, a projection that retains the correct lengths along the two axes, which are usually the parallels and meridians.

Equivalent: Another name for the *equal area* projection.

Loxodrome: Also known as a *rhumb line*. A line of constant bearing.

Map projection: A method for transforming the location of features on the curved surface of the Earth onto a flat or plane surface.

Mercator's projection: The *cylindrical orthomorphic* projection. It has the characteristic that straight lines on the map are lines of constant bearing.

Orthomorphic projection: Also known as a *conformal* projection. A projection that preserves the shape of small areas by maintaining scale along short distances around that point. Scale may, however, be different at different points.

Rhumb line: Also known as a *loxodrome*. A line of constant bearing.

Simple cylindrical: The simplest projection based on plotting north–south distances and east–west distances correctly. It introduces very large distortions east–west when nearing the North or South Pole.

Standard parallel: The parallel of latitude where the encircling cone touches the sphere or, in the case of two standard parallels, it intersects the sphere (or ellipsoid).

Stereographic projection: The *orthomorphic* version of the *azimuthal* projection. It has the characteristic that circles on the Earth plot as circles on the projection.

Transverse cylindrical projection: A projection based on a cylinder wrapped around a meridian of longitude.

Transverse Mercator: An *orthomorphic* projection based on a cylinder wrapped around a meridian of longitude. It is often used in topographic mapping.

UTM: The Universal Transverse Mercator projection. It is based on a series of central meridians at six-degree intervals around the globe, extending three degrees either side.

Zenithal projection: An *azimuthal projection* based on one or other pole. The meridians radiate out from the center as straight lines and the parallels are circles around that center.

12 Basic Statistics

12.1 PROBABILITIES

So far we have dealt with matters for which there is at least in theory an exact answer—curves and surfaces have been fitted through points and values such as sin ø have been calculated to as many significant figures as were necessary. In many circumstances, however, things are not exact. Every measure can be subject to a small amount of error, which means that we must look for a "best-fit" solution.

Many measures are of phenomena for which there are no precise answers, such as measures of human preference as in estimates of how people may vote in an election prior to the actual event. We need some way of estimating the reliability of any measurement and some way of handling inconsistencies. In other words, we need statistical measures where a *statistic* is some function of random variables that can be used as an estimate of a population. A *population* (sometimes referred to as a *universe*) is the complete set of individual components or events from which *samples* are drawn (Figure 12.1). The sample should be chosen so that its characteristics mimic those of the whole population. If it does not, then the sample will be biased.

Descriptive statistics are sets of numbers used to summarize a set of known data in a clear and concise way while *inferential statistics* result from the theory and practice of using statistical data to draw conclusions from random samples. Both processes are *heuristic,* meaning that they are guided by experience and experiment rather than by rigorous logical argument from precisely defined axioms.

The word *random* means that the value of the item under consideration cannot be predetermined other than in terms of the probability that it may occur; in particular, it cannot be determined by what has happened before. Thus, a *random number* is one that cannot be determined from any of the previous numbers that have been selected and, therefore, it does not follow any particular regular or repetitive pattern. In a series of random numbers, for example, each number is as likely to occur as often as any other in the series. If this does not happen, then the data are biased.

Processes that can be described by random variables are said to be *stochastic* and are described in terms of *probability,* which is a measure of the degree of confidence that can be had in any event. Thus, when tossing an unbiased coin, there is a 50% probability that it will come down head-side up and 50% that it will be "tails."

If we consider the toss of a coin, we can describe the probability of the outcome as $(ph + qt)$ where h is simply a dummy term meaning "heads" and t is a dummy meaning "tails"; p is the chances out of 1 that the main event will happen and q is the chances that it will not happen. $q = (1 - p)$ since it is 100% certain that either it will or will not happen. In the case of tossing an unbiased coin, $p = q = (1/2) = 0.5$.

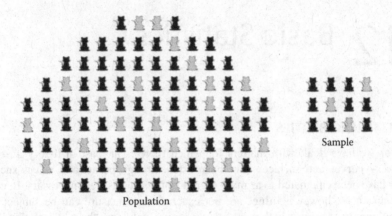

Sample

Population

FIGURE 12.1   Population and sample.

Thus, after one throw, we have $(0.5h + 0.5t) = (h + t)/2$. If we repeat the experiment a second time, we can multiply the probabilities and obtain

$$\left\{\frac{h+t}{2}\right\} * \left\{\frac{h+t}{2}\right\} = \left\{\frac{h^2 + 2ht + t^2}{4}\right\}$$

This tells us that out of four possible outcomes (the denominator) there is one chance that we have heads twice (h^2), two chances of having either a head and a tail or a tail and a head ($2ht$), and one chance of having two tails (t^2). If we repeat the process a third time, we obtain

$$\left(\frac{h+t}{2}\right)^3 = \left(\frac{h^3 + 3h^2t + 3ht^2 + t^3}{8}\right) = \left(\frac{h^3t^0 + 3h^2t^1 + 3h^1t^2 + h^0t^3}{8}\right)$$

Writing this as $(h^3t^0 + 3h^2t^1 + 3h^1t^2 + h^0t^3)/8$ tells us that after three throws, there are eight possible outcomes, there being one chance in eight of obtaining three heads (h^3) with no tails (t^0), three chances of there being two heads and one tail ($3 * h^2t^1$), three chances of there being one head and two tails ($3 * h^1t^2$), and one chance out of eight of obtaining three tails (t^3) with no heads (h^0).

Note that in the expression $(h^3t^0 + 3h^2t^1 + 3h^1t^2 + h^0t^3)/8$, the sum of the coefficients of the terms $h^m t^n$ is $(1 + 3 + 3 + 1) = 8$, which is the same as the denominator. This confirms that the total probability $= 8/8$ which is precisely one. In each component $h^m t^n$ the sum of the indices $= m + n$ is the same as the number of events under consideration; here, after three throws, it equals 3.

For four throws,

$$\left(\frac{h+t}{2}\right)^4 = (h^4t^0 + 4h^3t^1 + 6h^2t^2 + 4h^1t^3 + h^0t^4)/16$$

BOX 12.1   THE BINOMIAL EXPANSION

$$(p + q)^n = p^n + np^{(n-1)}q + \{n\,(n - 1)/2)\}p^{(n-2)}q^2$$
$$+ ... + {_nC_r}\,p^{(n-r)}q^r + \cdots + npq^{(n-1)} + q^n$$

where

$$_nC_r = \frac{n!}{(n-r)!\,r!}$$

---

and again $(1 + 4 + 6 + 4 + 1)/16 = 1$. The indices of h and t all add up to 4. There is one chance in 16 of obtaining four heads, four chances of obtaining three heads, six of obtaining two heads, four of obtaining only one head, and one chance in 16 of there being no heads at all.

The expansion of $\left(\frac{h+t}{2}\right)^n$ is known as the *binomial expansion* (Box 12.1) and results in

$$\{h^n t^0 + {_nC_1}h^{(n-1)}t^1 + {_nC_2}h^{(n-2)}t^2 + \cdots + {_nC_{(n-1)}}h^1 t^{(n-1)} + h^0 t^n\}/2^n$$

where $_nC_r$ is a shorthand way of writing the number $\frac{n!}{(n-r)!\,r!}$ and $n! = 1 * 2 * 3...*$ $(n - 1) * n$. For instance, $_5C_2 = \frac{5*4*3*2*1}{(3*2*1)*(2*1)} = 10$.

The sum of the indices of h and t or p and $q = n$. The sum of the coefficients in the numerator divided by the denominator $= 1$. If, for example, $n = 100$, we can calculate the chances that on a purely random basis all the 100 throws of the coin will be heads. The chances of this are $1/2^{100}$—pretty unlikely! The chances of getting 99 heads will be $100/2^{100}$ while for 98 heads they will be $4950/2^{100}$ where $4950 = \frac{100*99}{1*2}$.

We can extend the use of the binomial expansion where the probabilities are not 50–50. As an example, all other things being equal, the chances that a Wednesday will be the most rainy day in the week will be one in seven. The probability is $(pw + q)$ where $p = 1/7$ and $q = 6/7$ and w is a dummy to represent "Wednesday." For two weeks we have $(pw + q) * (pw + q)$ or for three weeks $(pw + q)^3$. Hence, the chance that a Wednesday is the most rainy day in the week for three weeks running is

$$\left(\frac{(w+6)}{7}\right)^3 = \left(\frac{w^3}{343} + \frac{18w^2}{343} + \frac{108w}{343} + \frac{216}{343}\right)$$

This means there is one chance in 343 that Wednesday will be the most rainy day of the week for three weeks, 18/343 chances that it happens twice, 108/343 chances that it happens once, and 216/343 chances that no Wednesdays are the most rainy. (Note that $1 + 18 + 108 + 216 = 343$.)

The coefficients in the expansion of $(p + q)^n$ are given in Box 12.2. For instance, the second line, (1 2 1) indicates $1 * p^2 + 2 * p^1 q^1 + 1 * q^2$ while in the fifth line, (1 5 10 10 5 1) means that

$$(p + q)^5 = p^5 + 5p^4 q^1 + 10p^3 q^2 + 10\,p^2 q^3 + 5p^1 q^4 + q^5$$

BOX 12.2   THE BINOMIAL COEFFICIENTS

$n =$	Coefficients in $(p + q)^n$
1	1 1
2	1 2 1
3	1 3 3 1
4	1 4 6 4 1
5	1 5 10 10 5 1
6	1 6 15 20 15 6 1
7	1 7 21 35 35 21 7 1
8	1 8 28 56 70 56 28 8 1
9	1 9 36 84 126 126 84 36 9 1
10	1 10 45 120 210 252 210 120 45 10 1
11	1 11 55 165 330 462 462 330 165 55 11 1
12	1 12 66 220 495 792 924 792 495 220 66 12 1
13	1 13 78 286 715 1287 1716 1716 1287 715 286 78 13 1
14	1 14 91 364 1001 2002 **3003 3432** 3003 2002 1001 364 91 14 1
15	1 15 105 455 1365 3003 5005 **6435** 6435 5005 3003 1365 455 105 15 1

There is, of course, a pattern in all this whereby the numbers in the lower row are the sum of the two numbers directly above them—for example, the 8th number in row 15 is 6435 = 3003 + 3432, the seventh and eighth numbers in row 14. The number 3003 is itself the sum of the two numbers above it, namely, 1287 + 1716; likewise, 3432 is the sum of 1716 + 1716, and so on. The triangular array of integers shown in Box 12.2 is known as a *Pascal triangle*, named after the French mathematician, Blaise Pascal, who lived in the 17th century.

To obtain the probabilities assuming $p = q = \frac{1}{2}$, for row 15 we must divide all the numbers by $2^{15} = 32768$ so that the whole probability adds up to 1. Hence, dividing through, we obtain 0.00003, 0.00047, 0.0032, 0.0139, 0.0417, 0.0916, 0.1527, 0.1964 with the numbers repeated in reverse order for the remaining eight coefficients, all adding up to 1.0.

12.2   MEASURES OF CENTRAL TENDENCY

In Figure 12.2, the 16 values of the coefficients of p and q have been plotted as a *histogram,* which is a set of contiguous rectangles with width proportional to the size of the class interval and the area proportional to its frequency. They have also been traced in the form of a continuous curve that interpolates the frequency. The chances of there being no heads or 15 heads in 15 throws are 1 in 32,768 or about 0.00003. This means that it is probable that if you carried out 32,768 experiments with tossing a coin 15 times, then you would expect on average one case where there were 15 consecutive heads. But it might not happen or it might happen several times. Nothing is certain. What you would expect is that if you carried out a set of 32,768 experiments a large number of times, then on average you would find 15 consecutive heads once in each set of experiments.

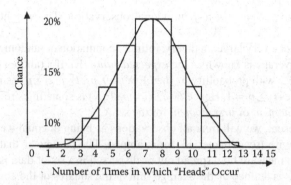

FIGURE 12.2   A plot of equal probability after 15 events.

In Figure 12.2, almost all the area under the curve lies between 3 and 12. The area under the whole curve is made up of all the separate probabilities and must equal 1 or 100%. We achieved this by dividing the original coefficients by 2^{15}. The areas represented by rectangles in Figure 12.2 are proportional to the overall area and hence represent graphically the probability of the number shown in the bottom line occurring; the total area of these boxes (and under the curve) must be exactly 1.

Furthermore, the middle or average value is 7.5. Of course, there cannot be 7.5 heads in a single experiment, since we have specified that the outcome is either heads or tails with a 50% probability of either. But with a series of experiments throwing 15 coins one would expect that in 7.5 cases there would be 8 or more heads and in 7.5 cases there would be 7 or less heads. In general, if there are n attempts at something for which the probability of it happening is p, then the average number of times that it will happen will be $np - n$ times p. Here, $n = 15$, $p = \frac{1}{2}$, so $np = 7.5$.

In producing a general description of the shape of the curve in Figure 12.2, there is a need to know not only the mean value, but also how spread out or dispersed the individual answers are likely to be. There are three parameters that are used to express the central point—the *mean*, the *median*, and the *mode*. The mean is the average value of a set of n observations x_i, calculated as $\sum x_i/n$, meaning "add all the x values together and divide by n." The median value is the middle number while the mode is the one that most frequently occurs.

For example, with two sets of numbers A (1, 5, 24) and B (1, 2, 2, 2, 8, 12, 15, 22), both have a median value of 5 since in the case of B where there is an even number of entities and 5 lies half way between 2 and 8. In B, the mode is 2, while the mean is 8. For data set A, the mean is 10. If the distribution were symmetrical about the middle, then all the values of the mean, median, and mode would coincide.

The average or mean value calculated from $\sum x_i/n$ assumes that each value of x is comparable with every other value. In practice this may not be the case—some observations may be dodgy while others may be more reliable. We may therefore need to introduce a weighting factor w_i to allow for the fact that, for example, one observation is twice as reliable as another. The weight is in effect a relative value of the importance of any individual observation. The weighted mean then becomes $\Sigma w_i x_i / \Sigma w_i$.

If x is part of a very large, in theory, infinite, population of random variables, then its weighted average is known as the *expected value* (E). If a random variable X can take the value x_i with probability p_i, then $E(X) = \Sigma p_i x_i / \Sigma p_i = \Sigma p_i x_i$ since the sum of all probabilities $= \Sigma p_i = 1$. Hence, $w_i / \Sigma w_i = p_i$. $E(X)$ is sometimes referred to as the *first product moment* or *first moment* for the set X.

For the present, we will treat all observations as being of equal weight unless we declare otherwise (see Chapter 13, for example). The parameter that is most commonly used to measure how spread out or dispersed is a set of data is known as the *variance*. This is defined as the average sum of the squares of the amount by which each observation differs from the mean. If an observation has a value x_i and if the Greek letter for m, which is "μ" (pronounced "mu") is used to denote the mean value of the population, then $(x_i - \mu)$ is said to be the *residual* value for x_i.

For n observations, the average value of the sum of the squares of all the residuals is equal to $\Sigma(x_i - \mu)^2/n$ where Σ (the Greek capital letter "sigma") means "the sum for all the values of x_i." The variance is usually represented by σ^2 where the symbol "σ" represents the Greek lower case letter "sigma." So, $\sigma^2 = \Sigma(x_i - \mu)^2/n$.

The square root of the variance, σ, is known as the *standard deviation*. Thus,

$$\sigma = \sqrt{(\Sigma(x_i - \mu)^2/n)} \text{ for the whole population.}$$

The amount by which any observation differs from the mean when measured in units of standard deviation is called the *Z-score* where $z_i = (x_i - \mu)/\sigma$. The mean of a set of z-scores $= 0$ while the variance and the standard deviation for the set of z-scores $= 1$.

We have summarized the important statistics in any given data set in Box 12.3. In Example 12.1 we give a simple example with two sets of measurements and their means and standard deviations.

BOX 12.3   VARIANCE AND STANDARD DEVIATION

For any set of n numbers of equal weight $x_1, x_2, \ldots x_i, \ldots x_n$: The mean value is

$$(x_1 + x_2 + \cdots + x_i + \cdots + x_n)/n = \Sigma x_i/n = \mu$$

The variance is

$$\sigma^2 = \Sigma(x_i - \mu)^2/n$$

The standard deviation is

$$\sigma = \sqrt{(\Sigma(x_i - \mu)^2/n)}$$

For each number x_i the Z value is

$$z_i = (x_i - \mu)/\sigma$$

EXAMPLE 12.1: EXAMPLE OF VARIANCES

Consider two groups of 10 people and their heights in meters.

Group A has heights of:

        1.66, 1.66, 1.67, 1.69, 1.70, 1.71, 1.71, 1.72, 1.73, 1.75

Group B has heights of:

        1.46, 1.52, 1.58, 1.60, 1.68, 1.74, 1.78, 1.84, 1.89, 1.91

The average or mean height in both groups is 1.70 meters. The residuals are:

For group A: −0.04, −0.04, −0.03, −0.01, 0, 0.01, 0.01, 0.02, 0.03, 0.05

For group B: −0.24, −0.18, −0.12, −0.10, −0.02, 0.04, 0.08, 0.14, 0.19, 0.21

For group A the variance = 0.00082 and the standard deviation ≈ 0.03. For group B the variance = 0.02226 and the standard deviation ≈ 0.15.

As demonstrated in Box 12.4, on average, there will be $A = np$ successful outcomes from n experiments where the probability of success $= p$. Put another way, the mean value $\mu = np$ as illustrated in Example 12.2, which demonstrates what commonsense should indicate, namely, that if each event has a probability of p, then thereafter n tries, there are likely to be np successful outcomes. The derivation of the standard deviation is perhaps less obvious and is shown in Box 12.5 to be $\sqrt{(npq)}$ where $q = 1 - p$.

To summarize, with a set of probabilities, $\mu = np$ and $\sigma = \sqrt{(npq)}$.

In Box 12.5 we show that the mean and variance and standard deviation can also be calculated for any set of numbers whatever they represent if we assume that the numbers are part of a binomial distribution. Normally, however, we simply add the numbers all together and divide by the number n to obtain the mean. Sum the squares of the differences between each number and the mean, and again divide by n and you have the variance σ^2; take the square root of the variance to obtain the standard deviation σ.

The smaller the standard deviation, the more bunched are the observations around the mean, and therefore the greater confidence there is that the mean represents the best possible estimate of the quantity being measured. The variance and standard deviation are sometimes referred to as *measures of central tendency*.

The binomial function allows precise probabilities that can be calculated, given certain basic assumptions such as heads occurring one time out of two, or rain falling more on a Wednesday one time out of seven. The function becomes difficult to handle when the numbers become large and, of course, it assumes a precise number of outcomes. Where there is more uncertainty, then a slightly different approach is needed.

BOX 12.4   MEAN OUTCOME FROM BINOMIAL EXPANSION

In Box 12.1 we quoted the binomial expansion in the generalized form

$$(q + p)^n = (q^n p^0 + {}_nC_1 q^{(n-1)}p^1 + ..{}_nC_r q^{(n-r)}p^r.. + q^0 p^n)$$

where $(p + q) = 1$, p being the probability that an event will happen, q that it will not happen. $p^0 = q^0 = 1$ while $p^1 = p$ and $q^1 = q$. In the expansion of $(q + p)^n$, we defined ${}_nC_r$ as $n!/\{r!(n - r)!\}$ where r is an integer between 0 and n. It follows that:

$$_{(n-1)}C_{(r-1)} = \frac{(n-1)!}{\{(n-1)-(r-1)\}!(r-1)!} = \frac{(n-1)!}{\{(n-1)-(r-1)\}!(r-1)!} * \frac{n*r}{n*r} = \frac{n!}{n} * \frac{r}{(n-r)!r!}$$

$$= (r/n)\, {}_nC_r$$

Hence,

$$_nC_r = (n/r)\, {}_{(n-1)}C_{(r-1)}$$

On average, the probability of there being no successes $= q^n p^0$. The probability of one successful outcome $= {}_nC_1 q^{(n-1)} p^1$. The probability of r successful outcomes $= {}_nC_r q^{(n-r)}p^r$.

   If the total number of expected successful outcomes after n attempts $= A$, then

$$A = \{q^n * 0 + {}_nC_1 q^{(n-1)} p * 1 + ... {}_nC_r q^{(n-r)} p^r * r ...... +{}_nC_n p^n\, n\}$$

with 0, 1, ...r.....n representing the number of possible outcomes. We can write this as:

$$A = \{q^n * 0 + {}_nC_1 q^{(n-1)}p^0 * (np/n) + \cdots +{}_nC_r q^{(n-r)}p^{(r-1)} * (npr/n) + \cdots + {}_nC_n p^{(n-1)} * (np)\}$$

and hence as:

$$A = \{{}_nC_1 q^{(n-1)}p^0 * (np/n) + \cdots +{}_nC_r q^{(n-r)}p^{(r-1)} * (npr/n) + \cdots + {}_nC_n p^{(n-1)} * (np)\}$$

Since ${}_nC_1 = n$ and ${}_nC_n = 1$, we have

$$A = (np) * \{q^{(n-1)}p^0 + \cdots +{}_nC_r q^{(n-r)}p^{(r-1)} * (r/n) + \cdots + p^{(n-1)}\}$$

But we have just shown that ${}_nC_r = (n/r)\, {}_{(n-1)}C_{(r-1)}$; hence,

$$A = np\{q^{(n-1)} + {}_{n-1}C_1 q^{(n-2)}p + ... {}_{n-1}C_r q^{(n-1-r)}p^r ...... p^{(n-1)}\}$$

$$= np(q + p)^{(n-1)} = np \text{ since } (q + p) = 1$$

Thus, on average, there will be $(A =) np$ successful outcomes or the mean value from the binomial distribution $= np$.

EXAMPLE 12.2: MEAN OUTCOME UNDER THE BINOMIAL DISTRIBUTION

We show in Box 12.4 that the mean value of a set of n observations with probability p of a particular outcome will on average $= np$. For instance, after 20 throws of a dice with $p = 1/2$ then one would expect $20 * 1/2 = 10$ heads. With rain on a Wednesday ($p = 1/7$) then over 14 weeks it is probable that there will be $14 * (1/7) = 2$ occasions on which Wednesday was the most rainy day, which is what one would expect.

BOX 12.5   THE BINOMIAL DISTRIBUTION— VARIANCE AND STANDARD DEVIATION

Given a mean of np then the original binomial equation shows the probability of the event occurring $(np - 1)$, $(np - 2)$, $(np - 3)$,....$(np - r)$,....$(np - n)$ times. These numbers will be positive or negative, depending on whether the event is less or more common than the average. The variance $\sigma^2 =$ the sum of the squares of these differences from the mean multiplied by their relative frequency, or

$$\sigma^2 = q^n p^0 (np - 0)^2 + {}_nC_1\, q^{n-1} p^1 (np - 1)^2 + \cdots\, {}_nC_r\, q^{(n-r)} p^r (np - r)^2 + \cdots + p^n (np - n)^2$$

where ${}_nC_r\, q^{(n-r)} p^r$ is the relative frequency of $(np - r)$.

This can be expanded to $L + M + N$ where

$$L = n^2 p^2 (q^n p^0 + {}_nC_1 q^{n-1} p^1 + {}_nC_2 q^{n-2} p^2 + \cdots + {}_nC_r q^{n-r} p^r + \cdots + p^n)$$
$$M = -2\, n\, p\, (q^n p^0 * 0 + {}_nC_1 q^{n-1} p^1 * 1 + {}_nC_2 q^{n-2} p^2 * 2 + \cdots + {}_nC_r q^{n-r} p^r * r + \cdots + p^n * n)$$
$$N = +(q^n p^0 * 0 + {}_nC_1 q^{n-1} p^1 * 1^2 + {}_nC_2 q^{n-2} p^2 * 2^2 + \cdots + {}_nC_r q^{n-r} p^r * r^2 + \cdots + p^n * n^2)$$

The first line, L, gives

$$n^2 p^2 (q^n p^0 + {}_nC_1 q^{n-1} p^1 + {}_nC_2 q^{n-2} p^2 + \cdots + {}_nC_r q^{n-r} p^r + \cdots + p^n) = n^2 p^2 (q + p)^n$$

But $(q + p) = 1$, hence the first line $= n^2 p^2$.

By expanding and rearranging the terms in the second and third lines M and N we obtain $M = -2n^2 p^2$ and $N = np + n(n - 1)p^2$. Hence,

$$L + M + N = \sigma^2 = n^2 p^2 - 2n^2 p^2 + np + n(n - 1)p^2 = np - np^2 = np(1 - p) = npq$$

Hence, $\sigma^2 = npq$. This represents the variance in the probability. The standard deviation $= \sigma = \sqrt{(npq)}$.

12.3   THE NORMAL DISTRIBUTION

The curve that has been drawn in Figure 12.2 passes through discrete points but implies intermediate values. If we accept the possibility of an observation taking any value and still retain the area under the curve to be 1 in order to give 100% probability, then we need a more complex function than the binomial distribution.

The fundamental assumption behind much of statistical analysis is known as the *central limit theorem*, which states that if a sequence of independent identically distributed random variables each has a finite variance, then as the number of observed values increases, they tend to mirror the probability distribution known as the *normal distribution*. Randomness means a lack of bias with the result of each observation being totally independent of any other observation.

In Figure 12.2, we showed a series of rectangles that represented the probabilities or proportion of events such as 1, 2, 3, and so forth, heads occurring after a given number of throws. The rectangles were 1 unit wide with heights derived from the binomial expansion. The total area made up by all the rectangles was 1.0 when every eventuality was considered. In the case shown, there were 15 throws. In Box 12.6 we

---

BOX 12.6   THE CURVE OF THE NORMAL DISTRIBUTION

Let us consider the proportions as being the result of n events as shown in Figure 12.3. Each proportional representation is separated by an amount w. Let the height of each rectangle be y_r so that the area of rectangle $r = w * y_r$. This represents the probability of an event occurring. According to the binomial equation, the frequency of an event at point r is $_nC_r p^r q^{n-r}$. To ensure that the total overall probabilities add up to 1, we must scale all the values y accordingly. Hence, we need to let $y_r = K * {}_nC_r p^r q^{n-r}$ where K is some appropriate scaling factor.

As shown in Boxes 12.4 and 12.5, the mean value $= n * pw$ and the variance $= \sigma^2 = n * pw * qw$. Let

$$x_r = \text{distance from the mean value to rectangle } r = (r - np)w$$

As the number n gets larger, the widths of the rectangles get narrower and n becomes infinite and w becomes zero and we finish up with a smooth curve. The probability of an error

$$x_r + \delta x_r \text{ is } y_r + \delta y_r = K * {}_nC_{r+1} q^{n-r-1} p^{r+1}$$

Hence,

$$\delta y_r = K * \{{}_nC_{r+1} q^{n-r-1} p^{r+1} - {}_nC_r q^{n-r} p^r\} = K * \{q^{n-r-1} p^r\}\{p * {}_nC_{r+1} - q * {}_nC_r\}$$

Since

$${}_nC_{r+1} = \frac{n!}{(n-r-1)!(r+1)!} = \frac{(n-r)}{(r+1)} {}_nC_r, \text{ then:}$$

$$\delta y_r = K\{{}_nC_r\}\{q^{n-r} p^r\}\{q^{-1}\}\{[(n - r)p - q]/(r + 1)\} = \{y_r\}\{(np - r)/q(r + 1)\}$$

or

$\delta y_r/y_r = (np - r)/\{q(r + 1)\} = (np - r)w/\{q(r + 1)w\}$ based on strips of width w

r represents the value $(npw + x_r)$ since $(r - np)w = x_r$

Thus,

$$(np - r)w = -x_r \text{ while } (r + 1)w = npw + x_r + w$$

Hence,

$$\delta y_r/y_r = -x_r/q(npw + x_r + w)$$

Again multiplying top and bottom by w

$$\delta y_r/y_r = -x_r w/\{npwqw + qw(x_r + w)\} = -x_r w/\{\sigma^2 + q(x_r + w)w\}$$

As the widths get smaller we can treat w as δx_r. Hence,

$$\delta y_r/y_r = -x_r \delta x_r/\{\sigma^2 + q(x_r + \delta x_r)\delta x_r\};$$

or

$$\delta y_r/\delta x_r = -y_r x_r/\{\sigma^2 + q(x_r + \delta x_r)\delta x_r\}$$

In the limit as we increase n and reduce w (= δx), then $q(x_r + \delta x_r)\delta x_r$ tends to zero, so

$$dy/dx = -xy/\sigma^2; \text{ and}$$
$$y = \int(-xy/\sigma^2)dx; \text{ or}$$
$$\int (1/y) \, dy = \int(-x/\sigma^2)dx; \text{ or}$$
$$\log_e y = -x^2/2\sigma^2 + c \text{ where } c \text{ is the constant of integration.}$$

If we measure x from the mean value "μ," we can rearrange this as

$$y = k \, e^{-\{(x-\mu)^2/2\sigma^2\}}$$

where k is some constant and "μ" is the mean value. The curve is symmetrical about the mean value "$x = \mu$."

In order that the total area under the curve is unity and since $\int_\infty^\infty e^{-t^2} \, dt$ can be shown (though is not demonstrated in this text) to equal $\sqrt{\pi}$, then $k = 1/\{\sigma\sqrt{(2\pi)}\}$. The equation of the smooth curve, called the *normal curve*, becomes

$$y = [1/\{\sigma\sqrt{(2\pi)}\}] \, e^{-\{(x-\mu)^2/2\sigma^2\}}$$

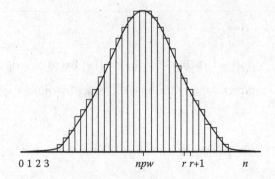

0 1 2 3　　　　　　　　*npw*　　　　*r r+1*　　　　*n*

FIGURE 12.3　A plot of probability for *n* events.

extend this further to *n* events and then to what in theory is an infinite number so that the curves in Figures 12.2 and 12.3 become smooth. Such a curve is then known as the *normal curve*.

Figure 12.3 shows the plot of probability of *n* events that are normally distributed. Figure 12.4 shows the percentage of a distribution that lies within 1, 2, 3, or more standard deviations of the mean.

For those who are interested, Box 12.6 outlines how the formula for representing the normal distribution is derived. What is important is that the normal distribution has the form

$$y = [1/\{\sigma\sqrt{(2\pi)}\}]\ e^{-\{x^2/2\sigma^2\}}$$

where e is the base of natural logarithms.

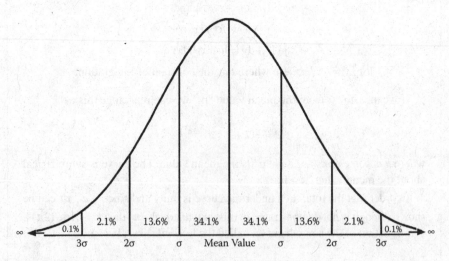

FIGURE 12.4　Percentage of distribution within one, two, and three standard deviations from the mean.

TABLE 12.1
Partial Areas under the Normal Curve

$Z = (x - \mu)/\sigma$	+0.0	0.2	0.4	0.6	0.8
0.0	0%	7.93%	15.54%	22.57%	28.81%
1.0	34.13%	38.49%	41.92%	44.52%	46.41%
2.0	47.72%	48.61%	49.18%	49.53%	49.74%
3.0	49.87%	49.93%	49.97%	49.98%	49.99%

Or, using the z statistics

$$y = \frac{1}{\sigma\sqrt{2\pi}} e^{-(z^2/2)}$$

Those who are prepared to take Box 12.6 as read should note that:

(a) The total area under the curve in each of the Figures 12.3 and 12.4 is 1.
(b) The curve is symmetrical about the value $x = \mu$.
(c) Although the curve goes off to infinity either side of the mean, in practice over 99% of the observations lie within about three standard deviations of the mean.

If we choose to record our measurements about a mean of zero, then $\mu = 0$ and we have

$$y = ke^{-\{x^2/2\sigma^2\}} \quad \text{with} \quad k = 1/\{\sigma\sqrt{(2\pi)}\}$$

This is symmetrical about $x = 0$. The area under the curve = 1 and the variance = σ^2. If we assume that x is a measure of the error in an observation, then the area under the curve between $x = 0$ and $x = r$ represents the total probability of an error equal to or less than r. This is given by $[1/\{\sigma\sqrt{(2\pi)}\}] \int_0^r e^{-\{x^2/2\sigma^2\}} dx$.

Unfortunately, there is no easy way in which to calculate this when using elementary techniques. Suffice it to say that, using various numerical methods, the sums have been worked out and are available in statistical tables. The ratio between the size of the residual and the standard deviation is illustrated in Table 12.1 where values of r/σ are tabulated against the area under the curve in Figure 12.3. A summary of results is shown in Figure 12.4.

12.4   LEVELS OF SIGNIFICANCE

The normal distribution shows that almost all observations should fall within the limits of three standard deviations away from the mean. If, for example, we take a series of measurements then it is easy enough to calculate the mean and standard deviation of the set of measurements. If we try to determine any quantity very precisely, then we would expect to have slight differences each time we make the measurement. Any observation that was more than three standard deviations away from the mean should be suspect and it is probably wise to reject it.

Table 12.1 shows that if we measure how far an observation is away from the mean value in units of σ (i.e., r/σ) then, if errors in the measurement are random, the chances are that 0.4987 out of 0.5 (half the area) or 99.74% of the observations will be within the range $-3 < r/\sigma < +3$ or that 99.74% of the observations will fall within three standard deviations of the mean. Hence, if we have an observation for which $r/\sigma = 4$, then it is outside the probable range. This suggests that, in practical terms, it is highly likely that there is a gross error and the observation does not belong to the group under investigation. Example 12.3 illustrates a set of observations where one member of the set may or may not be included.

Before carrying out any statistical analysis we need to create what is called the *null hypothesis*. A hypothesis is essentially a supposition or an unproved theory; the null hypothesis is the assumption that there is no significant difference between any particular measure and any other unless we can demonstrate to the contrary. The object of a statistical test is to prove that something is significantly different from what we would otherwise expect. Statistics cannot prove that anything is true, but they can indicate the probability that something is significantly different from what might otherwise be anticipated. The key questions then become how to phrase the null hypothesis and how to judge what is significant.

It is highly likely that if an observed quantity x is such that $(x - \mu)/\sigma$ is greater than 3, there has been an error and the observation can be rejected as very unlikely to have happened by chance. There is, of course, a very remote possibility that this was a random event since the normal curve extends to infinity. If we have a hypothesis

---

EXAMPLE 12.3: ERROR DETECTION

In a series of 10 observations, the measurement of an angle was found to be 28° 33′ plus 24.8″, 23.9″, 25.8″, 32.2″, 25.0″, 24.5″, 24.2″, 23.7″, 24.1″, and 23.8″.

The mean value = 28° 33′ 25.2″.
The residuals are 0.4, 1.3, −0.6, −7.0, 0.2, 0.7, 1.0, 1.5, 1.1, and 1.4.
The sum of the squares of the residuals = 58.16, so $\sigma = \sqrt{\Sigma(x_i - \mu)^2 / n} = 2.4$.

If we set a threshold of three standard deviations away from the mean as being our limit, (here being 3 * 2.4 or 7.2), then all the observations lie just within it. If we set two standard deviations as the limit, then the observation with 32.2″ should be rejected and our mean would become 28° 33′ 24.4″ and the standard deviation would be 0.6″.

that an event is so rare that it cannot belong to the population under consideration and an observation that really should be included is rejected, then a *Type I Error* is said to occur; if, however, an observation is included that should be rejected, then the reverse is the case and a *Type II Error* occurs.

Type I errors can be reduced by increasing the threshold for rejecting observations (known as the *significance level*) but in so doing the chances of making a Type II error are increased. In practice, it is usual to set significance levels at either 95% (the equivalent approximately of $r/\sigma = 2$) or 99% (r/σ is approximately equal to 3).

Consider a data set A that is a sample set of measurements of a particular phenomenon for which there is potentially an infinite possible number of observations. Let the mean and variance of the sample data set A be $\bar{x}$ and S_A^2. In practice it is very likely that any small sample will have a slightly different mean and variance from that derived from the infinite series.

Let us assume that we have n observations in data set A and let these be $x_1, x_2 \ldots x_n$ with a mean $\bar{x}$. Then for the variance of the sample set A, we have

$$S_A^2 = (1/n)\Sigma(x_i - \bar{x})^2 = (1/n)\Sigma(x_i^2 - 2\bar{x}x_i + \bar{x}^2)$$

$$= (1/n)\,\Sigma x_i^2 - (2\bar{x}/n)\Sigma x_i + (n/n)\,\bar{x}^2$$

$$= (1/n)\,\Sigma x_i^2 - \bar{x}^2 \text{ since } (1/n)\Sigma x_i = \bar{x}$$

If the whole population has a variance σ^2 and mean μ we would expect that

$$\sigma^2 = (1/n)\Sigma(x_i - \mu)^2 = (1/n)\,\Sigma x_i^2 - (2\mu/n)\Sigma x_i + \mu^2$$

$$= (1/n)\,\Sigma x_i^2 - 2\mu\bar{x} + \mu^2$$

Subtracting S_A^2 from this we obtain $(\sigma^2 - S_A^2) = (\bar{x} - \mu)^2$.

Thus, the sample variance is smaller than the variance of the infinite set. The mean values of the population and the sample will also differ. The question then arises as to how reliable is the value $\bar{x}$ as an estimate of μ, which is the mean of the whole population.

In fact, if we have N data sets A, B, C, and so on, each of size n_A, and so on, then they in turn form a set of values with means $\bar{x}_A$, $\bar{x}_B$, $\bar{x}_C$, and so on, the mean of all of which should be close to the mean of the overall population μ. The estimated variance for the combined N data sets will be

$$S^2 = (S_A^2 + S_B^2 + \ldots)/N = \sigma^2/N, \text{ so } S = \sigma/(\sqrt{N})$$

What this shows is that a sample of mean $\bar{x}$ is itself part of a wider distribution arising from many possible samples that could be taken. This larger distribution has the same mean as the total population (μ) and a standard deviation of $\sigma/(\sqrt{n})$ where σ is the standard deviation of the whole population and n is the total size of all the samples that have been taken. Put another way, the estimated standard deviation of the mean of a sample A of size n_A will be $(1/\sqrt{n_A})$ times the standard deviation of the sample itself.

We showed above that $(\sigma^2 - S_A^2) = (\bar{x} - \mu)^2$ where σ^2 and μ relate to the whole population and S_A^2 and $\bar{x}$ to the sample. The average value of $(\bar{x} - \mu)^2$ over n samples will be σ^2/n so for any general sample $(\sigma^2 - S^2) = \sigma^2/n$ or $S^2 = (n-1)\sigma^2/n$. Thus, if we have a sample with standard deviation S_A the better estimate for the standard deviation for the population as a whole is

$$S = S_A \sqrt{\left(\frac{n}{(n-1)}\right)}$$

Box 12.7 summarizes all these relationships.

Putting $\sigma^2 = \{1/(n-1)\}\Sigma(x_i - \bar{x})^2$ [rather than $\sigma^2 = \{1/n\}\Sigma(x_i - \bar{x})^2$] gives what is called the *unbiased estimate* of the population variance, although it makes little practical difference when the number n is large. The estimate is based on $(n-1)$ *degrees of freedom*, the term referring to the minimum number of parameters necessary to describe completely the state of a system. In statistics the number of degrees of freedom is equal to the number of independent unrestricted random variables that constitute a particular statistic. Hence, for example, if we have a mean value of 10 observations, the mean is fixed, and once nine of the 10 observations have been chosen, the final 10th observation is already determined. There are thus nine degrees of freedom.

BOX 12.7   MEAN, VARIANCE, AND STANDARD DEVIATION

For any sample set of n numbers $x_1, x_2, \ldots x_i, \ldots x_n$ taken from a large population with variance σ^2, standard error σ and mean μ:

The mean value of the sample is

$$(x_1 + x_2 + \cdots + x_i + \cdots + x_n)/n = \Sigma x_i/n = \bar{x}$$

The variance of the sample is

$$S^2 = \Sigma(x_i - \bar{x})^2/n = (1/n)\,\Sigma x_i^2 - \bar{x}^2 = \sigma^2 * (n-1)/n$$

where σ^2 is the estimate for the population as a whole.

The standard deviation of the sample $A = S_A = \sqrt{\frac{\Sigma(x_i - \bar{x})^2}{n}}$. A better estimate for the standard deviation of the population $= S = S_A \sqrt{\left(\frac{n}{(n-1)}\right)}$.

Also: The standard deviation of the mean $(\bar{x}) = \frac{\sigma}{\sqrt{n}} = \frac{S}{\sqrt{(n-1)}}$.

12.5   THE t-TEST

The normal curve is derived on the basis of a random selection of measures from an infinite population, although in practice it applies equally well so long as the size of the sample is large. When this is not the case, the normal distribution has to be modified, but we can still use units of standard deviation and calculate $z = (x - \bar{x})/\sigma$.

$(x - \bar{x})$ is the difference between an observation and the mean, and the division by σ means that we are measuring this difference in units of σ.

An English mathematician, William Gosset, writing under the pseudonym of "Student" developed a way in which to test whether a random sample of normally distributed observations of unknown parameters has the same mean as the population. His statistic is known as t and his test as the *Student's t-test* (Box 12.8).

The number $t = (\bar{x} - \mu)/\{S/\sqrt{n}\} = \{(\bar{x} - \mu)\sqrt{n}\}/S$, where n is the number of observations in a sample, $\bar{x}$ is the sample mean, S the standard deviation of the sample (and hence, $S/\sqrt{n}$ is the standard deviation of its mean). "μ" is the estimated population mean. Put another way, t is a measure of the difference in the means of two sets of data divided by the standard deviation of the mean. It is the measure of the disparity between the means of two data sets as measured in units of the standard deviation of their means.

The test can be used to compare a sample mean with the estimated mean of the whole population. Alternatively, it can be used to test two samples to see if their means differ significantly, in which case $\bar{x}$ would be the mean of one sample and μ the mean of the other, with S and n being taken either from the combination of the two samples or from one sample, depending on the hypothesis.

The various values of t can be found in statistical tables. These show the probability (p) or significance level that arises for various values of n and the calculated value of t, as illustrated in Example 12.4. Table 12.2 shows a sample set of values.

BOX 12.8   THE t-TEST

1. Tests the significance of the difference in means between two data sets.
2. Assumes that the population is normally distributed and that the samples are random.
3. Takes the form $t = \{(\bar{x} - \mu)\sqrt{n}\}/S$. It measures the difference in means by measuring $(\bar{x} - \mu)$ in units of the standard deviation of the mean $\{S/\sqrt{n}\}$.
4. The level of significance or probability can be found from special tables.

EXAMPLE 12.4: USING THE t-TEST

In Example 12.1 we had two groups A and B of 10 people whose heights had been measured. Both had mean values of 1.70 but their standard deviations were (A) 0.03 and (B) 0.15. Does either population differ significantly from an estimated population average of 1.75 meters?

$$\bar{x} = 1.70, \mu = 1.75, n = 10, \sqrt{n} = 3.16 \text{ in both cases.}$$

For Group A,

$$t = \frac{(\bar{x} - \mu)\sqrt{n}}{S} = 0.158/0.03 = 5.27$$

For Group B,

$$t = \frac{(\bar{x} - \mu)\sqrt{n}}{S} = 0.158/0.15 = 1.05$$

There are 9 degrees of freedom, and (from tables) the level of significance for group A is less than 1%. So one can expect this to happen in less than 1 out of 100 cases (see Table 12.2).

For group B, the result can be expected in slightly more than 0.3 out of 1 or more than 30% of cases and the result is insignificant. Thus, group A seems to have been a special case while group B is not unusual.

TABLE 12.2

Levels of Significance (p) for Values of t (Given 9 Degrees of Freedom)

p	0.9	0.8	0.7	0.6	0.5	0.4	0.3	0.2	0.1	0.05	0.01
t	.13	.26	.40	.54	.70	.88	1.10	1.38	1.83	2.26	3.25

12.6 ANALYSIS OF VARIANCE

The t-test allows one to compare sample means. The F-test (Box 12.9) may be used to compare the variances between two samples, where $F = S_1^2/S_2^2$. F is the ratio between the variance of sample set "1" and the variance of sample set "2." For very large data sets from the same population, the value of F would be expected to be $\approx$ 1 if the two samples were from the same population. It is usual to place the set with the larger variance as set "1" and the smaller as set "2." In Example 12.1 we had set A with a standard deviation of 0.03 and set B with 0.15. It is clear that since the two sets used in this example share the same mean, the smaller standard deviation fits easily within the range for set B. The reverse is not true.

For smaller data sets, the significance of the value of F will depend on the number of degrees of freedom in A and B, for instance, $(n_A - 1)$ and $(n_B - 1)$. It will also

---

BOX 12.9   THE F-TEST

1. Is applied to sample variances.
2. Assumes that the samples A and B are normally distributed.
3. Assumes that the variance of A (= S^2_A) and B (= S^2_B) are independent estimates of the population variance σ^2 and $S^2_A > S^2_B$.
4. Takes the form $F = S_A^2/S_B^2$.
5. If there are n_A measures in A and n_B in B, then the degrees of freedom are $(n_A - 1)$ and $(n_B - 1)$.

The F-distribution is of particular use in the tests known as the analysis of variance.

---

depend on whether one is looking for the variance in one of the data sets, such as A being greater than B or whether one is looking at the probability of their difference being significant regardless of whether A is greater or less than B. If the difference is only relevant in one direction, the test is said to be *one-tailed*; if the absolute difference is important regardless of whether one sample is greater or less than the other, then the test is *two-tailed*.

In practical terms, the importance lies in where the level of significance is set and the difference only affects observations that are around the limit of acceptability. Roughly speaking, if 95% of observations are to be within the limit, then for the one-tailed test, the 5% that are outside the limit all occur at one end of the left-hand, bell-shaped probability curve in Figure 12.5. For a two-tailed test (the right-hand, bell-shaped curve), the two black areas are each 2.5% of the total area below the curve.

The idea of comparing variances leads to a powerful series of tests known as *analysis of variance* or *ANOVA*. If a series of observations are taken, for instance, of the yield of some crop when treated with different types of fertilizer, and if there is no significant difference between any of the fertilizers, then one would expect that the variance of each sample would be much the same. If, however, one fertilizer gave rise to significantly greater yields, then the samples taken from the crop treated with it would consistently have a greater mean. The difference in the mean values could be tested using the t-test, while the analysis of variance could help to discriminate between possible factors that are less obvious in a simple test of the mean values. There may be a whole variety of variables and the analysis of variance can be used to identify which of these is significant.

For example, if the results of a series of c different experiments are recorded for r different samples (see Table 12.3), then they represent $n = r * c$ different experiments for which the overall mean ($\bar{x}$) and standard deviation can be calculated. Let us arrange these in a table with r rows and c columns and label each observation x_{ij} as in a matrix.

The sum of the squares of the residuals from the mean of all the observations

$$= \Sigma(x_{ij} - \bar{x})^2 = \Sigma\{x_{ij}^2 - 2\bar{x}\,x_{ij} + (\bar{x})^2\} = \Sigma x_{ij}^2 - 2\bar{x}\Sigma x_{ij} + n(\bar{x})^2\}$$

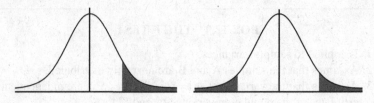

FIGURE 12.5   One- and two-tailed tests.

Since $\bar{x} = (\sum x_{ij})/n$ then the sum of the squares of the residuals $= \sum x_{ij}^2 - (\sum x_{ij})^2/n$. This is sometimes called the *variation*. It equals n times the variance.

$$\text{Sample variation} = \sum x^2 - (\sum x)^2/n$$
$$\text{Sample variance} = (\sum x^2)/n - \{(\sum x)/n\}^2$$

The unbiased estimate of the variance of the population as a whole from which the observations have been drawn is therefore

$$(\sum x^2)/(n-1) - (\sum x)^2/\{n(n-1)\}$$

The reason for calculating the variation and variance in this way is that it is no longer necessary to calculate each of the individual residuals. Rather than having to work out all the residuals from the mean and then squaring each one in turn, we can calculate the sum of each observation squared and subtract the total sum squared divided by n. This gives us the variation from which we can obtain the variance and standard deviation by applying the appropriate number of degrees of freedom.

TABLE 12.3
Data Classified into Rows and Columns

	Col 1	Col 2	...	Col j	...	Col c	$\sum$ = Row Sum	Number in Row	Sum of Squares
Row 1	x_{11}	x_{12}	...	x_{1j}	...	x_{1c}	T_{r1}	n_{r1}	SS_{r1}
Row 2	x_{21}	x_{22}	...	x_{2j}	...	x_{2c}	T_{r2}	n_{r2}	SS_{r2}
Row 3	x_{31}	x_{32}	...	x_{3j}	...	x_{3c}	T_{r3}	n_{r3}	SS_{r3}
...	...	...	...	...	...	...	...	...	...
Row i	x_{i1}	x_{i2}	...	x_{ij}	...	x_{ic}	T_{ri}	n_{ri}	SS_{ri}
...	...	...	...	...	...	...	...	...	...
Row r	x_{r1}	x_{r2}	...	x_{rj}	...	x_{rc}	T_{rr}	n_{rr}	SS_{rr}
Column Sum	T_{c1}	T_{c2}	...	T_{cj}	...	T_{cc}	$T = \sum x_{ij}$		
No. in Column	n_{c1}	n_{c2}	...	n_{cj}	...	n_{cc}		n	
Sum of Squares	SS_{c1}	SS_{c2}	...	SS_{cj}	...	SS_{cc}			

Referring to Table 12.3, we can calculate the overall variation or sum of the squares of the residuals as $\Sigma(x_{ij}^2) - (\Sigma x_{ij})^2/n$ where x_{ij} is the observed value in row i and column j, and n = the total number of observations. If all elements are filled, then $n = r * c$ where r is the number of rows and c is the number of columns. The unbiased estimate of the variance for the population as a whole is therefore $\Sigma(x_{ij}^2)/(n - 1) - (\Sigma x_{ij})^2/\{n(n - 1)\}$.

Let the sum of each column $j = T_{cj} = \Sigma\ x_{ij}$ for $i = 1$ to n_{cj} and the sum of each row $i = T_{ri} = \Sigma\ x_{ij}$ for $j = 1$ to n_{ri}. $\Sigma\ x_{ij}$ for the whole data set $= T = \Sigma T_{ri} = \Sigma T_{cj}$. If all elements are filled (that is, there are r observations in each column), then all $n_{ri} = c$ and all $n_{cj} = r$.

The mean value for each column $= T_{cj}/n_{cj}$. If we treat every element in the column as having the value of the mean for that column, then summing all c columns we have a total variation $= \Sigma\ \{n_{cj} * (T_{cj}/n_{cj})^2\} - T^2/n = \Sigma\ (T_{cj})^2/n_{cj} - T^2/n$. This is known as the *variation between columns*. If we assume that all elements were observed, then we can write this as $\Sigma\ (T_{cj})^2/r - T^2/n$. If we divide this by the number of degrees of freedom $(c - 1)$, we obtain an estimate of the overall population variance. Similarly, the variation between rows $= \Sigma\ (T_{ri})^2/c - T^2/n$. These variances are known as the *between-group variability* or *explained variance*.

Assuming that all the elements were observed, the unbiased estimate for the variance of the population obtained solely from column j is $SS_{cj}/(r - 1) - (T_{cj})^2/\{r(r - 1)\}$. Let us call this V_{cj}.

The pooled estimate for all c columns will be the weighted mean of each such unbiased estimate, the weights being the degrees of freedom in each column. Call the result V_c. Since there are r items in each column, there will be $(r - 1)$ degrees of freedom for each of the c columns. The sum of the weights with c lots of "$(r - 1)$" or $c * (r - 1)$ will be $= (n - c)$. Hence,

$$V_c = \{\Sigma V_{cj} * (r - 1)\}/(n - c) = \{\Sigma SS_{cj} - \Sigma(T_{cj})^2/r\}/(n - c)$$

This is known as the *overall variance within columns*. It is also known as the *unexplained variance* or *within-group variability*. It is an estimate of the overall population variance based on $(n - c)$ degrees of freedom. It can also be applied across the rows rather than across the columns, depending on the hypothesis being tested.

In order to determine whether there are any significant differences, we can apply the F-test, which in the context of analysis of variance takes the form:

F = (between-samples estimate of variance)/(within-samples estimate of variance)

or

F = (explained variance)/(unexplained variance)

The value of F can then be checked against the values given in statistical tables, based on the appropriate degrees of freedom.

The same calculations can be made with reference to rows rather than columns, resulting in different numbers. Their significance depends on the hypotheses being tested and whether one is attempting a one-way analysis of variance (in which only one result, for instance, the column differences, is being tested) or two-way analysis of variance in which more complex combinations are evaluated. The data are analyzed in terms of whether there is a significant difference between the circumstances giving rise to what is shown in rows or columns compared with the whole, or whether there are any anomalies.

In each case, the mean and standard deviation can be calculated and if there is no overall significant difference between any of the samples, both the variance between columns and the variance between rows should be about the same as the variance of the whole n experiments. This can then be checked by the F-test, see Example 12.5.

EXAMPLE 12.5: USING ANALYSIS OF VARIANCE

Consider a set of observations with five observations under four headings A, B, C, D with T referring to totals and SS to sums of squares.

	A	B	C	D	T_r	SS_r
	15	17	20	18	70	1238
	23	27	29	24	103	2675
	18	13	24	16	71	1325
	11	12	15	14	52	686
	17	16	21	14	68	1182
T_c	84	85	109	86	$\Sigma = 364$	
SS_c	1488	1587	2483	1548		$\Sigma = 7106$

$n = 4 * 5 = 20$. Total variation $= 7106 - 364^2/20 = 481.2$.

Between sample variation

$$= \Sigma (T_{cj})^2/r - T^2/n$$

$$= 84^2/5 + 85^2/5 + 109^2/5 + 86^2/5 - 364^2/20$$

$$= 6711.6 - 6624.8 = 86.8$$

With four columns there are three degrees of freedom; hence,

Between sample variance $= 86.8/3 = 28.87 =$ "explained variance"
Within sample variance $= \{\Sigma SS_{cj} - \Sigma(T_{cj})^2/r\}/(n - c)$

$$= (7106 - 6711.6)/16 = 394.4/16$$

$$= 24.65 =$$ "unexplained variance"

$F = 28.87/24.65 = 1.17$. The between column differences are not significant.

12.7   THE CHI-SQUARED TEST

Analysis of variance compares the frequencies of certain events under differing conditions. A somewhat similar test is known as the chi-square or χ^2 test (from the Greek letter "chi" or χ) (Box 12.10). The test considers how often something is observed to happen (f_o) compared with the expected or calculated frequency (f_e) in which $\chi^2 = \Sigma(f_o - f_e)^2/f_e$.

The χ^2 test may be used, for example, if there is a predicted distribution of points on a map (such as cases of cancer near some electrical transmitter) so that the expected numbers can be compared with the observed numbers using the χ^2 test, as in Example 12.6.

The χ^2 test provides a way in which an observed distribution can be examined to see if it is probably part of a normal distribution—the estimated frequency being provided by the normal function while the observed frequency is that which has been measured. The χ^2 test is an example of a *nonparametric test* in that it can be applied to ordinal and nominal data. All the other tests that have been discussed so far have been *parametric* tests in that they have had parameters such as the mean and standard deviation that can be subject to arithmetic operations.

Parametric tests are based on fairly stringent assumptions, for instance, about the nature of the distribution of the population, whereas nonparametric tests, sometimes referred to as *distribution free*, are much less demanding. For example, nonparametric tests can be used to compare two lists that have been ranked in order to see if they show similar characteristics, such as whether secondary or high school level examination results correlate well with final university degree awards. For further discussion on nonparametric tests the reader should consult specialist books on statistics.

BOX 12.10   THE χ^2 TEST

1. The test checks whether an observed distribution is probably consistent with estimated values based on a predictive model.
2. The data must be measured at least at a nominal scale or any higher level of measurement. There must be at least two mutually exclusive categories in which the data can be placed, each with a frequency greater than five.
3. If there are more than two categories, then at least 80% of these categories must have at least five expected outcomes.
4. If these restrictions cannot be met then a different test must be used or data categories amalgamated so that the criteria are met.
5. The test assumes an observed frequency (f_o) and expected frequency (f_e).
6. $\chi^2 = \Sigma(f_o - f_e)^2/f_e$.
7. The significance level for the value of χ^2 can be found from statistical tables.

EXAMPLE 12.6: χ^2 TEST

According to evidence, 15% of properties are sold each year, on average across the whole of a country. In one town with 20,000 properties, only 1000 properties were sold last year. Is that significantly below average?

The expected frequency f_e would be 3000 out of 20,000 or 3000 sold and 17,000 not sold. The observed quantities were 1000 sold and 19,000 not sold (we need to take the nonsales into account to ensure that the total probability {referred to before as $(p + q)$} is equal to 1 or 100%.

$$\chi^2 = (1000 - 3000)^2/3000 + (19{,}000 - 17{,}000)^2/17{,}000$$

$$= 4/3 + 4/17 = 1.57$$

The tables for χ^2 (not shown here) give the chances of this happening as greater than 20%; thus, at least one time in five this may happen. There is nothing particularly significant about the shortfall in house sales.

12.8 THE POISSON DISTRIBUTION

χ^2 can also be used to test other types of distribution, for example, that which bears the name of a French mathematician, Siméon Poisson. Although related to the normal distribution, it is particularly useful when dealing with small samples.

It takes the form

$$p = \sum \frac{m^r}{r!} e^{-m}$$

where p is the probability of an event, m is the mean value of a set of samples, r is the number of events, and e is the Euler number. Since

$$\frac{m^r}{r!} = \frac{m}{r} * \frac{m^{(r-1)}}{(r-1)!}$$

it means that each term except the first is $\frac{m}{r}$ times the previous term. An example is given in Examples 12.7 and 12.8.

Thus, if we lay a grid over an area and count the number of events that occur within each square, then we can calculate the probability of their being distributed with an expected frequency. This can be tested using the χ^2 test to see whether the predictions of the Poisson model are in conformity with what is found.

EXAMPLE 12.7: EXAMPLE OF POISSON DISTRIBUTION (1)

A grid with 10 * 10 squares was laid over an area and the numbers of events were counted (for instance, the number of plants or a particular type of worm in the quadrat). The Poisson prediction is calculated in Example 12.8. The results were as follows.

r		p
Number Found per Square	Number of Squares	Poisson Prediction
0	1	0.5
1	2	2.8
2	6	7.3
3	13	12.7
4	16	16.6
5	19	17.4
6	15	15.2
7	12	11.4
8	10	7.5
9	3	4.3
10	2	2.3
11 or more (= 11)	1	2.0
	Total	100

The significance of the differences between what was found and the Poisson estimate can be tested using the χ^2 test.

EXAMPLE 12.8: THE POISSON DISTRIBUTION (2)

Using the data in Example 12.7 that shows the number of squares that have 0, 1, 2, 3, and so forth, plants in them:

The total number of plants = $1 * 2 + 2 * 6 + 3 * 13 + \cdots + 11 * 1 = 524$

There are 100 squares, hence, the mean number per square = $m = 5.24$. The probabilities are then calculated from $p = \Sigma \frac{m^r}{r!} e^{-m}$

The value of $(1/e^{5.24}) = e^{-m} = 0.0053$

According to the Poisson formula, the probabilities for $P_0 = 1/(e^{5.24}) = 0.0053$, so out of 100 squares the expected frequency of nil finds = $100 * 0.0053 = 0.53$.

Thus, $P_0 = 0.53$. Hence, we can calculate:

$$P_1 = (m/1) * P_0 = 2.8$$
$$P_2 = (m/2) * P_1 = 7.3$$
$$P_3 = (m/3) * P_2 = 12.7$$
$$P_4 = (m/4) * P_3 = 16.6$$
$$P_5 = (m/5) * P_4 = 17.4$$
$$P_6 = (m/6) * P_5 = 15.2$$

and so on.

SUMMARY

Analysis of variance: Also known as ANOVA, tests the observed variance within and between samples to determine whether the differences are random or significant.

Binomial expansion: An expansion of the form

$$(p + q)^n = p^n + np^{(n-1)}q + \{n\,(n-1)/2)\}p^{(n-2)}q^2$$
$$+ \cdots + {}_nC_r\,p^{(n-r)}q^r + + \cdots + npq^{(n-1)} + q^n$$

This has a *mean* outcome of np with *standard deviation* $\sqrt{(npq)}$.

Central limit theorem: If a sequence of independent identically distributed random variables each has a finite variance, then as the number of observed values increases, they tend to mirror a *normal distribution*.

Chi-squared test: Compares the frequency of an event as observed with what might be expected in the form

$$\chi^2 = \Sigma(f_o - f_c)^2/f_c$$

Degrees of freedom: The minimum number of parameters necessary to describe completely the state of a system.

Descriptive statistics: Sets of numbers used to summarize a set of known data in a clear and concise way, such as the mean value.

Expected value: The value (E) in a set of observations that will most probably occur.

Explained variance: In *analysis of variance*, the overall *variance* between columns or rows. Also known as the *between-group variability*.

First moment: The *expected value* of the deviation from zero (*residual*) for all the values of a *random* variable (usually the weighted mean). The second moment is the expected value of the squares of the residuals.

F-test: A test that compares the variances between two samples in the form

$$F = S_A{}^2/S_B{}^2$$

Heuristic: Using processes that are guided by experience and experiment rather than by rigorous logical argument from precisely defined axioms.

Histogram: A graphical method for displaying frequencies.

Inferential statistics: The use of statistical data to draw conclusions from random samples.

Mean: The mean is the average value by summing all the terms x_i and then dividing by the number of terms (n). Often written as μ, its value is $\Sigma_{i=1}^{i=n} x_i / n$

Measures of central tendency: Measures of the spread of a set of observations around their mean value, for instance, *variance* and *standard deviation*.

Median: The median is the middle number in a frequency list of numbers.

Mode: The mode is the number that most often occurs, that is, it has the highest frequency in a list of numbers.

Nonparametric test: Test that can be applied to nominal and ordinal data, namely data that are distinguishable only by their class (nominal) or their place in a sequence (ordinal).

Normal curve: A curve that follows the normal distribution, sometimes described as being bell-shaped.

Normal distribution: Also known as a *Gaussian distribution*, it is symmetrical with the mean, median, and mode coinciding and has the form

$$y = [1/\{\sigma\sqrt{(2\pi)}\}]\, e^{-\{(x-\mu)^2/2\sigma^2\}}$$

or, using the Z statistic

$$y = \frac{1}{\sigma\sqrt{2\pi}}\, e^{-(z^2/2)}$$

Null hypothesis: The supposition that there is no significant difference between any particular measure and any other sample value.

One-tailed test: The difference is only significant if the quantity is exceptionally greater (or less) than anticipated but not significant if it is smaller (or greater)—unlike the *two-tailed test*.

Parametric test: Data that have parameters such as a mean and standard deviation.

Pascal triangle: A triangular array of integers starting with 1, in which each number is the sum of the two above it.

Poisson distribution: A method for predicting what might be expected at an average rate of distribution. It takes the form

$$p = \sum \frac{m^r}{r!}\, e^{-m}$$

Population: Sometimes referred to as a universe, it is the complete set of individual components or events from which *samples* are drawn.

Probability: A measure of confidence that can be had in any event, measured on a scale from 0 to 1 (which represents certainty).

Random number: A number that cannot be determined from any of the previous numbers that have been selected.

Residual: The difference between an observation (x_i) and the mean of all the observations (μ). The residual $= r = (x_i - \mu)$.

Residual variation: The difference between the sum of the variation between rows plus the variation between columns and the variation of the whole data set.

Sample: A set of individual observations of elements of a population that can be used to predict the nature of the population.

Significance level: The threshold at which there is a critical level of probability of wrongly rejecting the null hypothesis.

Standard deviation: The square root of the *variance*. $\sigma = \sqrt{(\sum(x_i - \mu)^2/n)}$

Statistic: A function of random variables that can be used as an estimator of a *population*.

Stochastic: Processes that can be described by random variables.

***t*-Test:** Also known as the *Student's t-test*, it tests the significance of the difference in means between two data sets, assuming that the population is normally distributed and the samples are random.

$$t = [(\overline{x} - \mu)\sqrt{n}]/S$$

Two-tailed test: A test that judges the significance of whether the quantity is exceptionally greater or less than the expected value for a random sample (unlike the *one-tailed test* where the significance is only tested in one direction).

Type I error: An observation that should be included is rejected.

Type II error: An observation that should be rejected is included.

Unbiased estimate: Using $\sigma^2 = \{1/(n-1)\}\sum(x_i - \overline{x})^2$ rather than $\sigma^2 = \{1/n\}\sum(x_i - \overline{x})^2$ as the estimate of the population *variance*.

Unexplained variance: In *analysis of variance*, the overall *variance* within columns or rows. Also known as the *within-group variability*.

Variance: The average sum of the squares of the *residuals*. Written as sigma squared, $\sigma^2 = \sum(x_i - \mu)^2/n$.

Variance between sample means: A term used in *analysis of variance*. It equals the *variance* of the means of samples and is approximately equal to σ^2/n.

Variance within samples: A term used in *analysis of variance*. It is the estimate of the overall population variance by considering the weighted average of the *variances* between columns (or rows).

Variation: *n* times the *variance*.

Weight: A factor by which an observation is multiplied to make it compatible with others in a data set.

Weighted mean: $\sum w_i x_i/\sum w_i$ where w_i is the weight of observation i.

Z score: The *residual* when measured in units of standard deviation.

$$z_i = (x_i - \mu)/\sigma$$

13 Correlation and Regression

13.1 CORRELATION

The statistical tests described so far apply primarily to circumstances in which there is one variable, albeit treated in different ways, giving rise to such techniques as the t-test and analysis of variance. The assumption is that the data come from the same population.

In practice, and especially in geomatics and GIS, at least two things are going on at the same time: for instance, movement in the x-direction and independent movement in the y-direction. The question then becomes, are these somehow interdependent or, put another way, are they correlated? Also, how can we find the best compromise between all the conflicting data?

Given two variables X and Y where $X = (x_1, x_2,... x_i ... x_n)$ and $Y = (y_1, y_2,... y_i ... y_n)$ and x_i is possibly related to y_i, then treating X and Y independently:

$$\text{Mean of } X = \bar{x} = (1/n)\Sigma x$$

$$\text{Mean of } Y = \bar{y} = (1/n)\Sigma y$$

$$\text{Variance of } X = S^2_x = (1/n)\Sigma(x_i - \bar{x})^2$$

$$\text{Variance of } Y = S^2_y = (1/n)\Sigma(y_i - \bar{y})^2$$

A measure of the relationship between X and Y is given by $(1/n)\Sigma\{(x_i - \bar{x})(y_i - \bar{y})\}$. This is called the *covariance* of X and Y. It is often written as "Cov (X,Y)." With smaller data sets the relationship is often given as $(1/(n-1))\Sigma\{(x_i - \bar{x})(y_i - \bar{y})\}$ as this gives a better estimate of the whole population. Here we will assume large samples and use:

$$\text{Cov }(X,Y) = (1/n)\sum_{i=1}^{i=n}(x_i - \bar{x})(y_i - \bar{y})$$

Now $\Sigma(x_i - \bar{x})(y_i - \bar{y}) = \Sigma(x_i y_i - x_i \bar{y} - y_i \bar{x} + \bar{x}\ \bar{y})$ for all $i = 1$ to n.

$$= \Sigma(x_i y_i) - (x_1 + x_2 + \cdots + x_n)\ \bar{y} - (y_1 + y_2 + \cdots + y_n)\ \bar{x} + n\bar{x}\ \bar{y}$$

$$= \Sigma(x_i y_i) - n\bar{x}\ \bar{y} - n\bar{x}\ \bar{y} + n\bar{x}\ \bar{y} = \Sigma(x_i y_i) - n\bar{x}\ \bar{y}$$

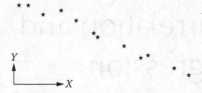

FIGURE 13.1 Correlation between X and Y.

Hence,

$$\text{Cov }(X,Y) = (1/n)\ \Sigma(x_iy_i) - \bar{x}\ \bar{y} = \frac{n\Sigma(x_iy_i) - n\bar{x}.n\bar{y}}{n^2} = \frac{n\Sigma(x_iy_i) - \Sigma x_i \Sigma y_i}{n^2}$$

The covariance has a similar form to the variance but unlike the latter, which is always positive (being the sum of squares), the covariance may be either positive or negative.

If we divide Cov (X, Y) by $\sqrt{S_x^2 S_y^2}$ (which is the *geometric mean* of the two variances), we obtain a ratio r that is known as the *coefficient of correlation*. An illustration of correlation can be seen in Figure 13.1, where there is a clear trend that relates in approximately linear form the relationship between X and Y, using the data in Example 13.1. Since

$$S_x^2 = \{\Sigma(x_i - \bar{x})^2\}/n = \{n\Sigma x_i^2 - (\Sigma x_i)^2\}/n^2$$

and

$$S_y^2 = \{\Sigma(y_i - \bar{y})^2\}/n = \{n\Sigma y_i^2 - (\Sigma y_i)^2\}/n^2$$

we can express the relationship between X and Y as r with

$$\text{Cov }(X,Y)/\sqrt{S_x^2 S_y^2} = r = \frac{n\Sigma x_i y_i - \Sigma x_i \Sigma y_i}{\sqrt{\{n\Sigma x_i^2 - (\Sigma x_i)^2\}}\sqrt{\{n\Sigma y_i^2 - (\Sigma u_i)^2\}}}$$

r is also known as the *Pearson correlation coefficient*.

Expressing r in this way makes it easier to program the calculation since, in tabular form, it is relatively easy to sum the raw observations without having first to calculate the means and then take the differences between the means and the observed quantities. We did this with the residuals in Chapter 12 when discussing analysis of variance. The process simplifies the way in which the residuals and hence the variance are calculated (Table 13.1).

It will be seen from the equation for r (Box 13.1), that if $x = y$, then the value of $r = 1$ and if $x = -y$ then $r = -1$. If we put $2x$ for x or any other multiple of x, then r remains the same; similarly with y. r will have a value of $+1$ for full agreement between x and y while $r = -1$ for full disagreement; $r = 0$ for no correlation between the two.

The significance level of r can be tested using the appropriate statistical tables. Alternatively, it can be shown by suitable rearrangement of the data that t from the t-test equates with $\frac{r\sqrt{(n-2)}}{\sqrt{(1-r^2)}}$ so by calculating this quantity, the t tables can be used.

EXAMPLE 13.1: EXAMPLE OF CORRELATION

In a set of 12 observations made of both x and y the following values were found:

Obs	x	y	x^2	y^2	xy
1	21	27	441	729	567
2	23	27	529	729	621
3	25	25	625	625	625
4	26	26	676	676	676
5	29	23	841	529	667
6	31	20	961	400	620
7	32	19	1024	361	608
8	35	17	1225	289	595
9	37	14	1369	196	518
10	38	15	1444	225	570
11	41	11	1681	121	451
12	43	9	1849	81	387
Total	$\Sigma x =$ 381	$\Sigma y =$ 233	$\Sigma x^2 =$ 12665	$\Sigma y^2 =$ 4961	$\Sigma xy =$ 6905

The coefficient of correlation $= r =$
$$r = \frac{n\sum x_i y_i - \sum x_i \sum y_i}{\sqrt{\left\{ n\sum x_i^2 - \left(\sum x_i\right)^2 \right\}}\sqrt{\left\{ n\sum y_i^2 - \left(\sum y_i\right)^2 \right\}}}$$

$$= (82860 - 88773)/\sqrt{\{(6819) * (5243)\}}$$

$$= -5913/5979 = -0.989$$

This shows a high level of negative correlation. Since there are two variables X and Y and 12 paired observations so there are $(n - 2) = 10$ degrees of freedom.

TABLE 13.1
Framework for Calculating r

Observation	x	y	x^2	y^2	xy
1	x_1	y_1	x_1^2	y_1^2	$x_1 y_1$
2	x_2	y_2	x_2^2	y_2^2	$x_2 y_2$
......	..	..	..	..	..
n	x_n	y_n	x_n^2	y_n^2	$x_n y_n$
Total	Σx	Σy	Σx^2	Σy^2	Σxy

BOX 13.1   COVARIANCE AND COEFFICIENT OF CORRELATION

For two independent variables X and Y:

$$\text{Covariance } (X, Y) = \text{Cov } (X, Y) = \frac{n\Sigma(x_i y_i) - \Sigma x_i \Sigma y_i}{n^2}$$

$$\text{Coefficient of Correlation} = r = \text{Cov } (X, Y)/\sqrt{\{S^2_x S^2_y\}}$$

$$r = \frac{n\sum x_i y_i - \sum x_i \sum y_i}{\sqrt{\left\{n\sum x_i^2 - \left(\sum x_i\right)^2\right\}}\sqrt{\left\{n\sum y_i^2 - \left(\sum y_i\right)^2\right\}}}$$

$$-1 \leq r \leq +1$$

We can present the relationships that we have so far pursued in this chapter in matrix form. In Chapter 12, we introduced the idea of *expectation* or what we would most likely expect for any set of samples x. We described this as $E(x)$. It is the mean value of all possible samples, assuming that there has been a very large, in effect infinite, number of observations. If we take the mean of different sets of samples, calling each mean m_i with an overall "mean of means" m, then the expected value $E(m)$ also equals the mean of the infinite population.

Using "ε," the Greek letter "epsilon" or "e" for error, we can define the residuals "ε_j" for the set with mean m_i as $\varepsilon_j = m_i - x_j$. Then, $E(\varepsilon_j) = 0 = E(m - m_i)$, because the sum of the residuals is, by the definition of the mean value, equal to zero.

For the population as a whole, we can also write $E(\varepsilon^2) = n\,\sigma^2$, that is, we would expect that if we made a sufficiently large number of observations of the whole population, then the mean of the sum of the squares of the residuals would equal the population variance. If we consider the individual values as ε_1, ε_2, and so on, we can regard this as a one row matrix $\mathbf{V}$. Its transpose is the column matrix $\mathbf{V^T}$ and $\mathbf{VV^T} = \Sigma\varepsilon^2_i =$ sum of the squares of the residuals = the variation = $n\,\sigma^2$.

We can then write $E(\varepsilon^2) = E(\mathbf{V}^2) = E(\mathbf{VV^T}) = n\,\sigma^2$.

Further, the product $(\mathbf{V^T V})$ is an $n * n$ matrix
$$\begin{bmatrix} \varepsilon_1^2 & \varepsilon_1\varepsilon_2 & \varepsilon_1\varepsilon_3 \dots \varepsilon_1\varepsilon_n \\ \varepsilon_2\varepsilon_1 & \varepsilon_2^2 & \varepsilon_2\varepsilon_3 \dots \varepsilon_2\varepsilon_n \\ \varepsilon_3\varepsilon_1 & \varepsilon_3\varepsilon_2 & \varepsilon_3^2 \dots \varepsilon_3\varepsilon_n \\ \dots \\ \varepsilon_n\varepsilon_1 & \varepsilon_n\varepsilon_2 & \varepsilon_n\varepsilon_3 \dots \varepsilon_n^2 \end{bmatrix}$$

where the leading diagonal is made up of the squares of the residuals and both the upper and lower triangular forms are made up of the elements of the covariance.

This matrix is known as the *dispersion matrix* or *variance–covariance matrix*.

We can write this as

$$
\begin{pmatrix}
\sigma_1^2 & \sigma_{12} & \sigma_{13} & ..... & \sigma_{1n} \\
\sigma_{21} & \sigma_2^2 & \sigma_{23} & ..... & \sigma_{2n} \\
\sigma_{31} & \sigma_{32} & \sigma_3^2 & ..... & \sigma_{3n} \\
.... & & & & \\
\sigma_{n1} & \sigma_{n2} & \sigma_{n3} ..... & \sigma_n^2
\end{pmatrix}
\quad \text{or as} \quad
\begin{pmatrix}
\sigma_1^2 & \sigma_{12} & \sigma_{13} & ..... & \sigma_{1n} \\
\sigma_{12} & \sigma_2^2 & \sigma_{23} & ..... & \sigma_{2n} \\
\sigma_{13} & \sigma_{23} & \sigma_3^2 & ..... & \sigma_{3n} \\
.... & & & & \\
\sigma_{1n} & \sigma_{2n} & \sigma_{3n} ..... & \sigma_n^2
\end{pmatrix}
$$

since $\varepsilon_1\varepsilon_2 = \varepsilon_2\varepsilon_1$ and therefore $\sigma_{12} = \sigma_{21}$, and so forth. If all the error or residual terms are truly random and independent, the covariance will be zero and the dispersion matrix reduces to the form

$$
\begin{pmatrix}
\sigma_1^2 & 0 & 0 & ..... & 0 \\
0 & \sigma_2^2 & 0 & ..... & 0 \\
0 & 0 & \sigma_3^2 & .... & 0 \\
.... & & & & \\
0 & 0 & 0 & ... & \sigma_n^2
\end{pmatrix}
$$

The smaller the variance σ^2, the smaller the errors are likely to be, and therefore the more likely our observations are to be accurate. Hence, we can weight our observations in inverse relation to their variance. This leads to the weight matrix

$$
\mathbf{W} =
\begin{pmatrix}
1/\sigma_1^2 & 0 & 0 & ..... & 0 \\
0 & 1/\sigma_2^2 & 0 & ..... & 0 \\
0 & 0 & 1/\sigma_3^2 & .... & 0 \\
.... & & & & \\
0 & 0 & 0 & ... & 1/\sigma_n^2
\end{pmatrix}
\quad \text{that we will refer to in Chapter 14.}
$$

13.2 REGRESSION

For certain data sets, it is probably easier to test levels of correlation by viewing them pictorially. Consider a set of points for which there are values of X and Y in pairs with (x_A, y_A), and (x_B, y_B), and so on, as shown in Figure 13.2. They appear to fall approximately along a straight line. There appears to be some degree of correlation between them and indeed some linear function, which gives a good approximation to the relationship between the set of points X and their partners Y.

The relationship could be expressed by calculating the coefficient of correlation between them (r). Expressing the relationship in the form of $y = mx + c$ provides additional information. The process of choosing the best-fit relationship between Y and X is called *regression*, in this case *simple linear regression*.

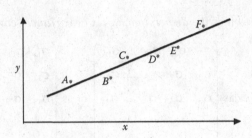

FIGURE 13.2   Regression line.

The line that has been drawn in Figure 13.2 does not actually pass through any of the points. It could have done but then it might not. Let us exaggerate the situation as in Figure 13.3.

Each point for which there has been a pair of observations (x_A, y_A), (x_B, y_B), and so on, there will be observed $(_o)$ quantities $(_ox_i, _oy_i)$ where i is any point. There will also be a set of computed $(_c)$ values such as $(_cx_i, _cy_i)$ that are derived from the equation for the straight line.

If we assume that the Y values are dependent on the X values and that $_ox_i = _cx_i$, implying that the X values are correctly determined, then $_oy_i$ will approximately equal $_cy_i$. In practice, there may be small differences as shown by the short vertical lines above or below points A, B, C, and so forth, in Figure 13.3. If the line has the form

$$y = mx + c$$

then

$$_oy_i = m * {_o}x_i + c + \varepsilon_i$$

where "ε" represents a small amount and is sometimes called the *error term*.

$$\varepsilon_i = \text{observed } y - \text{computed } y = {_o}y_i - {_c}y_i = {_o}y_i - c - m * {_o}x_i$$

As can be seen from Figure 13.3, ε will sometimes be positive and sometimes negative and in general will be small if there is good correlation between X and Y.

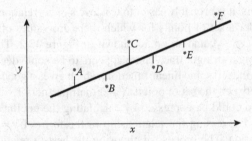

FIGURE 13.3   Residuals from the regression line.

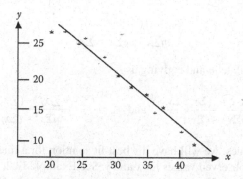

FIGURE 13.4 Example of a regression line.

The value of ε will depend not only upon the various values of X and Y but also upon the choice of the two constants m and c for the regression line (Figure 13.4).

If we assume that ε is normally distributed, the most likely values for m and c will be those that ensure the $\Sigma\varepsilon^2$ is a minimum. This is known as the *least square solution*. This fundamental assumption occurs in all sorts of problems in measurement science. In Chapter 12, we introduced the idea of residuals, variances, and standard deviations. ε is a residual and by minimizing $\Sigma\varepsilon^2$, we are maximizing the probability of achieving the best answer to whatever problem we may be addressing. As can be seen from the formula for the coefficient r, the maximum correlation is achieved when the denominator in the expression for r (i.e., the value of $\sqrt{S_x^2 S_y^2}$) is as small as possible. This occurs when the variance is a minimum.

In the general case, we need to choose m and c so that $\Sigma\varepsilon^2$ and therefore $\Sigma(y - c - mx)^2$ is a minimum. If there are n observations, this means that

$$\Sigma(y^2 + c^2 + m^2x^2 - 2cy - 2mxy + 2cmx)$$

is a minimum or if we call this relationship F, then

$$F = \Sigma y^2 + nc^2 + m^2\Sigma x^2 - 2c\Sigma y - 2m\Sigma xy + 2cm\Sigma x$$

is a minimum.

We can choose any values for m and c, but the minimum value for F will occur when the partial differentials with respect to m and c are zero. This will happen when:

(Differentiating with respect to c) $\partial F/\partial c = 2nc - 2\Sigma y + 2m\Sigma x = 0$

(Differentiating with respect to m) $\partial F/\partial m = 2m\Sigma x^2 - 2\Sigma xy + 2c\Sigma x = 0$

Hence,

$$m\Sigma x + cn = \Sigma y$$

and

$$m\Sigma x^2 + c\Sigma x = \Sigma xy$$

Thus, we can calculate m and c giving us

$$m = \frac{n\Sigma(x_i y_i) - (\Sigma x_i)(\Sigma y_i)}{n\Sigma x_i^2 - (\Sigma x_i)^2} \quad \text{and} \quad c = \frac{(\Sigma y_i)(\Sigma x_i^2) - (\Sigma x_i)(\Sigma x_i y_i)}{n\Sigma x_i^2 - (\Sigma x_i)^2}$$

Using these values, we will have the best-fit solution for a line that runs as close as possible to our observed values for x and y. See Example 13.2.

It may be that Y and X are not linear functions of each other. For example, Y may increase in the form of y^k where k is some constant (for example, 2 or 3 so that X is related to Y^2 or Y^3). We discussed aspects of linearization in Chapter 6, Section 6.4. Here, we note a relatively simple solution.

If we consider G rather than Y as the function that is dependent on X and if we let $G = \log(y^k) = k \log y$, then we can still plot the regression line between X and G by

---

EXAMPLE 13.2: LINEAR REGRESSION

For two variables X and Y where Y is dependent on X, the best-fit straight line takes the form $y = mx + c$, where

$$m = \frac{n\Sigma(x_i y_i) - (\Sigma x_i)(\Sigma y_i)}{n\Sigma x_i^2 - (\Sigma x_i)^2}$$

and

$$c = \frac{(\Sigma y_i)(\Sigma x_i^2) - (\Sigma x_i)(\Sigma x_i y_i)}{n\Sigma x_i^2 - (\Sigma x_i)^2}$$

For example, using the data in Example 13.1,

$$\Sigma x = 381; \quad \Sigma y = 233$$

$$\Sigma x^2 = 12665; \quad \Sigma y^2 = 4961; \quad \Sigma xy = 6905$$

$$m = (82860 - 88773)/(151980 - 145161)$$

$$= -5913/6819 = -0.87$$

$$c = (2950945 - 2630805)/(151980 - 145161)$$

$$= 320140/6819 = 46.95$$

Thus,

$$y = -0.87x + 46.45$$

using (log y_i) instead of (y_i), the value k merely affecting the slope of the line. Plotting one set of values on a logarithmic scale does not invalidate the process as long as the relationship is correctly interpreted.

The principles enunciated above also apply in three dimensions where X and Y are freely independent variables and Z is dependent on X and Y. If Z is some function of X and Y in which $Z = f(x, y)$, then we have a surface. In particular, if the relationship is linear, then the surface is a plane. In general, the surface representing Z is known as the *trend surface*. Trend surface analysis is an extension of linear regression into three or more dimensions.

The technique of finding the best fit by minimizing the variance of the residual values can be extended to the combination of different measurements that are subject to different conditions. This is possible because as shown in Boxes 13.2 and 13.3, if

$$X = ax + by + cz + ...$$

BOX 13.2   THE MEAN OF LINKED INDEPENDENT VARIABLES

Consider m observations of x and n of y that are combined to form X such that

$$X = ax + by$$

There are $m * n$ possible values for X since there are m ways of choosing x and n ways of choosing y. Let

$$x_i = \bar{x} + r_i \text{ and } \qquad y_i = \bar{y} + \rho_i$$

Thus, $\sum r_i = 0$ and $\sum \rho_i = 0$ (because $(1/m)\sum x_i = \bar{x}$ and $(1/n)\sum y_i = \bar{y}$). Also,

$$(1/m) \sum r^2_i = S^2_x \quad \text{and} \quad (1/n)\sum \rho^2_i = S^2_y$$

(since r_i and ρ_i are the residuals of x and y about their individual means). Thus,

$$\bar{X} = \frac{1}{mn} \sum_{i=1}^{m} \sum_{j=1}^{n} (ax_i + by_j)$$

$$= \frac{1}{mn} \sum_{i=1}^{m} \sum_{j=1}^{n} (a\bar{x} + b\bar{y} + ar_i + b\rho_j)$$

$$= \frac{1}{mn} \sum_{i=1}^{m} \sum_{j=1}^{n} (a\bar{x} + b\bar{y}) \text{ since both } \sum r_i \text{ and } \sum \rho_i = 0$$

$$= a\bar{x} + b\bar{y}$$

BOX 13.3   THE VARIANCE OF LINKED INDEPENDENT VARIABLES

Following on from Box 13.2, the variance of $X = S^2$

$$= \frac{1}{mn} \sum_{i=1}^{m} \sum_{j=1}^{n} (ax_i + by_j - a\bar{x} - b\bar{y})^2$$

$$= \frac{1}{mn} \sum_{i=1}^{m} \sum_{j=1}^{n} (ar_i + b\rho_j)^2$$

With a little manipulation this gives us $S^2 = a^2 S^2_x + b^2 S^2_y$. Once again, what is true for x and y will also be true for x, y, z, and so forth. When summing independent variables, the variances are additive. Thus, if $X = ax + by + cz + \ldots$ where a, b, c are numbers and x, y, and z are related variables and subject to all possible combinations, then

Mean of X is $\bar{X} = a\bar{x} + b\bar{y} + c\bar{z} + \ldots$

The variance of X is $S^2 = a^2 S^2_x + b^2 S^2_y + c^2 S^2_z + \ldots$

where a, b, c are any numbers, and if the means of x, y, z are $\bar{x}$, $\bar{y}$, $\bar{z}$, and so on, and the standard deviations are S_x, S_y, S_z, and so on, then:

Mean of $X = \bar{X} = a\bar{x} + b\bar{y} + c\bar{z} + \ldots$

Variance of $X = \sigma^2_X = a^2 S^2_x + b^2 S^2_y + c^2 S^2_z + \ldots$

In the derivations given in Boxes 13.2 and 13.3, if $a = b = c$, and so on, $= 1$, then for all the possible combinations of x, y, and z, and so on,

Mean of $X = \bar{x} + \bar{y} + \bar{z} \ldots$

Variance $= S^2_x + S^2_y + S^2_z + \ldots$

13.3   WEIGHTS

In what we have explored above, the assumption has been that all the observations are equally reliable and therefore the errors in each quantity have been distributed normally with the same value for their standard deviations σ. Sometimes, the observations are of different reliability and therefore we may wish to give them different *weight*, in effect leaning more heavily toward one set of observations than another.

Thus, if we give twice as much weight to y than to x and three times as much to z, then the weighted mean will be $(x + 2y + 3z)/6$. With weights w_1, w_2, and w_3, the weighted mean is given by

$$\frac{w_1 x + w_2 y + w_3 z}{w_1 + w_2 + w_3}$$

Example 13.3 illustrates the use of weights. Weights are relative values, not absolute values. From a theoretical perspective, according to the normal distribution discussed in Chapter 12, the probability of an error of unit weight $(x_i - \bar{x})$ is

EXAMPLE 13.3: EXAMPLE OF A WEIGHTED MEAN

A distance is measured by three different methods and the results are estimated as:

$$a \; 19.412 \pm 0.005 \qquad b \; 19.417 \pm 0.008 \qquad c \; 19.419 \pm 0.010$$

The estimated standard deviation for observation a was 0.005 and the variance was 0.000025 and its inverse is $40{,}000 = w_a$. For b the inverse $= w_b = 15{,}625$ and for $c = w_c = 10{,}000$.

Allocating weights according to these inverse variances and reducing the calculations by starting at 19.410, so that

$$a = \{19.410\} + 0.002 \pm 0.005, \; b = \{19.410\} + 0.007 \pm 0.008,$$
$$c = \{19.410\} + 0.009 \pm 0.010$$

The weighted mean is

$$\{19.410\} + \frac{(0.002 * w_a + 0.007 * w_b + 0.009 * w_c)}{(w_a + w_b + w_c)}$$

$$= \{19.410\} + (80 + 109.375 + 90)/(65625) = \{19.410\} + 0.004 = 19.414$$

The new residuals are -0.002, 0.003, and 0.005 while $(n - 1) = 2$. Hence,

$$\sigma^2 = \frac{1}{n-1} \sum_{i=1}^{i=n} w_i (x_i - \mu)^2$$

$$= (1/2) \frac{(0.002^2 * w_a + 0.003^2 * w_b + 0.005^2 * w_c)}{(w_a + w_b + w_c)}$$

This gives $\sigma = 0.002$. Thus, a better estimate is that the distance $= 19.414 \pm 0.002$.

proportional to $e^{-(x_i-\bar{x})^2/2\sigma^2}$. If the weight of observation $i = w_i$, then the probability of the error $(x_i - \bar{x})$ will be $e^{-w_i(x_i-\bar{x})^2/2\sigma^2}$. For the whole set of n values $x_1, x_2 \dots x_i \dots x_n$ the probability will be the sum of all these weighted values, namely,

$$e^{-\Sigma w_i(x_i-m)^2/2\sigma^2}$$

Since we may not know the true population mean, we can replace $\bar{x}$ by m. The quantity $e^{-\Sigma w_i(x_i-m)^2/2\sigma^2}$ will be a maximum (to give the maximum probability) when we choose m so that $\Sigma w_i(x_i-m)^2$ is a minimum. Differentiating with respect to m, we obtain $-2\Sigma w_i(x_i - m)$ and this must equal zero for a minimum value. Hence,

$$2\Sigma w_i x_i - 2m\Sigma w_i = 0 \quad \text{or} \quad m = \Sigma w_i x_i / \Sigma w_i$$

Let m have the value μ as the weighted mean of x, namely, $\mu = \Sigma w_i x_i / \Sigma w_i$. This is the most probable mean value for observations, assuming that all the weights are known. Note that if all observations are of the same weight, $\Sigma w_i = n$ and the mean $= \Sigma x_i / n$.

For weighted observations the variance becomes $\sigma^2 = \frac{1}{n} \sum_{i=1}^{i=n} w_i(x_i - \mu)^2$ or, more strictly speaking for a smaller data set, $\frac{1}{n-1} \sum_{i=1}^{i=n} w_i(x_i - \mu)^2$.

13.4   SPATIAL AUTOCORRELATION

The assumption behind regression is that each error term or residual is a random number that is independent of neighboring errors or residuals. The only assumption is that the sum of the squares of the residuals should be a minimum with the errors, weighted if necessary, forming part of a normal distribution. We illustrated a linear relationship between two data sets x and y in Example 13.1 where we derived the best-fit line assuming that, although the values of x and y were correlated, the residuals were not. If successive observations do not appear to be independent, then either we have fitted the wrong type of curve or there is some sequential connection between the residuals, in which case there is said to be *autocorrelation*. The former will happen if, for instance, we try to fit a straight-line $y = mx + c$ to a relationship of the form $y = ax^2 + c$ (see Figure 13.5, which shows the apparent correlation between residuals resulting from fitting a straight line to part of an elliptical curve). If the relationship is correctly modeled, then autocorrelation will occur when the residuals (as distinct from the original observations) have a correlation that is not zero. This means that they are not independent. This will be shown when the covariance is not

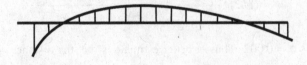

FIGURE 13.5   Residuals from fitting a straight line to an elliptical curve.

(approximately) zero, the condition being known as *autocovariance*. To be able to draw valid conclusions from many forms of statistical analysis, the residuals must be independent.

In the study of time series, the question can arise as to what are the chances of an event affecting a subsequent event, for example, of it being sunny tomorrow or the day after if it is sunny today. Consider two sets of observations x_s and x_t taken at time s and time t. Let their means be $E(x_s) = \bar{x}_s$ and $E(x_t) = \bar{x}_t$, where E means the "expected value" of x, which in this case is the mean value.

Similarly, the expected value of the variance $\sigma^2 = E[(x - E(x))^2] = E(x^2) - [E(x)]^2$. The autocorrelation between the two sets x_s and x_t is expressed as

$$r_{s,t} = \frac{E[(x_s - \bar{x}_s)(x_t - \bar{x}_t)]}{\sigma_s \sigma_t} = \frac{\sum_{i=1}^{n}(x_{s,i} - \bar{x}_s)(x_{t,i} - \bar{x}_t)}{(n-1)\sigma_s \sigma_t} = \frac{\sum_{i=1}^{n}(x_{s,i} - \bar{x}_s)(x_{t,i} - \bar{x}_t)}{\sqrt{\sum_{i=1}^{n}(x_{s,i} - \bar{x}_s)^2 \sum_{i=1}^{n}(x_{t,i} - \bar{x}_t)^2}}$$

If we combine the two sets into one, then $r_{s,t} = \frac{E[(x_s - \bar{x}_s)(x_t - \bar{x}_t)]}{\sigma^2}$ with, for instance, $t = s + 1$ or $s + 2$ for a one-day or two-day delay.

As we have seen in Section 13.1, the value r is often referred to as the *Pearson correlation coefficient* or more simply as the population correlation coefficient. This relationship is based on different values of x at different instances, with time here being taken as a linear or one-dimensional function.

In much geographical analysis, there is an assumption that neighbors influence what is going on around them and this means at different points in two- or three-dimensional space, as well as at different points in time. Geographers are interested in whether there is any apparent spatial correlation between quantities measured at different points. If there is, there is said to be *spatial autocorrelation*. This correlation can arise in three-dimensional space but here we will only consider two dimensions x and y.

A traditional test for spatial autocorrelation is known as *Moran's statistic I*, where

$$I = N\frac{\sum_{i=1}^{i=n}\sum_{j=1}^{j=n}W_{i,j}(x_i - \bar{x})(x_j - \bar{x})}{\left(\sum_{i=1}^{i=n}\sum_{j=1}^{j=n}W_{i,j}\right)\sum_{i=1}^{i=n}(x_i - \bar{x})^2} = \frac{N}{S} * \frac{\sum_{i=1}^{i=n}\sum_{j=1}^{j=n}W_{i,j}(x_i - \bar{x})(x_j - \bar{x})}{\sum_{i=1}^{i=n}(x_i - \bar{x})^2}$$

In this equation, N is the number of cases; x_i is the value of the variable at a particular location I, and x_j is its value at another location j; $\bar{x}$ is the mean of the N values of the variable x. W_{ij} is a weight matrix applied to the comparison between location i and location j and $S = \left(\sum_{i=1}^{i=n}\sum_{j=1}^{j=n}W_{i,j}\right)$ is the sum of all the weights so that the overall probability can remain 1.

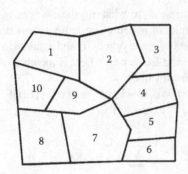

Area	1	2	3	4	5	6	7	8	9	10
1	0	1	0	0	0	0	0	0	1	1
2	1	0	1	1	0	0	0	0	1	0
3	0	1	0	1	0	0	0	0	0	0
4	0	1	1	0	1	0	1	0	0	0
5	0	0	0	1	0	1	1	0	0	0
6	0	0	0	0	1	0	1	0	0	0
7	0	0	0	1	1	1	0	1	1	0
8	0	0	0	0	0	0	1	0	1	1
9	1	1	0	0	0	0	1	1	0	1
10	1	0	0	0	0	0	0	1	1	0

FIGURE 13.6   A set of polygons and the contiguity matrix.

W_{ij} is a contiguity matrix that in the simple case of a rectangular grid would be made up of 0s and 1s; if sample j is in a polygon next to sample i the weight is 1 while if it does not share a boundary with i, then the weight is zero (see Figure 13.6). In a more complex case the weight could be based on the distance between the sample points i and j—for instance, the inverse $(1/d_{ij})$ or even the inverse square $(1/d_{ij}^2)$.

With a regular grid it could be based on only those squares north, south, east, or west of the chosen point (known from the game of chess as the *Rook's Case*), those on the diagonals (the *Bishop's Case*) or a combination of the two (the *Queen's* or *King's Case*)—see Figure 13.7.

Although we do not prove it here, if $E(I)$ is the expected value of the Moran statistic I then, if there is no autocorrelation,

$$E(I) = -1/(N-1)$$

If $I > E(I)$ there is positive spatial autocorrelation; if $I < E(I)$ there is negative spatial autocorrelation. If the value of N is large, then effectively $-1/(N-1)$ can be treated as zero.

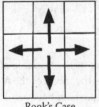
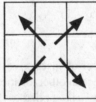
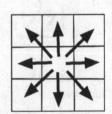

Rook's Case   Bishop's Case   Queen's or King's Case

FIGURE 13.7   Contiguity cases.

The z statistic for the distribution of I is given by

$$Z(I) = \{I - E(I)\}/\sigma_I$$

where σ_I is the standard deviation of I. The derivation of σ_I is complicated and we will not show it here. Various approximations appear in the literature. In one version, it can be shown to be:

$$\sigma_I^2 = \text{variance of } I =$$

$$\frac{\left\{n\left[(n^2 - 3n + 3)S_1 - nS_2 + 3S_0^2\right]\right\} - \left\{nk\left[(n^2 - n)S_1 - 2nS_2 + 6S_0^2\right]\right\}}{(n-1)(n-2)(n-3)S_0^2} - \frac{1}{(n-1)^2}$$

where n = number of observations, $S_0 = \sum_{i=1}^{i=n}\sum_{j=1}^{j=n} W_{i,j}$ = the sum of the elements in the spatial weight matrix

$$S_1 = \frac{\sum_{i=1}^{i=n}\sum_{j=1}^{j=n}(W_{i,j} + W_{j,i})^2}{2}$$

If the weight matrix is symmetric,

$$S_1 = 2\sum_{i=1}^{i=n}\sum_{j=1}^{j=n}(W_{i,j})^2$$

then

$$S_2 = \sum_{i=1}^{i=n}(W_{i,m} + W_{m,i})^2$$

W_{im} is the sum of ith column of the weight matrix $= \sum_{j=1}^{j=n} W_{ij}$; W_{mi} is the sum of the ith row $= \sum_{j=1}^{j=n} W_{ji}$. If the matrix is symmetric, then

$$S_2 = 4\sum_{i=1}^{i=n}\left(\sum_{j=1}^{j=n} W_{i,j}\right)^2$$

Finally,

$$k = \frac{\sum_{i=1}^{n}(x_i - \bar{x})^4}{\left(\sum_{i=1}^{n}(x_i - \bar{x})^2\right)^2}$$

This is based on the sum of each value in the data matrix minus the mean value.

By calculating the variance, the standard deviation of I can be derived. $Z(I)$ or $\{I - E(I)\}/\sigma_I$ then follows the normal distribution and its significance can therefore be tested using standard statistical tables.

An alternative measure of spatial autocorrelation is known as *Geary's C statistic*. It is calculated from

$$C = \frac{(n-1)}{2} * \frac{\sum\limits_{i=1}^{i=n}\sum\limits_{j=1}^{j=n} W_{i,j}(x_i - x_j)^2}{\left(\sum\limits_{i=1}^{i=n}\sum\limits_{j=1}^{j=n} W_{i,j}\right)\sum\limits_{i=1}^{i=n}(x_i - \bar{x})^2}$$

The theoretical mean for $C = 1$ and the value of C will lie between 0 and 2. If its value is 1 then there is no spatial autocorrelation. If $0 < C < 1$ there is positive correlation while if $1 < C < 2$, there will be negative correlation.

The two statistics—Geary's C and Moran's I—are similar and inversely related but are not the same. Moran's I depends on the difference between each value and the mean for the data set and is similar to the Pearson correlation coefficient and is a covariance test. Geary's C is more concerned with neighboring values and is in essence a variance test.

SUMMARY

Autocorrelation: A condition occurring when successive items in a sequence are correlated.

Autocovariance: The condition occurring when adjacent observations in a sequence are not independent.

Coefficient of correlation: The extent of correspondence between two variables, expressed by the coefficient r in the form

$$r = \frac{n\sum x_i y_i - \sum x_i \sum y_i}{\sqrt{\left\{n\sum x_i^2 - \left(\sum x_i\right)^2\right\}}\sqrt{\left\{n\sum y_i^2 - \left(\sum y_i\right)^2\right\}}}$$

Covariance: A measure of the association between two variables X and Y

$$\text{Cov}(X, Y) = (1/n)\sum_{i=1}^{i=n}(x_i - \bar{x})(y_i - \bar{y})$$

Error term: The difference between the observed and computed quantity.

Expectation E: A term for the probability that an event will occur.

Geary's statistic C: A measure of spatial autocorrelation in which

$$C = \frac{(n-1)}{2} * \frac{\sum\limits_{i=1}^{i=n}\sum\limits_{j=1}^{j=n} W_{i,j}(x_i - x_j)^2}{\left(\sum\limits_{i=1}^{i=n}\sum\limits_{j=1}^{j=n} W_{i,j}\right)\sum\limits_{i=1}^{i=n}(x_i - \bar{x})^2}$$

Geometric mean: The nth root of the product of n numbers, for example, for three numbers a, b, c, it would be the cube root of $a * b * c$.

Moran's statistic I: A measure of spatial autocorrelation in which

$$I = N \frac{\sum\limits_{i=1}^{i=n}\sum\limits_{j=1}^{j=n} W_{i,j}(x_i - \bar{x})(x_j - \bar{x})}{\left(\sum\limits_{i=1}^{i=n}\sum\limits_{j=1}^{j=n} W_{i,j}\right)\sum\limits_{i=1}^{i=n}(x_i - \bar{x})^2}$$

$$= \frac{N}{S} * \frac{\sum\limits_{i=1}^{i=n}\sum\limits_{j=1}^{j=n} W_{i,j}(x_i - \bar{x})(x_j - \bar{x})}{\sum\limits_{i=1}^{i=n}(x_i - \bar{x})^2}$$

Regression: A measure of association between a dependent variable and one or more independent variables.

Residual: The difference between an observed and a predicted value. Also referred to as the *error term*.

Simple linear regression: A linear measure of association between one dependent and one independent variable.

Spatial autocorrelation: The property that mapped data possess when because of their location they are not independent.

Trend surface: A form of regression in which the third dimension Z is dependent on X and Y.

Weight: A relative value of the importance of an observation.

14 Best-Fit Solutions

14.1 LEAST SQUARE SOLUTIONS

Returning to the ideas behind regression we have seen that, given a set of observations in which each observation or measurement may contain a small amount of error, we can obtain the most likely value of a quantity by distributing the residuals on the basis of minimum variance, which means that the sum of the squares of the residuals is a minimum. This is called the principle of *least squares*. Although the process of calculating the minimum sum of the squares can be applied to any set of residuals, the interpretation of the results assumes that the errors are normally distributed and of equal weight or that appropriate weights have been applied.

The weighted mean gives the overall most likely value of any data set while the probable errors are, in accordance with the normal distribution, of the form $e^{-w_i(x_i-\bar{x})^2/2\sigma^2}$. The maximum probability will occur when this is a maximum, which will happen when $\sum w_i(x_i-\bar{x})^2$ is a minimum.

We can extend this idea not just to fitting lines or surfaces, but when dealing with a set of observations, various combinations of which are subject to a set of conditions that can be determined. In the general case, if we have n observed quantities x_1, x_2, x_3,.. x_n that are linearly related and m independent relationships, we can express this in the form

$$ax_1 + bx_2 + cx_3 + \cdots + nx_n + l = 0$$

where l is a numerical value. We can write such relationships in longhand as shown in Table 14.1. Alternatively, we can use the matrix form $\mathbf{MX} + \mathbf{L} = 0$ where $\mathbf{M}$ is the $(m * n)$ matrix made up from the as, bs, and cs, and so forth, $\mathbf{X}$ is the column vector of unknowns, and $\mathbf{L}$ is the column vector of constants. In geomatics, the observed quantities are normally angles, distances, or time; but the technique can be used in all sorts of applications from optimizing financial balance sheets through to quality control procedures. The key point is that linear relationships can be established between the independent variables; where this is not immediately so, the relationships will need to be "linearized," an example of which is discussed later in the case of a set of angles. If $m = n$, that is, if we have the same number of independent equations as there are unknown, then we can solve for all the values of x_i precisely. If $m \neq n$, then we have to establish the best-fit solution.

Assuming that the observations are of equal weight, and that there are small errors in each of the observations x_i, then for equation j in Table 14.1, let

$$v_j = a_jx_1 + b_jx_2 + c_jx_3 + \cdots + n_jx_n + l_j$$

TABLE 14.1

Conditions to Be Satisfied: MX + L = 0

1:     $a_1x_1 + b_1x_2 + c_1x_3 + \cdots f_1x_i + \cdots + n_1x_n + l_1 = 0$

2:     $a_2x_1 + b_2x_2 + c_2x_3 + \cdots f_2x_i + \cdots + n_2x_n + l_2 = 0$

$\vdots$

j: $a_jx_1 + b_jx_2 + c_jx_3 + \cdots f_jx_j + \cdots + n_jx_n + l_j = 0$

$\vdots$

m: $a_mx_1 + b_mx_2 + c_mx_3 + \cdots f_mx_i \quad \cdots + n_mx_n + l_m = 0$

---

where j takes the values $1, 2,\ldots m$. v_j is known as the residual value. If we assume that we want to minimize the errors in each of the m equations, then we have to make the sum of the squares of the quantities v_j as small as possible by suitably amending each of the values of x_i. This will give us the most likely consistent solution.

v_j has $(n + 1)$ terms so that v_j^2 would have $(n + 1)^2$ terms and the sum of all these, Σv_j^2, would have $m * (n + 1)^2$ terms. However, we can simplify matters.

To minimize Σv_j^2 for x_1, we need to make the partial derivative $\partial(\Sigma v_j^2)/\partial x_1 = 0$. To minimize Σv_j^2 for x_2, we need to make $\partial(\Sigma v_j^2)/\partial x_2 = 0$, and so forth.

$$\frac{\partial(\Sigma v_j^2)}{\partial x_1} = \frac{\partial(v_1^2 + v_2^2 + v_3^2 + \ldots)}{\partial x_1} = 2\left\{ v_1 \frac{\partial v_1}{\partial x_1} + v_2 \frac{\partial v_2}{\partial x_1} + \cdots + v_m \frac{\partial v_m}{\partial x_1} \right\}$$

$= 0$ when things reach a minimum.

$$\frac{\partial v_1}{\partial x_1} = a_1, \text{ and so on, hence,}$$

$$v_1 \frac{\partial v_1}{\partial x_1} = a_1\{a_1x_1 + b_1x_2 + c_1x_3 + \cdots f_1 x_i + \cdots + n_1x_n + l_1\}$$

$$v_2 \frac{\partial v_2}{\partial x_1} = a_2\{a_2x_1 + b_2x_2 + c_2x_3 + \cdots f_2x_i + \cdots + n_2x_n + l_2\}, \text{ and so on,}$$

for all the m equations. If we add all these together, we obtain for x_1

$$\frac{\partial(\Sigma v_j^2)}{\partial x_1} = 2\{(\Sigma a_j^2)x_1 + (\Sigma a_jb_j)x_2 + \cdots (\Sigma a_jf_j)x_i + \cdots + (\Sigma a_jn_j)x_n + (\Sigma a_jl_j)\} = 0$$

for minimum variation and therefore maximum probability. (Σa_jf_j means "add all the m values $a_1f_1 + a_2f_2 + \cdots + a_mf_m$.") Similarly,

$$\frac{\partial(\Sigma v_j^2)}{\partial x_2} = 2\{(\Sigma a_jb_j)x_1 + (\Sigma b_j^2)x_2 + \cdots (\Sigma b_jf_j)x_i + \cdots + (\Sigma b_jn_j)x_n + (\Sigma b_jl_j)\} = 0$$

TABLE 14.2

The Normal Equations: $M^T(MX + L) = 0$

$[aa]x_1 + [ab]x_2 + [ac]x_3 + \cdots + [af]x_i + \cdots + [an]x_n + [al] = 0$

$[ab]x_1 + [bb]x_2 + [bc]x_3 + \cdots + [bf]x_i + \cdots + [bn]x_n + [bl] = 0$

$[ac]x_1 + [bc]x_2 + [cc]x_3 + \cdots + [cf]x_i + \cdots + [cn]x_n + [cl] = 0$

$\vdots$

$[an]x_1 + [bn]x_2 + [cn]x_3 + \cdots + [fn]x_i + \cdots + [nn]x_n + [nl] = 0$

This gives us n equations for the n unknown adjusted values of x.

If we write $(\sum a_j^2)$ as $[aa]$, $(\sum a_j b_j)$ as $[ab]$, and so forth, for the sake of simplicity, then we can express these as in Table 14.2. Thus,

$$[aa] = a_1^2 + a_2^2 + a_3^2 + \cdots + a_m^2$$

$$[ab] = a_1 b_1 + a_2 b_2 + a_3 b_3 + \cdots + a_m b_m, \text{ and so forth.}$$

The equations in Table 14.2 are known as the *normal equations*. Note that in these n equations, the coefficients for x are symmetrical about the diagonal, which itself is made up from the sum of the squares of the relevant coefficients.

We wrote the condition equations in Table 14.1 as $MX + L = 0$, where

$$X = \begin{pmatrix} x_1 \\ x_2 \\ x_3 \\ \cdots \\ x_n \end{pmatrix} ; \quad L = \begin{pmatrix} l_1 \\ l_2 \\ l_3 \\ \cdots \\ l_m \end{pmatrix}$$

and

$$M = \begin{pmatrix} a_1 & b_1 & c_1 & \cdots & n_1 \\ a_2 & b_2 & c_2 & \cdots & n_2 \\ a_3 & b_3 & c_3 & \cdots & n_3 \\ \cdots & \cdots & \cdots & \cdots & \cdots \\ a_m & b_m & c_m & \cdots & n_m \end{pmatrix}$$

Note that

$$M^T = \begin{pmatrix} a_1 & a_2 & a_3 & \cdots & a_m \\ b_1 & b_2 & b_3 & \cdots & b_m \\ c_1 & c_2 & c_3 & \cdots & c_m \\ \cdots & \cdots & \cdots & \cdots & \cdots \\ n_1 & n_2 & n_3 & \cdots & n_m \end{pmatrix}$$

so that

$$\mathbf{M^TM} = \begin{pmatrix} [aa] & [ab] & [ac] & \cdots & [an] \\ [ab] & [bb] & [bc] & \cdots & [bn] \\ [ac] & [bc] & [cc] & \cdots & [cn] \\ \cdots & \cdots & \cdots & \cdots & \cdots \\ [an] & [bn] & [cn] & \cdots & [nn] \end{pmatrix}$$

or

$$\mathbf{M^TM} = \begin{pmatrix} a_1a_1 + a_2a_2 + a_3a_3 + \cdots + a_ma_m & a_1b_1 + a_2b_2 + a_3b_3 + \cdots + a_mb_m & \cdots & a_1n_1 + a_2n_2 + a_3n_3 + \cdots + a_mn_m \\ b_1a_1 + b_2a_2 + b_3a_3 + \cdots + b_ma_m & b_1b_1 + b_2b_2 + c_2c_2 + \cdots + b_mb_m & \cdots & b_1n_1 + b_2n_2 + c_3n_3 + \cdots + b_mn_m \\ \cdots\cdots & & & \\ n_1a_1 + n_2a_2 + n_3a_3 + \cdots + n_ma_m & n_1b_1 + n_2b_2 + n_3b_3 + \cdots + n_mb_m & \cdots & n_1n_1 + n_2n_2 + n_3n_3 + \cdots + n_mn_m \end{pmatrix}$$

and

$$\mathbf{MM^T} = \begin{pmatrix} a_1a_1 + b_1b_1 + c_1c_1 + \cdots + n_1n_1 & a_1a_2 + b_1b_2 + c_1c_2 + \cdots + n_1n_2 & \cdots & a_1a_m + b_1b_m + c_1c_m + \cdots + n_1n_m \\ a_2a_1 + b_2b_1 + c_2c_1 + \cdots + n_2n_1 & a_2a_2 + b_2b_2 + c_2c_2 + \cdots + n_2n_2 & \cdots & a_2a_m + b_2b_m + c_2c_m + \cdots + n_2n_m \\ \cdots\cdots & & & \\ a_ma_1 + b_mb_1 + c_mc_1 + \cdots + n_mn_1 & a_ma_2 + b_mb_2 + c_mc_2 + \cdots + n_mn_2 & \cdots & a_ma_m + b_mb_m + c_mc_m + \cdots + n_mn_m \end{pmatrix}$$

Also,

$$\mathbf{M^TL} = \begin{pmatrix} a_1 & a_2 & a_3 & \cdots & a_m \\ b_1 & b_2 & b_3 & \cdots & b_m \\ c_1 & c_2 & c_3 & \cdots & c_m \\ \cdots & \cdots & \cdots & \cdots & \cdots \\ n_1 & n_2 & n_3 & \cdots & n_m \end{pmatrix} * \begin{pmatrix} l_1 \\ l_2 \\ l_3 \\ \cdots \\ l_m \end{pmatrix} = \begin{pmatrix} [al] \\ [bl] \\ [cl] \\ \cdots \\ [nl] \end{pmatrix}$$

TABLE 14.3
Weighted Normal Equations: $M^TWMX + M^TWL = 0$

$[waa]x_1 + [wab]x_2 + \cdots + [waf]x_i + \cdots + [wan]x_n + [wal] = 0$

$[wab]x_1 + [wbb]x_2 + \cdots + [wbf]x_i + \cdots + [wbn]x_n + [wbl] = 0$

$[wac]x_1 + [wbc]x_2 + \cdots + [wcf]x_i + \cdots + [wcn]x_n + [wcl] = 0$

$\vdots$

$[wan]x_1 + [wbn]x_2 + \cdots + [wfn]x_i + \cdots + [wnn]x_n + [wnl] = 0$

M^T has n rows and m columns, $\dot{M}$ has m rows and n columns so that M^TM is an $n * n$ matrix and MM^T is $m * m$. Thus, Table 14.2 can be written in shorthand as $M^T(MX + L) = 0$.

Once we have formulated the necessary conditions, it becomes a matter of routine to form and transform the relevant matrices and solve for the n unknowns. An example of how this can be used is given in Example 14.1.

If the m equations are of unequal weight, then instead of minimizing Σv_j^2, we need to minimize $\Sigma w_j v_j^2$ and the normal equations become as shown in Table 14.3 where W is the weight matrix. Assuming that there is no correlation between the residuals, the variance/covariance matrix will reduce to a diagonal matrix as will the weight matrix W.

In Example 14.1 we have $m = 5$ conditions affecting $n = 3$ unknowns and we reduced these to three independent equations linking these three unknowns. We may, however, have more unknowns than conditions, that is, $m < n$. This will arise when we make redundant observations, for instance, when we measure the three angles of a triangle where the only condition is that the sum of the three observations must add up to $180°$. Here we will have $x_1 + x_2 + x_3 + r = 0$ where $x_1, x_2,$ and x_3 are the corrections to be applied to each angle and r is the amount by which our observed angles fail to add to $180°$. In this case we have:

$$M = (1, 1, 1), \quad M^TM = \begin{pmatrix} 1 & 1 & 1 \\ 1 & 1 & 1 \\ 1 & 1 & 1 \end{pmatrix}$$

and hence the three equations "$x_1 + x_2 + x_3 + r = 0$" that result from $M^TMX + M^TL$ are identical and therefore offer no solution. We must therefore take a different approach, as outlined in Section 14.2.

EXAMPLE 14.1: SOLVING FOR MORE EQUATIONS THAN THERE ARE UNKNOWNS

Let

$$x + y + z - 9.2 = 0$$

$$x + y - z - 0.9 = 0$$

$$2x + 3y + 2z - 20.6 = 0$$

$$3x - 4y + 5z - 14.1 = 0$$

$$x + 5y - 3z - 4.9 = 0$$

We have three values (x, y, z) with five equations, assumed to be of equal weight. In matrix form, $\mathbf{MX} + \mathbf{L} = 0$, where

$$\mathbf{M} = \begin{pmatrix} 1 & 1 & 1 \\ 1 & 1 & -1 \\ 2 & 3 & 2 \\ 3 & -4 & 5 \\ 1 & 5 & -3 \end{pmatrix} \quad \mathbf{X} = \begin{pmatrix} x \\ y \\ z \end{pmatrix} \quad \mathbf{L} = \begin{pmatrix} -9.2 \\ -0.9 \\ -20.6 \\ -14.1 \\ -4.9 \end{pmatrix}$$

$$\mathbf{M^T M} = \begin{pmatrix} 1 & 1 & 2 & 3 & 1 \\ 1 & 1 & 3 & -4 & 5 \\ 1 & -1 & 2 & 5 & -3 \end{pmatrix} * \begin{pmatrix} 1 & 1 & 1 \\ 1 & 1 & -1 \\ 2 & 3 & 2 \\ 3 & -4 & 5 \\ 1 & 5 & -3 \end{pmatrix} = \begin{pmatrix} 16 & 1 & 16 \\ 1 & 52 & -29 \\ 16 & -29 & 40 \end{pmatrix}$$

$$\mathbf{M^T L} = \begin{pmatrix} 1 & 1 & 2 & 3 & 1 \\ 1 & 1 & 3 & -4 & 5 \\ 1 & -1 & 2 & 5 & -3 \end{pmatrix} * \begin{pmatrix} -9.2 \\ -0.9 \\ -20.6 \\ -14.1 \\ -4.9 \end{pmatrix} = \begin{pmatrix} -98.5 \\ -40 \\ -105.3 \end{pmatrix}$$

Thus, $\mathbf{M^T(MX + L)}$ or $\mathbf{M^T MX + M^T L}$ gives three equations, namely,

$$16x + 1y + 16z - 98.5 = 0$$

$$1x + 52y - 29z - 40 = 0$$

$$16x - 29y + 40z - 105.3 = 0$$

from which

$$x = 2.02, y = 2.93, z = 3.95$$

The residuals are $-0.3, 0.1, 0.13, -0.01,$ and -0.08. Their sum of squares $= 0.1234$.

14.2   SURVEY ADJUSTMENTS

In much of geomatics, especially in position fixing, we deliberately make redundant observations in order to improve the accuracy of our work, both by identifying and thereby being able to eliminate any gross errors in measurement, but more particularly, by reducing the overall standard deviations of the residual errors in our measurements. This increases the probability of obtaining a better answer.

Consider then the case where we have more observations than there are conditions to satisfy. We must satisfy the basic condition while also ensuring that $\sum w_i \varepsilon_i^2$ is a minimum where w_i is the weight of each observation and ε_i is the correction to the observed quantity (o_i). i is an observation that takes the value from one to n where n is the total number of observed quantities.

If the best answer for each observation is x_i where $x_i = o_i + \varepsilon_i$, then for $\sum w_i \varepsilon_i^2$ to be a minimum, the partial differential with respect to ε must be zero. Thus,

$$2w_1\varepsilon_1\delta\varepsilon_1 + 2w_2\varepsilon_2\delta\varepsilon_2 + \cdots + 2w_n\varepsilon_n\delta\varepsilon_n = 0$$

or

$$w_1\varepsilon_1\delta\varepsilon_1 + w_2\varepsilon_2\delta\varepsilon_2 + \cdots + w_n\varepsilon_n\delta\varepsilon_n = 0$$

The weights can be expressed as a diagonal matrix $\mathbf{W}$ in the form

$$\begin{pmatrix} w_1 & 0 & 0......0 \\ 0 & w_2 & 0......0 \\ & ....0 \\ 0 & 0 & 0.....w_n \end{pmatrix}$$

or, as shown at the end of Section 13.1 in Chapter 13,

$$\mathbf{W} = \begin{pmatrix} 1/\sigma_1^2 & 0 & 0 ...... & 0 \\ 0 & 1/\sigma_2^2 & 0 ...... & 0 \\ 0 & 0 & 1/\sigma_3^2.... & 0 \\ .... & & & \\ 0 & 0 & 0 ... & 1/\sigma_n^2 \end{pmatrix}$$

In the general case, we have n observations of x and m conditions to satisfy a set of equations as shown in Table 14.4. Here, $x_i = o_i + \varepsilon_i$ where x is the adjusted value and o the observed value, such as one of a set of angles, the difference between them being the correction terms ε. $\mathbf{M}$ is the $m * n$ matrix of conditions, $\mathbf{O}$ and ε are $n * 1$ column matrices.

TABLE 14.4

Observations and Conditions: M(O + ε) + L = 0

$$a_1(o_1 + \varepsilon_1) + b_1(o_2 + \varepsilon_2) + \dots f_1(o_i + \varepsilon_i)\dots + n_1(o_n + \varepsilon_n) + l_1 = 0$$
$$a_2(o_1 + \varepsilon_1) + b_2(o_2 + \varepsilon_2) + \dots f_2(o_i + \varepsilon_i)\dots + n_2(o_n + \varepsilon_n) + l_2 = 0$$
$$\vdots$$
$$a_i(o_1 + \varepsilon_1) + b_i(o_2 + \varepsilon_2) + \dots f_i(o_i + \varepsilon_i)\dots + n_i(o_n + \varepsilon_n) + l_i = 0$$
$$\vdots$$
$$a_m(o_1 + \varepsilon_1) + b_m(o_2 + \varepsilon_2) + \dots f_m(o_i + \varepsilon_i)\dots + n_m(o_n + \varepsilon_n) + l_m = 0$$

The m relationships in Table 14.4 must be fulfilled while ensuring that $\Sigma\varepsilon^2$ is a minimum. There are n unknown errors or corrections that we must apply to satisfy the m relationships. If we consider the uncorrected observations, then we will have m equations of the form

$$a_i o_1 + b_i o_2 + c_i o_3 + \cdots f_i o_i + \cdots + n_i o_n + l_i = r_i$$

Here, r_i is the residual for condition i. The equations are of the form $\mathbf{MO + L = R}$ and the least square solution requires both $\Sigma\varepsilon^2$ and Σr^2 to be a minimum.

We can write the m equations in Table 14.4 as Table 14.5 in which ε is the column vector of corrections to the observed quantities, and $\mathbf{R}$ is the column vector of the residuals.

If we partially differentiate the equations in Table 14.5, assuming that ε is the variable, then we derive the m equations with n unknowns shown in Table 14.6, which we call E_1, E_2, and so forth.

Next we introduce m constants in the form of the column vector

$$\mathbf{K} = \begin{pmatrix} K_1 \\ K_2 \\ \dots \\ K_i \\ \dots \\ K_m \end{pmatrix}$$

TABLE 14.5

The Relationships to Be Optimized: Mε + R = 0

$$a_1\varepsilon_1 + b_1\varepsilon_2 + c_1\varepsilon_3 + \cdots f_1\varepsilon_i + \cdots + n_1\varepsilon_n + r_1 = 0$$
$$a_2\varepsilon_1 + b_2\varepsilon_2 + c_2\varepsilon_3 + \cdots f_2\varepsilon_i + \cdots + n_2\varepsilon_n + r_2 = 0$$
$$\vdots$$
$$a_i\varepsilon_1 + b_i\varepsilon_2 + c_i\varepsilon_3 + \cdots f_i\varepsilon_i + \cdots + n_i\varepsilon_n + r_i = 0$$
$$\vdots$$
$$a_m\varepsilon_1 + b_m\varepsilon_2 + c_m\varepsilon_3 + \cdots f_m\varepsilon_i \cdots + n_m\varepsilon_n + r_m = 0$$

TABLE 14.6
The Differential Equations: $M(\delta\varepsilon) = 0$

$E_1 \qquad a_1\,\delta\varepsilon_1 + b_1\,\delta\varepsilon_2 + c_1\,\delta\varepsilon_3 + ...\,f_1\,\delta\varepsilon_i + \cdots + n_1\,\delta\varepsilon_n = 0$

$E_2 \qquad a_2\,\delta\varepsilon_1 + b_2\,\delta\varepsilon_2 + c_2\,\delta\varepsilon_3 + ...\,f_2\,\delta\varepsilon_i + \cdots + n_2\,\delta\varepsilon_n = 0$

$\vdots$

$E_i \qquad a_i\,\delta\varepsilon_1 + b_i\,\delta\varepsilon_2 + c_i\,\delta\varepsilon_3 + ...\,f_i\,\delta\varepsilon_i + \cdots + n_i\,\delta\varepsilon_n = 0$

$\vdots$

$E_m \qquad a_m\,\delta\varepsilon_1 + b_m\,\delta\varepsilon_2 + c_m\,\delta\varepsilon_3 + ...\,f_m\,\delta\varepsilon_{i...} + n_m\,\delta\varepsilon_n = 0$

Initially, the numerical values of K_1, K_2, and so forth, are unknown. We apply each of these to its equivalent differentiated equation in Table 14.6 and combine all the equations together by adding them in the form $M(\delta\varepsilon)K$. The equations then become $(E_1) * K_1 + (E_2) * K_2 + ...$, and so on, where by E_1 we mean $(a_1\,\delta\varepsilon_1 + b_1\,\delta\varepsilon_2 + c_1\,\delta\varepsilon_3 + \cdots f_1\,\delta\varepsilon_i + \cdots + n_1\,\delta\varepsilon_n)$. This gives us a single equation:

$$K_1(a_1\,\delta\varepsilon_1 + b_1\,\delta\varepsilon_2 + c_1\,\delta\varepsilon_3 + \cdots f_1\,\delta\varepsilon_i + \cdots + n_1\,\delta\varepsilon_n)$$

$$+ K_2(a_2\,\delta\varepsilon_1 + b_2\,\delta\varepsilon_2 + c_2\,\delta\varepsilon_3 + \cdots f_2\,\delta\varepsilon_i + \cdots + n_2\,\delta\varepsilon_n) +$$

$$+ ......$$

$$+ K_m(\alpha_\mu\,\delta\varepsilon_1 + b_m\,\delta\varepsilon_2 + c_m\,\delta\varepsilon_3 + \cdots f_m\,\delta\varepsilon_i \cdots + n_m\,\delta\varepsilon_n) = 0$$

Hence, on rearranging terms

$$(a_1K_1 + a_2K_2 + \cdots + a_mK_m)\,\delta\varepsilon_1$$

$$+ (b_1K_1 + b_2K_2 + \cdots + b_mK_m)\,\delta\varepsilon_2 +$$

$$+ ......$$

$$+ (n_1K_1 + n_2K_2 + \cdots + n_mK_m)\,\delta\varepsilon_n = 0$$

For this to be identical with the condition that the weighted sum of the squares of the residuals is a minimum, namely, as we showed earlier:

$$w_1\varepsilon_1\,\delta\varepsilon_1 + w_2\varepsilon_2\,\delta\varepsilon_2 + \cdots + w_n\varepsilon_n\,\delta\varepsilon_n = 0$$

then we must have the n relationships shown in Table 14.7.

TABLE 14.7
The Relationships between the Correlatives: $M^\mathsf{T}K = W\varepsilon$

$(a_1K_1 + a_2K_2 + \cdots + a_mK_m) = w_1\varepsilon_1$

$(b_1K_1 + b_2K_2 + \cdots + b_mK_m) = w_2\varepsilon_2$

$\vdots$

$(n_1K_1 + n_2K_2 + \cdots + n_mK_m) = w_n\varepsilon_n$

TABLE 14.8

The Equations for the Correlatives

1. $a_1(a_1K_1 + a_2K_2 + \cdots + a_mK_m)/w_1$
   $+ b_1(b_1K_1 + b_2K_2 + \cdots + b_mK_m)/w_2 + \cdots + r_1 = 0$

2. $a_2(a_1K_1 + a_2K_2 + \cdots + a_mK_m)/w_1$
   $+ b_2(b_1K_1 + b_2K_2 + \cdots + b_mK_m)/w_2 + \cdots + r_2 = 0$

   $\vdots$

m. $a_m(a_1K_1 + a_2K_2 + \cdots + a_mK_m)/w_1$
   $+ b_m(b_1K_1 + b_2K_2 + \cdots + b_mK_m)/w_2 + \cdots + r_m = 0$

This gives us n relationships linking the m unknown quantities K with the n unknown quantities ε. Put another way, $\varepsilon_1 = (a_1K_1 + a_2K_2 + \cdots + a_mK_m)/w_1$, and so on. The quantities K are called the *correlatives*. Substituting the values of ε in our original equations (Table 14.5), we obtain the m equations for the correlatives as given in Table 14.8, which we can rearrange as shown in Table 14.9. We can then solve these m linear equations to give us the values K, which we can then substitute back into the equations given in Table 14.7 to find the numerical values ε.

To summarize, if the column vector of n corrections that must be applied to meet the conditions is

$$\varepsilon = (\varepsilon_1, \varepsilon_2, \dots \varepsilon_i, \dots \varepsilon_m)^T$$

and if **K** is the column vector

$$(k_1, k_2, \dots k_i, \dots k_m)^T$$

TABLE 14.9

Solving for the Correlatives

1. $K_1(a_1a_1/w_1 + b_1b_1/w_2 + \cdots + n_1n_1/w_n)$
   $+ K_2(a_1a_2/w_1 + b_1b_2/w_2 + \cdots + n_1n_2/w_n)$
   $+ \cdots$
   $+ K_m(a_1a_m/w_1 + b_1b_m/w_2 + \cdots + n_1n_m/w_n) + r_1 = 0$

2. $K_1(a_1a_2/w_1 + b_1b_2/w_2 + \cdots + n_1n_2/w_n)$
   $+ K_2(a_2a_2/w_1 + b_2b_2/w_2 + \cdots + n_2n_2/w_n)$
   $+ \cdots$
   $+ K_m(a_2a_m/w_1 + b_2b_m/w_2 + \cdots + n_2n_m/w_n) + r_2 = 0$

   $\vdots$

m. $K_1(a_1a_m/w_1 + b_1b_m/w_2 + \cdots + n_1n_m/w_n)$
   $+ K_2(a_2a_m/w_1 + b_2b_m/w_2 + \cdots + n_2n_m/w_n)$
   $+ \cdots$
   $+ K_m(a_ma_m/w_1 + b_mb_m/w_2 + \cdots + n_mn_m/w_n) + r_m = 0$

---

of correlatives and $\mathbf{M}$ is the set of m conditions that link n unknown corrections ε so that $\mathbf{M}\varepsilon + \mathbf{R} = 0$ and $\sum r^2$ and is a minimum, then assuming unit weight, if $\mathbf{M}\mathbf{M}^T\mathbf{K} + \mathbf{R} = 0$, we can obtain the values in $\mathbf{K}$ and thus $\varepsilon = \mathbf{M}^T\mathbf{K}$.

Since this is very abstract, let us consider an example. If all the eight internal angles are measured in Figure 14.1 (a figure known to surveyors as a *braced quadrilateral*), then there will be redundant measurements. In fact, if A and B are known then only two measures are required to fix C and two to fix D, hence, if we have eight measurements there are four extra observations. These can be used to improve the accuracy of the determinations of C and D from A and B.

If we take the braced quadrilateral in Figure 14.1a, then obviously:

Angles

$$1 + 2 + 3 + 8 = 180$$

$$4 + 5 + 6 + 7 = 180$$

$$2 + 3 + 4 + 5 = 180$$

$$6 + 7 + 8 + 1 = 180$$

$$1 + 2 + 3 + 4 + 5 + 6 + 7 + 8 = 360$$

We have five equations and four redundant observations, since if A and B are known, then angles 1, 2, 7, and 8 are sufficient to fix C and D; angles 3, 4, 5, and 6 are redundant. However, these five equations are not independent since the fifth equation is the sum of the first and second, and also of the third and fourth. Furthermore, $(\angle 4 + \angle 5) = (\angle 1 + \angle 8)$ while $(\angle 2 + \angle 3) = (\angle 6 + \angle 7)$ and we cannot distinguish between the pairs. The angles in Figure 14.1b are the same as in 14.1a, but the figure is geometrically different.

The conclusion from this is that care must be taken in defining the conditions that must be satisfied. In the case of the braced quadrilateral, in order to satisfy the geometry, we need to make sure that the scale is consistent. Since

$$AB/AC = \sin 3/\sin 8; \qquad AC/CD = \sin 5/\sin 2$$

$$CD/DB = \sin 7/\sin 4; \qquad DB/AB = \sin 1/\sin 6$$

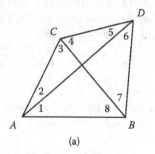 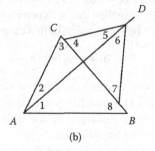

(a)   (b)

FIGURE 14.1   The braced quadrilateral.

Then on multiplying all these together, we obtain

$$\frac{AB}{AC} * \frac{AC}{AD} * \frac{CD}{DB} * \frac{DB}{AB} = \frac{AB*AC*CD*DB}{AC*CD*DB*AB} = 1$$

Hence,

$$\frac{\sin 3 * \sin 5 * \sin 7 * \sin 1}{\sin 8 * \sin 2 * \sin 4 * \sin 6} = 1$$

or

$$\sin 3 * \sin 5 * \sin 7 * \sin 1 = \sin 8 * \sin 2 * \sin 4 * \sin 6$$

So now we have four independent equations that must be satisfied, namely,

$$(4 + 5) = (1 + 8)$$

$$(2 + 3) = (6 + 7)$$

$$(1 + 2 + 3 + 4 + 5 + 6 + 7 + 8) = 360$$

$$\sin 3 * \sin 5 * \sin 7 * \sin 1 = \sin 8 * \sin 2 * \sin 4 * \sin 6$$

Satisfying these four equations will ensure that the geometry is robust. These equations are the condition equations and must be satisfied exactly.

Three of the equations are of linear form, but at present the relationship between the sine values of the angles is not linear. We can, however, express this relationship in the form:

$$\log \sin 3 + \log \sin 5 + \log \sin 7 + \log \sin 1$$

$$= \log \sin 8 + \log \sin 2 + \log \sin 4 + \log \sin 6$$

This is a linear relationship between the log sin values. Let each observed angle have a correction ε_i seconds where $i = 1$ to 8. From the log sine tables we can find the difference for one second of arc. Let us call this ϕ_i; for instance, for the angle 52° 48′ 12″, the log sine = −0.0987787 while for 52° 48′ 13″ it is −0.0987771 and hence $\phi_i =$ 0.0000016. Provided that the corrections are small, the difference in the logarithmic values will be $\varepsilon_i \phi_i$, which in this case would be 0.0000016 ε_i. Thus,

$$\varepsilon_3 \phi_3 + \varepsilon_5 \phi_5 + \varepsilon_7 \phi_7 + \varepsilon_1 \phi_1 = \varepsilon_8 \phi_8 + \varepsilon_2 \phi_2 + \varepsilon_4 \phi_4 + \varepsilon_6 \phi_6$$

This gives us a linear equation between the error terms.

Example 14.2 establishes a relationship based on the log sine conditions by reducing these to a set of linear relationships. Example 14.3 then summarizes the corrections that must be optimized.

In Example 14.2 the logarithm of the sine of each of the eight angles has been tabulated together with the differences in each logarithm for one second of arc $\Delta 1''$.

EXAMPLE 14.2: ANGLES IN A BRACED QUADRILATERAL

Angle	Observed	Log Sine (Odd angles)	Δ1"	Log Sine (Even angles)	Δ1"
1.	52° 48′ 12.3″	−0.0987783	16		
2.	89 33 15.8			−0.0000131	0
3.	15 13 58.9	−0.5804649	77		
4.	17 23 14.0			−0.5245787	67
5.	57 49 31.7	−0.0724090	13		
6.	68 09 09.5			−0.0323685	8
7.	36 38 19.6	−0.2241944	28		
8.	22 24 31.0			−0.4188365	51
Σ =	360° 00′ 12.8″	−0.9758466		−0.9757968	

If we adjust each angle by an amount to make the sine values consistent, then

$$16\varepsilon_1 + 77\varepsilon_3 + 13\varepsilon_5 + 28\,\varepsilon_7 - 9758466 = 67\varepsilon_4 + 8\varepsilon_6 + 51\varepsilon_8 - 9757968$$

or

$$16\varepsilon_1 + 77\varepsilon_3 - 67\varepsilon_4 + 13\varepsilon_5 - 8\varepsilon_6 + 28\varepsilon_7 - 51\varepsilon_8 - 498 = 0$$

This allows us to express the errors in terms of linear combinations derived from a relationship that was expressed as a series of sine values that were multiplied together.

Example 14.3 shows how the process is carried out, leading to a simple linear relationship between the corrections to be applied to each of the angles. The box also shows the simple relationships between the sums of the angles. There are four independent equations relating to eight unknowns. In addition, we want $\Sigma\varepsilon^2$ to be a minimum. Using the formulae derived above, $n = 8$, the number of observed angles and $m = 4$, the number of conditions. We can tabulate the four condition equations (Eqn1), (Eqn2), (Eqn3), and (Eqn4), and eight unknowns as in Example 14.4.

In Example 14.4, the rows are the condition equations and the columns the corrections. Here the row called Eqn1 shows that from Eqn1 in Example 14.3

$$1 * \varepsilon_1 - 1 * \varepsilon_4 - 1 * \varepsilon_5 + 1 * \varepsilon_8 - 2.4 = 0.$$

Using the equations in Example 14.4 and assuming the weights all = 1, we can substitute numbers for all the expressions such as

$$K_1(a_1a_1/w_1 + b_1b_1/w_2 + \cdots + n_1n_1/w_n) + K_2(a_1a_2/w_1 + b_1b_2/w_2 + \cdots + n_1n_2/w_n),$$ and so forth.

EXAMPLE 14.3: AN EXAMPLE OF A SURVEY ADJUSTMENT

Using the data in Example 14.2:

$$\text{Angles } 1 + 8 = 75° \; 12' \; 43.3'' \quad \text{and} \quad 4 + 5 = 75° \; 12' \; 45.7''$$

Hence,

$$\varepsilon_1 + \varepsilon_8 + 75° \; 12' \; 43.3'' = \varepsilon_4 + \varepsilon_5 + 75° \; 12' \; 45.7''$$

$$\varepsilon_1 - \varepsilon_4 - \varepsilon_5 + \varepsilon_8 - 2.4 = 0 \qquad\qquad (Eqn1)$$

Also, angles $2 + 3 = 104° \; 47' \; 14.7''$ and $6 + 7 = 104° \; 47' \; 29.1''$

$$\varepsilon_2 + \varepsilon_3 - \varepsilon_6 - \varepsilon_7 - 14.4 = 0 \qquad\qquad (Eqn2)$$

$$\varepsilon_1 + \varepsilon_2 + \varepsilon_3 + \varepsilon_4 + \varepsilon_5 + \varepsilon_6 + \varepsilon_7 + \varepsilon_8 + 12.8 = 0 \qquad\qquad (Eqn3)$$

Also, $\sin1 * \sin3 * \sin5 * \sin7 = \sin2 * \sin4 * \sin6 * \sin8$. Hence,

$\log \sin1 + \log \sin3 + \log \sin5 + \log \sin7 = \log \sin2 + \log \sin4 + \log \sin6 + \log \sin8$

The log (sine) values of the four odd numbered angles add up to -0.9758466 and the evens to -0.9757968 (see Example 14.2). The change in the log sine that arises from a change of $1''$ can be obtained from tables or using a pocket calculator.

This gives us, as shown in Example 14.2

$$16\varepsilon_1 + 77\varepsilon_3 - 67\varepsilon_4 + 13\varepsilon_5 - 8\varepsilon_6 + 28\varepsilon_7 - 51\varepsilon_8 - 498 = 0 \qquad (Eqn4)$$

Hence, we have four linear equations, *Eqn1*, *Eqn2*, *Eqn3*, and *Eqn4* above linking the eight unknowns ε_i each of which represents a correction to one of the angles.

EXAMPLE 14.4: FORMING THE CORRELATIVES

	$a = \varepsilon_1$	$b = \varepsilon_2$	$c = \varepsilon_3$	$d = \varepsilon_4$	$e = \varepsilon_5$	$f = \varepsilon_6$	$g = \varepsilon_7$	$h = \varepsilon_8$	r
Eqn1	1	0	0	-1	-1	0	0	1	-2.4
Eqn2	0	1	1	0	0	-1	-1	0	-14.4
Eqn3	1	1	1	1	1	1	1	1	12.8
Eqn4	16	0	77	-67	13	-8	28	-51	-498

Thus, in the first equation the coefficient of

$$K_1 \text{ from } Eqn1^2 = 1^2 + 0^2 + 0^2 + (-1)^2 + (-1)^2 + 0^2 + 0^2 + 1^2 = 4$$
$$K_2 \text{ from } Eqn1 * Eqn2 = 1 * 0 + 0 * 1 + 0 * 1 - 1 * 0 - 1 * 0 + 0 *$$
$$(-1) + 0 * (-1) + 1 * 0 = 0$$
$$K_3 \text{ from } Eqn1 * Eqn3 = 1 + 0 + 0 - 1 - 1 + 0 + 0 + 1 = 0$$
$$K_4 \text{ from } Eqn1 * Eqn4 = 16 + 0 + 0 + 67 - 13 + 0 + 0 - 51 = 19$$

This gives us

$$4 * K_1 + 0 * K_2 + 0 * K_3 + 19 * K_4 - 2.4 = 0$$

Similarly, for the second equation the coefficients are derived from

$$(Eqn1 * Eqn2), (Eqn2)^2, (Eqn2 * Eqn3), (Eqn2 * Eqn4)$$

For the third,

$$(Eqn1 * Eqn3), (Eqn2 * Eqn3), (Eqn3)^2, (Eqn3 * Eqn4)$$

For the fourth,

$$(Eqn1 * Eqn4), (Eqn2 * Eqn4), (Eqn3 * Eqn4), (Eqn4)^2$$

Hence, using the numbers in Example 14.4 and substituting them in the equations given in Table 14.9 we obtain the following:

1. $4 * K_1 + 0 * K_2 + 0 * K_3 + 19 * K_4 - 2.4 = 0$
2. $0 * K_1 + 4 * K_2 + 0 * K_3 + 57 * K_4 - 14.4 = 0$
3. $0 * K_1 + 0 * K_2 + 8 * K_3 + 8 * K_4 + 12.8 = 0$
4. $19 * K_1 + 57 * K_2 + 8 * K_3 + 14292 * K_4 - 498 = 0$

From 1,

$$K_1 = 0.6 - 4.75K_4$$

From 2,

$$K_2 = 3.6 - 14.25K_4$$

From 3,

$$K_3 = -1.6 - K_4$$

From 4,

$$K_4 = (294.2)/(13381.5)$$
$$= 0.022$$

So,

$$K_1 = 0.496; \qquad K_2 = 3.287; \qquad K_3 = -1.622; \qquad K_4 = 0.022$$

Substituting back, using Example 14.4 and reading the columns downward so that under the column headed "$a = \varepsilon_1$," we have:

$$\varepsilon_1 = 1 * Eqn1 + 0 * Eqn2 + 1 * Eqn3 + 16 * Eqn4$$

$$\varepsilon_2 = 0 * Eqn1 + 1 * Eqn2 + 1 * Eqn3 + 0 * Eqn4$$

$$\varepsilon_3 = 0 * Eqn1 + 1 * Eqn2 + 1 * Eqn3 + 77 * Eqn4, \text{ and so on.}$$

The results are shown in Example 14.5.

EXAMPLE 14.5: THE FINISHED SOLUTION

$$\varepsilon_1 = 1 * K_1 + 0 * K_2 + 1 * K_3 + 16 * K_4 = -0.774 = -0.8$$

Angle 1 = 52° 48′ 11.5″

$$\varepsilon_2 = 0 * K_1 + 1 * K_2 + 1 * K_3 + 0 * K_4 = 1.665 = +1.7$$

Angle 2 = 89° 33′ 17.5″

$$\varepsilon_3 = 0 * K_1 + 1 * K_2 + 1 * K_3 + 77 * K_4 = 3.358 = +3.4$$

Angle 3 = 15° 14′ 02.3″

$$\varepsilon_4 = -1 * K_1 + 0 * K_2 + 1 * K_3 - 67 * K_4 = -3.591 = -3.6$$

Angle 4 = 17° 23′ 10.4″

$$\varepsilon_5 = -1 * K_1 + 0 * K_2 + 1 * K_3 + 13 * K_4 = -1.832 = -1.8$$

Angle 5 = 57° 49′ 29.9″

$$\varepsilon_6 = 0 * K_1 + -1 * K_2 + 1 * K_3 - 8 * K_4 = -5.085 = -5.1$$

Angle 6 = 68° 09′ 04.4″

$$\varepsilon_7 = 0 * K_1 + -1 * K_2 + 1 * K_3 + 28 * K_4 = -4.293 = -4.3$$

Angle 7 = 36° 38′ 15.3″

$$\varepsilon_8 = 1 * K_1 + 0 * K_2 + 1 * K_3 - 51 * K_4 = -2.248 = -2.2$$

Angle 8 = 22° 24′ 28.8″

In matrix form this can be more elegantly expressed. To summarize:

$$\text{If } \mathbf{M} = \begin{pmatrix} 1 & 0 & 0 & -1 & -1 & 0 & 0 & 1 \\ 0 & 1 & 1 & 0 & 0 & -1 & -1 & 0 \\ 1 & 1 & 1 & 1 & 1 & 1 & 1 & 1 \\ 16 & 0 & 77 & 67 & 13 & -8 & 28 & -51 \end{pmatrix}; \mathbf{R} = \begin{pmatrix} -2.4 \\ -14.4 \\ 12.8 \\ -498 \end{pmatrix}; \mathbf{M}^T = \begin{pmatrix} 1 & 0 & 1 & 16 \\ 0 & 1 & 1 & 0 \\ 0 & 1 & 1 & 77 \\ -1 & 0 & 1 & -67 \\ -1 & 0 & 1 & 13 \\ 0 & -1 & 1 & -8 \\ 0 & -1 & 1 & 28 \\ 1 & 0 & 1 & -51 \end{pmatrix} \text{ then }$$

$$\mathbf{M} * \mathbf{M}^T * \mathbf{K} = \begin{pmatrix} 4 & 0 & 0 & 19 \\ 0 & 4 & 0 & 57 \\ 0 & 0 & 8 & 8 \\ 19 & 57 & 8 & 14292 \end{pmatrix} * \mathbf{K} = -\mathbf{R} = - \begin{pmatrix} -2.4 \\ -14.4 \\ 12.8 \\ -498 \end{pmatrix} \text{ where } \mathbf{K} = \begin{pmatrix} K_1 \\ K_2 \\ K_3 \\ K_4 \end{pmatrix}$$

This gives

$$\mathbf{K} = - \begin{pmatrix} 4 & 0 & 0 & 19 \\ 0 & 4 & 0 & 57 \\ 0 & 0 & 8 & 8 \\ 19 & 57 & 8 & 14292 \end{pmatrix}^{-1} * \begin{pmatrix} -2.4 \\ -14.4 \\ 12.8 \\ -498 \end{pmatrix} = - \begin{pmatrix} 0.2517 & 0.0051 & 0.0004 & -0.0004 \\ 0.0051 & 0.2652 & 0.0011 & -0.0011 \\ 0.0004 & 0.0011 & 0.1251 & -0.0001 \\ -0.0004 & -0.0011 & -0.0001 & 0.0001 \end{pmatrix} * \begin{pmatrix} -2.4 \\ -14.4 \\ 12.8 \\ -498 \end{pmatrix}$$

$$= \begin{pmatrix} 0.496 \\ 3.287 \\ -1.622 \\ 0.022 \end{pmatrix}$$

Hence,

$$\varepsilon = \mathbf{M}^T * \mathbf{K} = \begin{pmatrix} 1 & 0 & 1 & 16 \\ 0 & 1 & 1 & 0 \\ 0 & 1 & 1 & 77 \\ -1 & 0 & 1 & -67 \\ -1 & 0 & 1 & 13 \\ 0 & -1 & 1 & -8 \\ 0 & -1 & 1 & 28 \\ 1 & 0 & 1 & -51 \end{pmatrix} * \begin{pmatrix} 0.496 \\ 3.287 \\ -1.622 \\ 0.022 \end{pmatrix} = \begin{pmatrix} -0.774 \\ 1.665 \\ 3.358 \\ -3.591 \\ -1.832 \\ -5.085 \\ -4.293 \\ -2.248 \end{pmatrix}$$

The somewhat tortuous calculations that have led to Example 14.5 provide a solution in which, rounding off to the nearest 0.1″, gives $\Sigma \varepsilon^2 = 80.6$. This is the minimum least square value.

Returning to the problem posed at the end of Section 14.1 where we had three angles of a triangle with corrections $x_1 + x_2 + x_3 + r = 0$, and $\mathbf{M} = (1, 1, 1)$ then $\mathbf{M} * \mathbf{M}^T = 3$ and $\mathbf{M} * \mathbf{M}^T * \mathbf{K} = 3 * \mathbf{K} = -r$. Thus, $\mathbf{K} = (-r/3)$ and

$$X = \begin{pmatrix} -r/3 \\ -r/3 \\ -r/3 \end{pmatrix} \quad \text{and} \quad x_1 = x_2 = x_3 = -r/3$$

meaning that we apply equal corrections to each angle as the optimum least square solution.

Very many types of observation can be processed in this way, provided that we know the conditions that must be fulfilled and the relative weighting of the different observations. Although tedious to go through by longhand, once the equations are formed, the electronic computer can take over and provide a most probable value for each of the observations.

The technique of least square adjustments has many applications in surveying and mapping, and indeed other areas where a best-fit solution is required. The algorithms are ideally suited to electronic data processing but the quality of the results will depend on the quality of the observations and the assumptions underlying the adjustment procedures. If, for example, each observation is not truly independent and is somehow correlated with other observations, then the underlying assumptions of the least square process are not fully valid. Techniques are available to test for autocorrelation using variance and covariance analysis and were touched on in Chapter 13.

14.3   SATELLITE POSITION FIXING

Global positioning systems (GPS) make use of several mathematical processes to determine the coordinates of a point on the Earth's surface. A series of satellites circling the Earth transmit data about their position and time, the signals being picked up by a receiver. In essence, if satellite i has position (x_i, y_i, z_i) at time t_i and if its signal is received at a ground station R in position (x, y, z) at time t_r according to the local ground station clock, then the distance between the satellite and point R will be $\sqrt{\{(x-x_i)^2 +(y-y_i)^2 +(z-z_i)^2\}}$. The coordinates for this system are rectangular Cartesian and geocentric in that they are expressed relative to the notional center of the Earth, the z-axis lying through the North and South Poles, the direction of the x-axis being determined by the meridian through Greenwich (G in Figure 14.2) and the y-axis being at right angles thereto.

The time taken for the signal to travel at the speed of light will be $(t_r - e - t_i)$ where e is the error between the two clocks. The distance between the two is therefore $(t_r - e - t_i) * c$ where c = the speed of light. This is known precisely and = 299,792,458 meters per second.

We therefore have a distance (derived from a measurement of time) from a known point in space. We can imagine a sphere of radius $(t_r - e - t_i)$ around each satellite with the receiver R being the point where these spheres all intersect with the Earth's surface.

Using Pythagoras, we have $\sqrt{\{(x-x_i)^2 +(y-y_i)^2 +(z-z_i)^2\}} = (t_r - e - t_i) * c$.

In this equation there are four unknowns; namely x, y, z, and e. We therefore need four sets of observations to solve for the unknowns so that we must take observations to at least four satellites. This assumes that the clock error e is the same for each set of observations. In practice, the signals sent out by the satellites are based on identically the same time reference system. e is solely dependent on the receiver station and will be the same for all observations that are taken at the same time from the same place with the same equipment. The assumption that all we need is a minimum of four satellites is therefore valid.

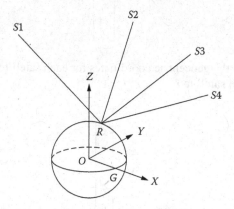

FIGURE 14.2   Four satellites and a receiver.

We can express these four relationships as

$$\sqrt{\{(x-x_i)^2 + (y-y_i)^2 + (z-z_i)^2\}} + ec = (t_r - t_i) * c$$

where $i = 1, 2, 3, 4$. The value $(t_r - t_i) * c$ is known as the *pseudo-range* and is often designated ρ_i.

The four equations are not linear but can be solved iteratively using Newton's method, as discussed later. Alternatively, if we write ec as l then the equation above can be expressed in the form

$$(x - x_i)^2 + (y - y_i)^2 + (z - z_i)^2 = (\rho_i - l)^2 = \rho_i^2 - 2l\rho_i + l^2$$

or

$$F_i = (x_i^2 + y_i^2 + z_i^2 - \rho_i^2) - 2(x_i x + y_i y + z_i z - l\rho_i) + (x^2 + y^2 + z^2 - l^2) = 0$$

Let us call this relationship F_i. The term $(x^2 + y^2 + z^2 - l^2)$ is the same for all observations when the receiver is at a given point. If we have five or more satellite observations then we can eliminate the term by working with pairs of observations giving us four or more linear relationships between x, y, z, and l. We can also use the method of least squares to determine the optimum solution by treating F_i not as zero but as the residual for which we must minimize the sum of its squares. The determination of our position should become more accurate the more observations that we make.

The equations F_i are quadratic implying that there will be two possible solutions, although only one of these will be point R where the receiver is located. Let us assume that there are n satellites.

Let

$$S_i = \begin{bmatrix} x_i \\ y_i \\ z_i \\ \rho_i \end{bmatrix}$$

where (x_i, y_i, z_i) are the geocentric coordinates for each satellite, $i = 1, 2, \ldots n$. ρ_i is the pseudo-range to satellite i.

Let

$$X = \begin{bmatrix} x \\ y \\ z \\ l \end{bmatrix}$$

represent the ground receiver coordinates (x, y, z) and l is the distance represented by the clock error multiplied by the speed of light.

$$\text{If } N = \begin{bmatrix} 1 & 0 & 0 & 0 \\ 0 & 1 & 0 & 0 \\ 0 & 0 & 1 & 0 \\ 0 & 0 & 0 & -1 \end{bmatrix} \quad \text{then} \quad X^T * N = [x \ y \ z - l]$$

$(X^TN)X = (x^2 + y^2 + z^2 - l^2)$. This is a constant for all observations from the ground station. Let us call it k so that $k = (X^TN)X = K$.

Let A = column matrix of terms $(x_i^2 + y_i^2 + z_i^2 - \rho_i^2)$ of size n rows and one column; also let B = matrix with n rows and four columns made of the elements of $[x_j \ y_j \ z_j, -\rho_j]$ where $j = 1, 2, \ldots n$

$$A = \begin{bmatrix} x_1^2 + y_1^2 + z_1^2 - \rho_1^2 \\ x_2^2 + y_2^2 + z_2^2 - \rho_2^2 \\ x_3^2 + y_3^2 + z_3^2 - \rho_3^2 \\ x_4^2 + y_4^2 + z_4^2 - \rho_4^2 \\ \ldots \\ x_n^2 + y_n^2 + z_n^2 - \rho_n^2 \end{bmatrix} \quad \text{and} \quad B = \begin{bmatrix} x_1 \ y_1 \ z_1 - \rho_1 \\ x_2 \ y_2 \ z_2 - \rho_2 \\ x_3 \ y_3 \ z_3 - \rho_3 \\ x_4 \ y_4 \ z_4 - \rho_4 \\ \ldots\ldots \\ x_n \ y_n \ z_n - \rho_n \end{bmatrix} \quad BX = \begin{bmatrix} x_1x + y_1y + z_1z - \rho_1 l \\ x_2x + y_2y + z_2z - \rho_2 l \\ x_3x + y_3y + z_3z - \rho_3 l \\ x_4x + y_4y + z_4z - \rho_4 l \\ \ldots\ldots \\ x_nx + y_ny + z_nz - \rho_n l \end{bmatrix}$$

The equations F_i show that we need to satisfy the condition that $A - 2BX + K = 0$. Because of potential for error in all the measurements, slightly different answers will be obtained for each set of four satellites observed. For now, let us assume that we have four satellites, hence $n = 4$ and B is a square matrix. Let B^{-1} be the inverse of B, and finally let K be the column vector with $n = 4$ rows, each row being equal to k where, as before, $k = (x^2 + y^2 + z^2 - l^2)$.

Then the equations F_i can be written in the form

$$A - 2BX + K = 0 \text{ or}$$

$$BX = 0.5(A + K) \text{ or}$$

$$X = 0.5B^{-1}(A + K)$$

If, for convenience, we write C for $0.5B^{-1}$ where C is a $4 * 4$ matrix, then

$$X = C(A + K)$$

and

$$X^T = (A + K)^T C^T$$

The elements of C and A are all observed quantities and hence are all known.

Each element of K, namely, k, is a scalar constant made up from x, y, z, and l. If we knew k we would have a series of at least four linear equations linking x, y, z, and l.

Now $k = (\mathbf{X}^T * \mathbf{N}) * \mathbf{X} = \{(\mathbf{A} + \mathbf{K})^T * \mathbf{C}^T * \mathbf{N} * \mathbf{C} * (\mathbf{A} + \mathbf{K})\}$. If in $\mathbf{A}$ we write $a_j = x_j^2 + y_j^2 + z_j^2 - \rho_j^2$, all of which are observed quantities, then

$$\mathbf{A} + \mathbf{K} = \begin{bmatrix} a_1 + k \\ a_2 + k \\ a_3 + k \\ a_4 + k \end{bmatrix} \quad \text{and} \quad (\mathbf{A} + \mathbf{K})^T = [a_1 + k, a_2 + k, a_3 + k, a_4 + k]$$

Let the elements of $\mathbf{C}$ be such that

$$\mathbf{C} = \begin{bmatrix} c_{11} & c_{12} & c_{13} & c_{14} \\ c_{21} & c_{22} & c_{23} & c_{24} \\ c_{31} & c_{32} & c_{33} & c_{34} \\ c_{41} & c_{42} & c_{43} & c_{44} \end{bmatrix}$$

Therefore,

$$(\mathbf{A} + \mathbf{K})^T * \mathbf{C}^T = [a_1 + k, a_2 + k, a_3 + k, a_4 + k] * \begin{bmatrix} c_{11} & c_{21} & c_{31} & c_{41} \\ c_{12} & c_{22} & c_{32} & c_{42} \\ c_{13} & c_{23} & c_{33} & c_{43} \\ c_{14} & c_{24} & c_{34} & c_{44} \end{bmatrix}$$

Hence,

$$(\mathbf{A} + \mathbf{K})^T * \mathbf{C}^T * \mathbf{N} \text{ is a } 1 * 4 \text{ row vector } [v_1, v_2, v_3, v_4]$$

where
$$v_1 = c_{11}(a_1 + k) + c_{12}(a_2 + k) + c_{13}(a_3 + k) + c_{14}(a_4 + k)$$
$$v_2 = c_{21}(a_1 + k) + c_{22}(a_2 + k) + c_{23}(a_3 + k) + c_{24}(a_4 + k)$$
$$v_3 = c_{31}(a_1 + k) + c_{32}(a_2 + k) + c_{33}(a_3 + k) + c_{34}(a_4 + k)$$
$$v_4 = -[c_{41}(a_1 + k) + c_{42}(a_2 + k) + c_{43}(a_3 + k) + c_{44}(a_4 + k)]$$

Also,

$$\mathbf{C} * (\mathbf{A} + \mathbf{K}) = \begin{bmatrix} c_{11} & c_{12} & c_{13} & c_{14} \\ c_{21} & c_{22} & c_{23} & c_{24} \\ c_{31} & c_{32} & c_{33} & c_{34} \\ c_{41} & c_{42} & c_{43} & c_{44} \end{bmatrix} * \begin{bmatrix} a_1 + k \\ a_2 + k \\ a_3 + k \\ a_4 + k \end{bmatrix} \quad \text{which is a } 4*1 \text{ column vector} \quad \begin{bmatrix} v_1 \\ v_2 \\ v_3 \\ v_4 \end{bmatrix}$$

If we call the row vector $[v_1, v_2, v_3, v_4]$ as $[p_1 + q_1k, p_2 + q_2k, p_3 + q_3k, p_4 + q_4k]$, then the elements of the column vector are the same, except for a change of sign at the end. We therefore have

$$(\mathbf{X}^T * \mathbf{N}) * \mathbf{X} = k = (p_1 + q_1k)^2 + (p_2 + q_2k)^2 + (p_3 + q_3k)^2 - (p_4 + q_4k)^2$$

or

$$(q_1^2 + q_2^2 + q_3^2 - q_4^2)k^2 + (2p_1q_1 + 2p_2q_2 + 2p_3q_3 - 2p_4q_4 - 1)k + (p_1^2 + p_2^2 + p_3^2 - p_4^2) = 0$$

This is a quadratic in k, which will give two possible values for k only one of which will be the ground receiver, the other being somewhere out in space.

We can now write $F_i = (x_i^2 + y_i^2 + z_i^2 - \rho_i^2) - 2(x_ix + y_iy + z_iz - l\rho_i) + k$, giving us four linear equations relating x, y, z, and l. With four satellites we can solve the equations directly, as we do in Example 14.6. If there are more than four then

EXAMPLE 14.6: SATELLITE POSITIONING

Consider four satellites where their (x, y, z) is known, along with the pseudo-ranges:

Satellite x	Satellite y	Satellite z	$t_r - t_s$	$(t_r - t_s) * c$	A
15629930.79	17229314.77	12797138.77	71,410,508.00	21,408,331.72	2.46594E + 14
1198579.24	15672887.77	21381561.43	76,622,350.95	22,970,802.93	1.76589E + 14
4927877.54	24742040.32	8352009.51	80,056,010.16	24,000,188.06	1.302E + 14
18740047.66	–4918293.33	18270380.04	70,918,574.34	21,260,853.72	2.57162E + 14

Column A is made up of terms $x_i^2 + y_i^2 + z_i^2 - \rho_i^2$. The notation E + 14 indicates an exponential value or scientific notation showing that the number is large with 14 terms before the decimal point. Here we have rounded off all the numbers.

The matrix **B** with elements $[x_j \, y_j \, z_j, \, -\rho_j] =$

$$\begin{bmatrix} 15629930.79 & 17229314.77 & 12797138.77 & -21,408,331.72 \\ 1198579.24 & 15672887.77 & 21381561.43 & -22,970,802.92 \\ 4927877.54 & 24742040.32 & 8352009.51 & -24,000,188.06 \\ 18740047.66 & -4918293.33 & 18270380.04 & -21,260,853.72 \end{bmatrix}$$

$$0.5\mathbf{B}^{-1} =$$

$$\begin{bmatrix} 0.000000044454 & -0.00000001484 & -0.00000002227 & -0.00000000359 \\ 0.000000037673 & 0.000000007393 & -0.00000001620 & -0.00000002763 \\ 0.000000034215 & 0.000000037870 & -0.00000005129 & -0.00000001747 \\ 0.000000059872 & 0.000000017754 & -0.00000005996 & -0.00000003530 \end{bmatrix}$$

We defined p_1 as $c_{11}a_1 + c_{12}a_2 + c_{13}a_3 + c_{14}a_4$ and $q_1 = c_{11} + c_{12} + c_{13} + c_{14}$, and so on. This gives values for p_1, q_1, and so on, as

	1	2	3	4
p	4,518,807.6085	1,379,841.7715	3,953,880.9441	−1,013,796.077
q	0.00000000375	0.00000000124	0.00000000333	0.00000001763

This in turn gives $k = 4.0865\text{E} + 13$. The coordinates and error term for the receiver then become

$$(x_i^2 + y_i^2 + z_i^2 - \rho_i^2) - 2(x_ix + y_iy + z_iz - l\rho_i) + k = 0, \text{ or}$$

$$2x_ix + 2y_iy + 2z_iz - 2l\rho_i = (x_i^2 + y_i^2 + z_i^2 - \rho_i^2) + k \text{ as shown below}$$

x	y	z	l	Equals
31259861.58	34458629.54	25594277.54	−42,816,663.44	2.87081E + 14
2397158.48	31345775.54	42763122.86	−45,941,605.86	2.17076E + 14
9855755.08	49484080.64	16704019.02	−48,000,376.13	1.70686E + 14
37480095.32	−9836586.66	36540760.08	−42,521,707.44	2.97648E + 14

On solving these four quadratics:

$x = 4670813.04$; $y = 1429701.03$; $z = 4088501.02$, and the time error $l = 299792.477$

(l suggests that the receiver clock was almost exactly one second out.)

we will need to derive a least squares solution using the principles outlined in Sections 14.1 and 14.2 to obtain a more accurate solution for the coordinates of point R.

A word of caution should, however, be given. The orbit of the satellites is around 26.5 million meters from the center of the Earth. The determinant for **B** in the above equations is a number in excess of 1,000,000,000,000,000,000,000,000,000. Even when working in kilometers rather than meters it is still in excess of 1,000,000,000,000,000. Any computation system must be capable of handling such large numbers. In Example 14.6 we show part of the calculations but with rounding up of some of the numbers in the printout.

It should also be noted that the description above is a simplification as, for example, it assumes perfect measurement conditions and ignores the theory of relativity.

An alternative approach makes use of a series of iterations to come up with a solution where there are four satellites and the approximate position of the receiver station and the time delay is known. If the "o" for "old" coordinates are

$\mathbf{X_o} = (x_o, y_o, z_o, l_o)$, and a better estimate is a new set $\mathbf{X_n} = (x_n, y_n, z_n, l_n)$, let

$$\mathbf{F} = \begin{bmatrix} f_1 \\ f_2 \\ f_3 \\ f_4 \end{bmatrix}$$

where

$$f_i = \sqrt{\{(x_o - x_i)^2 + (y_o - y_i)^2 + (z_o - z_i)^2\}} - \rho_i + l \text{ and } i = 1, 2, 3, 4$$

The partial derivatives of these equations are:

$$\partial f_i/\partial x = (x - x_i)/(\sqrt{\{(x - x_i)^2 + (y - y_i)^2 + (z - z_i)^2\}})$$

$$\partial f_i/\partial y = (y - y_i)/(\sqrt{\{(x - x_i)^2 + (y - y_i)^2 + (z - z_i)^2\}})$$

$$\partial f_i/\partial z = (z - z_i)/(\sqrt{\{(x - x_i)^2 + (y - y_i)^2 + (z - z_i)^2\}})$$

$$\partial f_i/\partial l = 1$$

We then use a version of Newton's method (sometimes referred to as Newton–Raphson), which is an extension to what was discussed in Chapter 6. We form the matrix $\mathbf{J}$ from these partial derivatives so that

$$\mathbf{J} = \begin{bmatrix} \partial f_1/\partial x & \partial f_1/\partial y & \partial f_1/\partial z & \partial f_1/\partial l \\ \partial f_2/\partial x & \partial f_2/\partial y & \partial f_2/\partial z & \partial f_2/\partial l \\ \partial f_3/\partial x & \partial f_3/\partial y & \partial f_3/\partial z & \partial f_3/\partial l \\ \partial f_4/\partial x & \partial f_4/\partial y & \partial f_4/\partial z & \partial f_4/\partial l \end{bmatrix}$$

This is known as the *Jacobian* of $\mathbf{F}$. It can be evaluated from the known coordinates of the satellites and the assumed value for the receiver station, for instance,

$$\partial f_1/\partial x = (x_o - x_1)/(\sqrt{\{(x_o - x_1)^2 + (y_o - y_1)^2 + (z_o - z_1)^2\}})$$

We then calculate the inverse of $\mathbf{J} = \mathbf{J}^{-1}$. Then we derive a new set of coordinates by calculating

$$X_{new} = X_{old} - J^{-1}F_{old} \text{ where } \mathbf{J}^{-1} \text{ is a } 4 * 4 \text{ square matrix}$$

or

$$
\begin{bmatrix} x_n \\ y_n \\ z_n \\ l_n \end{bmatrix} = \begin{bmatrix} x_o \\ y_o \\ z_o \\ l_o \end{bmatrix} - \mathbf{J}^{-1} \begin{bmatrix} f_1 \\ f_2 \\ f_3 \\ f_4 \end{bmatrix}
$$

We can then replace x_o by x_n, and so forth, and repeat the process. If the original estimate is reasonably close to the true value, the iteration should converge quite rapidly. See Example 14.7.

EXAMPLE 14.7: POSITIONING FROM SATELLITES USING ITERATION

Consider four satellites with geocentric coordinates (x, y, z), an apparent time difference $t_r - t_s$.

Satellite x	Satellite y	Satellite z	$t_r - t_s$	$(t_r - t_s) * c$
15629930.79	17229314.77	12797138.77	71,410,508.00	21,408,331.72
1198579.24	15672887.77	21381561.43	76,622,350.95	22,970,802.93
4927877.54	24742040.32	8352009.51	80,056,010.16	24,000,188.06
18740047.66	−4918293.33	18270380.04	70,918,574.34	21,260,853.72

Let us assume an approximate position for the receiver station as

(4,671,000; 1,430,000; 4,089,000) with zero clock error.

We can calculate the distances to each satellite by Pythagoras. These are 21,108,012.509, 22,670,470.636, 23,700,009.711, and 20,960,688.688 from the trial point. The pseudo-ranges are 21,408,331.720; 22,970,802.929; 24,000,188.064; and 21,260,853.719 but these distances include the clock error.

Using the partial derivative equations based on the estimated position

$$\partial f_i/\partial x = (x - x_i)/(\sqrt{\{(x - x_i)^2 + (y - y_i)^2 + (z - z_i)^2\}}), \text{ and so on,}$$

we can obtain the Jacobian with elements as shown in the following tabular form:

diff x/old dist	diff y/old dist	diff z/old dist	$\partial f_i/\partial l$
−0.519183451549186	−0.748498455879073	−0.412551336422655	1
0.1531693283319	−0.628257260285913	−0.762779110723671	1
−0.010838710329384	−0.983629990443701	−0.188042391302602	1
−0.671211135864479	0.302866638866031	−0.676570329051984	1

The inverse of the Jacobian becomes:

$$\begin{bmatrix} -2.023620428365 & 0.6221991481610 & 1.203438354395 & 0.1979829258089 \\ -1.644050579201 & -0.357954526276 & 0.8253402169906 & 1.176664888486 \\ -1.594018146643 & -1.783706488009 & 2.593669274997 & 0.7840553596546 \\ -1.938813875222 & -0.6807634043134 & 2.312592882192 & 1.306984397343 \end{bmatrix}$$

Applying $\mathbf{X}_{new} = \mathbf{X}_{old} - \mathbf{J}^{-1}\mathbf{F}_{old}$ the new approximate coordinates become:

(4,670,808.109; 1,429,697.652; 4,088,490.454) with $l = 299783.064$

The next iteration gives

(4,670,813.115; 1,429,701.081; 4,088,501.240) with $l = 299792.677$

A further iteration gives

(4,670,813.009; 1,429,701.009; 4,088,501.012) with $l = 299,792.473$

And once more gives

(4,670,813.011; 1,429,701.010; 4,088,501.016) with $l = 299,792.477$

Thus, quite quickly, we have honed in on the probable position and clock error using four satellites. Any difference with Example 14.6 is due to rounding errors in the computation.

SUMMARY

Correlatives: Unknown constants introduced in the solution by least squares.

Dispersion matrix: Also known as the *variance–covariance matrix*. If $\mathbf{V}$ is a row matrix of all the error terms and its transpose is $\mathbf{V}^T$ then the dispersion matrix $= \mathbf{V}^T\mathbf{V}$.

Error term: The difference between the observed and computed quantity.

Jacobian: A matrix made from the partial derivatives of given functions.

Least squares: A solution that makes the sum of the squares of the *residuals* a minimum.

Normal equations: The equations that, when solved, give the best estimate of a set of values by the principle of *least squares*. They take the form $\mathbf{M}^T(\mathbf{MX} + \mathbf{L}) = 0$ where $\mathbf{MX} + \mathbf{L} = 0$ represents the *condition equations*.

Pseudo-range: The distance to a satellite based on the time for light to travel from the satellite to the receiver, assuming the clocks are synchronized. The speed of light equals 299,792,458 meters per second.

Residual: The difference between an observed and a predicted value. Also referred to as the *error term*.

Variance–covariance matrix: Another name for the *dispersion* matrix.

Further Reading

There are many excellent books covering basic mathematics and more specific techniques in topics such as algebra, calculus, geometry, and statistical analysis. For the more advanced, CRC Press/Taylor & Francis publishes a wide range of texts on both mathematics and geoscience—see www.crcpress.com. The following is a small selection of some others that are available.

GIS

Heywood I., Cornelius C., and Carver S. (2011) *An Introduction to Geographical Information Systems*, 3rd ed. New York: Pearson.

Longley P.A., Goodchild M.F., Maguire D.J., and Rhind D.W. (2011) *Geographical Information Systems and Science*, 3rd ed. Hoboken, NJ: Wiley.

MATHEMATICS

Allan A.L. (2006) *Maths for Map Makers,* 2nd ed. Scotland, UK: Whittles Publishing.

Borowski E.J. and Borwein J.M. (2005) *Collins Internet-Linked Dictionary of Mathematics*. Collins.

de Smith M.J., Goodchild M.F., and Longley P.A. (2008) *Geospatial Analysis: A Comprehensive Guide to Principles, Techniques and Software Tools*, 2nd ed. Matador.

Ghilani G.D. and Wolf P.R. (2006) *Adjustment Computations—Spatial Data Analysis,* 4th ed. Hoboken, NJ: John Wiley & Sons.

SPECIFIC TOPICS

Ghilani G.D. and Wolf P.R. (2008) *Elementary Surveying. An Introduction to Geomatics*. New York: Pearson.

Iliffe J.C. and Lott R. (2008) *Datums and Map Projections: For Remote Sensing, GIS, and Surveying*, 2nd ed. Scotland, UK: Whittles Publishing.

Mikhail E., Bethel J., and McGlone J.C. (2001) *Introduction to Modern Photogrammetry*. Hoboken, NJ: John Wiley and Sons.

Salomon D. (2013) *Curves and Surfaces for Computer Graphics*. New York: Springer.

Vince J. (2010) *Mathematics for Computer Graphics*. New York: Springer.

Wolf P.R. and Dewitt B.A. (2000) *Elements of Photogrammetry with Applications in GIS*. Thomas Casson.

Index

Printed in the United States
by Baker & Taylor Publisher Services